W9-BXN-447

CONSCIOUSNESS BEYOND LIFE

The Science of the Near-Death Experience

Pim van Lommel

HarperOne
An Imprint of HarperCollins*Publishers*

HarperOne

Translation copyright © Laura Vroomen.
Originally published as *Eindeloos Bewustzijn* in 2007 by Uitgeverij Ten Have.
© Pim van Lommel, chapter 10 written by Monique Hennequin.

HarperCollins Web site: http://www.harpercollins.com
HarperCollins®, ⚒ ®, and HarperOne™ are
trademarks of HarperCollins Publishers

FIRST EDITION

Library of Congress Cataloging-in-Publication Data is available upon request.

ISBN 978–0–06–1777257

10 11 12 13 14   RRD(H)   10 9 8 7 6 5 4 3 2 1

Contents

Introduction

All science is empirical science, all theory is subordinate to perception; a single fact can overturn an entire system.
—FREDERIK VAN EEDEN

It is 1969. At the coronary care unit the alarm suddenly goes off. The monitor shows that the electrocardiogram of a patient with a myocardial infarction (heart attack) has flatlined. The man has suffered a cardiac arrest. Two nurses hurry over to the patient, who is no longer responsive, and quickly draw the curtains around his bed. One of the nurses starts CPR while the other places a mask over his mouth and administers oxygen. A third nurse rushes over with the resuscitation trolley that contains the defibrillator. The defibrillator is charged, the paddles are covered in gel, the patient's chest is bared, the medical staff let go of the patient and the bed, and the patient is defibrillated. He receives an electric shock to the chest. It has no effect. Heart massage and artificial respiration are resumed, and, in consultation with the doctor, extra medication is injected into the IV drip. Then the patient is defibrillated for the second time. This time his cardiac rhythm is re-established, and more than a minute later, after a spell of unconsciousness that lasted about four minutes, the patient regains consciousness, to the great relief of the nursing staff and attendant doctor.

That attendant doctor was me. I had started my cardiology training that year.

Following the successful resuscitation, everybody was pleased—

everybody except the patient. He had been successfully revived, yet to everybody's surprise he was extremely disappointed. He spoke of a tunnel, colors, a light, a beautiful landscape, and music. He was extremely emotional. The term *near-death experience* (NDE) did not yet exist, and I had never heard of people remembering the period of their cardiac arrest. While studying for my degree, I had learned that such a thing is in fact impossible: being unconscious means being unaware—and the same applies to people suffering a cardiac arrest or patients in a coma. At such a moment it is simply impossible to be conscious or to have memories because all brain function has ceased. In the event of a cardiac arrest, a patient is unconscious, is no longer breathing, and has no palpable pulse or blood pressure.

Near Death in the Hospital

The first coronary care units in Dutch hospitals opened in 1966, when massaging the heart, administering oxygen, and defibrillation were found to be effective in treating cardiac arrest patients. Cardiac arrest was and remains the most common cause of death for people with an acute myocardial infarction—in the United States, about one death each minute, and in the UK about one death every two minutes. Since the introduction of modern techniques of resuscitation and the establishment of coronary care units, mortality rates as a result of cardiac arrest have fallen sharply, and these days it is not uncommon for patients to survive cardiac arrest.

When I was working as a cardiologist, I was confronted with death on an almost daily basis. As a doctor, you are all but forced to reflect on the emotional, philosophical, and physiological aspects of life and death. But often such reflection does not actually take on any urgency until you are personally affected by the death of a family member. In my case, this happened when my mother died at the age of sixty-two and my brother at the age of forty-one.

Although I had never forgotten the successfully resuscitated patient in 1969, with his memories of the period of his cardiac arrest, I had not done anything with the experience. This changed in 1986 when I read a book about near-death experiences by George Ritchie with the title *Return from Tomorrow.*[1] When he had double pneumonia as a medical student in 1943, Ritchie experienced a period of clinical death. At the time antibiotics such as penicillin were not yet widely used. Following an episode of very high fever and extreme tightness of the chest, he passed away: he ceased breathing and his pulse also stopped. He was pronounced dead by a doctor and covered with a sheet. But a male nurse was so upset by the death of this medical student that he managed to persuade the attendant doctor to administer an adrenalin injection in the chest near the heart—a most unusual procedure in those days. Having been "dead" for more than nine minutes, George Ritchie regained consciousness, to the immense surprise of the doctor and nurse. It emerged that during his spell of unconsciousness, the period in which he had been pronounced dead, he had had an extremely powerful experience of which he could recollect a great many details. At first he was unable and afraid to talk about it. Later he wrote a book about what happened to him in those nine minutes. And after graduating as a psychiatrist, he began to share his experiences in lectures to medical students. One of the students attending these lectures was Raymond Moody, who was so intrigued by this story that he began to look into experiences that may occur during life-threatening situations. In 1975 he wrote the book *Life After Life,* which became a global best-seller. In this book Moody first coined the term *near-death experience* (NDE).[2]

After reading Ritchie's book, I kept wondering how somebody can possibly experience consciousness during cardiac arrest and whether this is a common occurrence. So in 1986 I began to systematically ask all the patients at my outpatient clinic who had undergone resuscitation whether they had any recollection of the period of their cardiac arrest. I was more than a little surprised to hear, within the space of two years, twelve reports of such a near-death experience among just over fifty cardiac arrest survivors. Since that first time in 1969, I had not heard any other such reports. I had not inquired after these experiences either because I had not been open to them. But all the reports I was hearing now roused my curiosity. After all, according to current medical knowledge it is impossible to experience consciousness when the heart has stopped beating.

During cardiac arrest patients are clinically dead. Clinical death is defined as a period of unconsciousness caused by a lack of oxygen to the brain because either circulation or breathing or both have stopped. If no resuscitation takes place, the brain cells will suffer irreparable damage within five to ten minutes and the patient will nearly always die, even if the cardiac rhythm is reestablished later.

Questions About Brain Function and Consciousness

For me it all started with curiosity—with asking questions, with seeking to explain certain objective findings and subjective experiences. Learning about near-death experience raised a number of fundamental questions for me. An NDE is a special state of consciousness that occurs during an imminent or actual period of physical, psychological, or emotional death. How and why does an NDE occur? How does the content of an NDE come about? Why does an NDE bring about such profound changes in someone's life? I was unable to accept some of the answers to these questions because they seemed incomplete, incorrect, or unsubstantiated. I grew up in an academic environment where I was taught that there is a reductionist and materialist explanation for everything. And up until that point, I had always accepted this as indisputably true.

After immersing myself in the personal, psychological, social, and scientific aspects of near-death experience, I found other frequently asked questions becoming important to me too: Who am I? Why am I here? What is the origin of my life? When and how will my life end? And what does death mean to me? Will my life go on after death? In all times and all cultures and during every phase of life—among them the birth of a child or grandchild and confrontations with death and other serious crises—these essential questions are asked again and again. You may have asked them yourself. Yet we seldom receive satisfactory answers. Whatever happens in our lives—whether we meet with success or disappointment, no matter how much fame, power, or wealth we acquire—death is inescapable. Everything we gather around us will perish within the not-too-distant future. Birth and death are realities during every single second of our lives because our bodies undergo a constant process of death and renewal.

Some scientists do not believe in questions that cannot be answered, but they do believe in wrongly formulated questions. In 2005 the journal *Science* published a special anniversary issue featuring 125 questions that scientists have so far failed to answer.[3] The most important unanswered question, What is the universe made of? was followed by, What is the biological basis of consciousness? I would like to reformulate this second question as follows: Does consciousness have a biological basis at all? We can also distinguish between temporary and timeless aspects of our consciousness. This prompts the following question: Is it possible to speak of a beginning of our consciousness, and will our consciousness ever end?

In order to answer these questions, we need a better understanding of the relationship between brain function and consciousness. We will have to find out if there is any indication that consciousness can be experienced during sleep, general anesthesia, coma, brain death, clinical death, the process of dying and, finally, after confirmed death. If the answer to any of these questions is yes, we must try to find scientific explanations and analyze the relationship between brain function and consciousness in these situations. This raises a series of other questions that will be addressed in this book:

- Where am I when I *sleep?* Can I be aware of anything during sleep?
- Sometimes there are indications of consciousness under *general anesthesia*. How is it possible that some patients under general anesthesia can later describe exactly what was being said or even done, usually at the moment when they suffered complications during surgery.
- Can we speak of consciousness when a person is in a *coma?* A recent article in *Science* looked at the scientific evidence of awareness in a patient in a *vegetative state*.[4] This is a form of coma with spontaneous breathing and brain-stem reflexes. Brain tests showed that when this patient was instructed to imagine certain activities like playing tennis or moving around her home, the

monitors recorded changes identical to those in healthy volunteers who carried out the same instructions. This means that the identified changes can be explained only by assuming that this patient, despite her vegetative state, not only understood the verbal instructions but also carried them out. The research demonstrated that this coma patient was aware of both herself and her surroundings but that her brain damage prevented her from communicating her thoughts and emotions directly to the outside world. In her book *Uit coma* (Out of Coma), Alison Korthals Altes also describes seeing staff and family in and around the intensive care unit during her three-week coma following a serious traffic accident.[5]

- Can we still speak of consciousness when a person has been pronounced *brain-dead*? In his book *Droomvlucht in coma* (Dream Flight in Coma), Jan Kerkhoffs tells us about his conscious experiences after neurologists declared him brain-dead following complications during brain surgery. Only because his family refused organ donation was he able to write about his experiences because, much to everybody's surprise, he regained consciousness after three weeks in a coma.[6]
- Does brain death really equal death, or does it mark the start of a *process of dying* that can last anywhere between hours and days? What happens to our consciousness during this process of dying?
- Does *clinical death* equal loss of consciousness? Many of the reports of near-death experience covered in this book suggest that during a cardiac arrest, that is, during a period of clinical death, people may experience an exceptionally lucid consciousness.
- Can we still speak of consciousness when a person is *confirmed dead* and the body is cold? I will look more closely at this question below.

Is There Consciousness After Death?

Can research into near-death experiences give us any indication of what happens to consciousness when a person is confirmed dead? We must start by exploring answers to the question whether and how consciousness may be experienced after death. How can we surmise what happens to our consciousness when we are dead? And where do our ideas about death come from? Why would we want to learn more about death, about the meaning of being dead?

The confrontation with death raises urgent questions because death remains a taboo in our society. Yet it is normal for people to die every day. Today, as you are reading this, approximately 6,925 people die in the United States (375 in the Netherlands and 1,400 in the United Kingdom). This means that more than 2,530,000 people die in the United States each year (155,000 people in the Netherlands and 509,000 people in the United Kingdom). Worldwide, more than 70 million people die every year. However, because global birth rates exceed mortality rates, the global population continues to grow. On average, every day in the United States about 11,000 babies are born (515 babies in the Netherlands and 1,600 in the United Kingdom). Dying is just as normal as being born. And yet death has been banished from our society. People increasingly die in hospitals and care homes although there is a growing preference for dying at home or in a hospice.

What is death, what is life, and what happens when I am dead? Why are most people so afraid of death? Surely death can be a release after a difficult illness? Why do doctors often perceive the death of a patient as a failure on their part? Because the patient lost his or her life? Why are people no longer allowed to "just" die of a serious, terminal illness but instead are put on a ventilator and given artificial feeding through tubes and drips? Why do some people in the final stages of a malignant disease opt for chemotherapy, which may prolong life for a short while but certainly does not always improve the quality of their remaining life? Why is our first impulse to prolong life and delay death at all costs? Is fear of death the reason why? And does this fear stem from ignorance of what death might be? Are our ideas about death accurate at all? Is death really the end of everything?

Even medical training pays scant attention to what death might be. By the time they graduate, most doctors have not given death much thought. Throughout life 500,000 cells in the body die every second, 30 million every minute, and 50 billion every day. These cells are all replaced again on a daily basis, giving a person an almost entirely new body every couple of years. Cell death is therefore not the same as physical death. In life, our bodies change constantly from one second to the next. Yet we neither feel nor realize it. How do we explain the continuity of this constantly changing body? Cells are building blocks comparable to the building blocks of a house, but who designs, plans, and coordinates the construction of a house? Not the building blocks themselves. So the obvious question is: What explains the construction and coordination of the ever-changing body from one second to the next?

All bodies function the same on a biochemical and physiological level, yet all people are different. The cause of this difference is not just physical. People have different characters, feelings, moods, levels of intelligence, interests, ideas, and needs. Consciousness plays a major role in this difference. This raises the question: do we human beings *equal* our bodies, or do we *have* bodies?

Just over 50 percent of the population of the Netherlands is relatively confident that death is the end of everything. These people believe that the death of our bodies marks the end of our identities, our thoughts, and our memories, and that death is the end of our consciousness. In contrast, approximately 40 to 50 percent of Dutch people believe in some form of afterlife. In the United States between 72 percent (male 67 percent and female 76 percent) and 74 percent of people believe in life after death. In the United Kingdom about 58 percent believe in an afterlife.[7] Yet many people never ask themselves whether their ideas about death are actually correct—until they are confronted with their own mortality after a death, serious accident, or life-threatening illness in their family or close circle of friends.

By studying everything that has been thought and written about death throughout history—in all times, cultures, and religions—we may be able to form a different, better picture of death. But the same can be achieved by studying recent scientific research into near-death experience. Evidence has shown that most people lose all fear of death after an NDE. Their experience tells them that death is not the end of everything and that life goes on in one way or another. One patient wrote to me after his NDE,

I'm not qualified to discuss something that can only be proven by death. However, for me personally this experience was decisive in convincing me that consciousness endures beyond the grave. Dead turned out to be not dead, but another form of life.

According to people who have had an NDE, death is nothing other than a different way of being with an enhanced and broadened consciousness, which is everywhere at once because it is no longer tied to a body.

The Role of Science in the Study of Consciousness

According to the philosopher of science Ilja Maso, most scientists employ the scientific method based on materialist, mechanistic, and reductionist assumptions. It attracts most of the funding, achieves the most striking results, and is thought to employ the brightest minds. The more a vision deviates from this materialist paradigm, the lower its status and the less money it receives. Indeed, experience shows us that the upper echelons of the research hierarchy receive a disproportionate percentage of funding, whereas the lower echelons actually address the condition, needs, and problems of people. True science does not restrict itself to materialist and therefore restrictive hypotheses but is open to new and initially inexplicable findings and welcomes the challenge of finding explanatory theories. Maso speaks of an *inclusive science,* which can accommodate ideas that are more compatible with our attempts to learn about subjective aspects of the world and ourselves than the materialist demarcation currently allows.[8]

The psychologist Abraham H. Maslow offered a fine definition of what such an inclusive science should entail:

The acceptance of the obligation to acknowledge and describe all of reality, all that exists, everything that is the case. Before all else science must be comprehensive and all-inclusive. It must accept within its jurisdiction even that which it cannot understand or explain, that for which no theory exists, that which cannot be measured, predicted, controlled, or ordered. It must accept even

*contradictions and illogicalities and mysteries, the vague, the am-
biguous, the archaic, the unconscious, and all other aspects of existence that are difficult to communicate. At best it is completely
open and excludes nothing. It has no "entrance requirements."*[9]

The American philosopher of science Thomas Kuhn claimed that most scientists are still trying to reconcile theory and facts within the routinely accepted (materialist) paradigm, which he describes as essentially a collection of articles of faith shared by scientists.[10] All research results that cannot be accounted for by the prevailing worldview are labeled "anomalies" because they threaten the existing paradigm and challenge the expectations raised by this paradigm. Needless to say, such anomalies are initially overlooked, ignored, rejected as aberrations, or even ridiculed. Near-death experiences are such anomalies. Anomalies offer the chance of modifying existing scientific theories or replacing them with new concepts that do offer an explanation. But it is rare for new concepts to be received and accepted with enthusiasm when they do not fit the prevailing materialist paradigm. The words of psychiatrist Ian Stevenson still ring true: "It's been said that there's nothing so troublesome as a new idea, and I think that's particularly true in science."

Most of the people who specialize in consciousness research, including neuroscientists, psychologists, psychiatrists, and philosophers, are still of the opinion that there is a materialist and reductionist explanation for consciousness. The well-known philosopher Daniel Dennett believes, and many with him, that consciousness is nothing other than matter and that our subjective experience of our consciousness as something purely personal and different from somebody else's consciousness is merely an illusion.[11] According to these scientists, consciousness arises entirely from the matter that constitutes our brain. If this were true, then everything we experience in our consciousness would be nothing other than the expression of a machine controlled by classical physics and chemistry, and our behavior would be the inexorable outcome of nerve-cell activity in our brain. Of course the notion that all subjective thoughts and feelings are produced by nothing other than the brain's activity also means that free will is an illusion. This viewpoint has enormous implications for concepts such as moral responsibility and personal freedom.

The Need for a New Approach

If you wish to upset the law that all crows are black . . . it is enough if you prove one single crow to be white.
—WILLIAM JAMES

When empirical scientific studies discover phenomena or facts that are *not* consistent with current scientific theories, these new facts must not be denied, suppressed, or even ridiculed, as is still quite common. In the event of new findings, the existing theories ought to be elaborated or modified and if necessary rejected and replaced. We need new ways of thinking and new forms of science to study consciousness and acquire a better understanding of the effects of consciousness. Some scientists, such as the philosopher David Chalmers, are more receptive and take consciousness seriously: "Consciousness poses the most baffling problems in the science of the mind. There is nothing that we know more intimately than conscious experience, but there is nothing that is harder to explain." Chalmers specializes in the problem of consciousness and has written an excellent overview of the various theories on the relationship between consciousness and the brain.[12] I will look at this overview in more detail in a later chapter.

In the past new forms of science emerged when prevailing scientific ideas could no longer explain certain phenomena. At the start of the twentieth century, for instance, quantum physics emerged because certain findings could not be accounted for with classical physics. Quantum physics overturned the established view of our material world. The fact that the new insights provided by quantum physics are being accepted only slowly can be attributed to the materialist worldview with which most of us grew up. According to some quantum physicists, quantum physics accords our consciousness a decisive role in creating and experiencing perceptible reality. This not-yet-commonly-accepted interpretation posits that our picture of reality is based on the information received by our consciousness. This transforms modern science into a subjective science in which consciousness plays a fundamental role. The quantum physicist Werner Heisenberg formulated it as follows:

> *Science no longer is in the position of observer of nature, but rather recognizes itself as part of the interplay between man and nature. The scientific method . . . changes and transforms its object: the procedure can no longer keep its distance from the object.*[13]

The experience of certain aspects of consciousness during an NDE is comparable, or analogous, to concepts from quantum physics. Of course quantum theory cannot explain consciousness, but in conjunction with the results and conclusions from NDE research it can contribute to a better understanding of the transition or interface between consciousness and the brain.

Science Equals Asking Questions with an Open Mind

In my opinion, current science must reconsider its assumptions about the nature of perceptible reality because these ideas have led to the neglect or denial of important areas of consciousness. Current science usually starts from a reality that is based solely on perceptible phenomena. Yet at the same time we can (intuitively) sense that besides objective, sensory perception there is a role for subjective aspects such as feelings, inspiration, and intuition. Current scientific techniques cannot measure or demonstrate the content of consciousness. It is impossible to produce scientific evidence that somebody is in love or that somebody appreciates a certain piece of music or a particular painting. The things that can be measured are the chemical, electric, or magnetic changes in brain activity; the content of thoughts, feelings, and emotions cannot be measured. If we had no direct experience of our consciousness through our feelings, emotions, and thoughts, we would not be able to perceive it.

Moreover, people must appreciate that their picture of the material world is derived from and constructed solely on the basis of their own perception. There is simply no other way. All of us create our own reality on the basis of our consciousness. When we are in love the world is beautiful, and when we are depressed that very same world is a torment. In other words, the material, "objective" world is merely the picture constructed in our consciousness. People thus preserve their own worldview. This is precisely the kind of idea that a large part of the scientific community has difficulty accepting.

Endless Consciousness

On the basis of prospective studies of near-death experience, recent results from neurophysiological research, and concepts from quantum physics, I strongly believe that consciousness cannot be located in a particular time and place. This is known as nonlocality. Complete and endless consciousness is everywhere in a dimension that is not tied to time or place, where past, present, and future all exist and are accessible at the same time. This endless consciousness is always in and around us. We have no theories to prove or measure nonlocal space and nonlocal consciousness in the material world. The brain and the body merely function as an interface or relay station to receive part of our total consciousness and part of our memories into our waking consciousness. Nonlocal consciousness encompasses much more than our waking consciousness. Our brain may be compared both to a television set, receiving information from electromagnetic fields and decoding this into sound and vision, and to a television camera, converting or encoding sound and vision into electromagnetic waves. Our consciousness transmits information to the brain and via the brain receives information from the body and senses. The function of the brain can be compared to a transceiver; our brain has a facilitating rather than a producing role: it enables the experience of consciousness. There is also increasing evidence that consciousness has a direct effect on the function and anatomy of the brain and the body, with DNA likely to play an important role.

Near-death experience prompted the concept of a nonlocal and endless consciousness, which allows us to understand a wide range of special states of consciousness, such as mystical and religious experiences, deathbed visions (end-of-life experiences), perimortem and postmortem experiences (nonlocal communication), heightened intuitive feelings (nonlocal information exchange), prognostic dreams, remote viewing (nonlocal perception), and the mind's influence on matter (nonlocal perturbation). Ultimately, we cannot avoid the conclusion that endless consciousness has always been and always will be, independently of the body. There is no beginning and there will never be an end to our consciousness. For this reason we ought to seriously consider the possibility that death, like birth, may be a mere passing from one state of consciousness into another and that during life the body functions as an interface or place of resonance.

The Near-Death Experience: Bridging Science and Spirituality

I hope that readers will approach this book with empathy and without prejudice. By making a scientific case for consciousness as a nonlocal and thus ubiquitous phenomenon, this book can contribute to new ideas about consciousness in relation to the brain. I am aware that this book can be little more than a springboard for further study and debate because we still lack definitive answers to the many important questions about our consciousness and the relationship between consciousness and the brain. No doubt many questions about consciousness and the mystery of life and death will remain unanswered. Nevertheless, when faced with exceptional or abnormal findings, we must question a purely materialist paradigm in science. A near-death experience is such an exceptional finding. Although consciousness remains a huge mystery, new scientific theories based on NDE research appear to be making a major contribution to the search for answers. It looks as if a single anomalous finding that defies explanation with commonly accepted concepts and ideas is capable of bringing about a fundamental change in science.

I suspect that reading this book will raise many questions. I am aware that some topics in this book may be new or even unimaginable to many readers, especially those who have never heard or read anything about near-death experiences. But the hundreds of thousands of people who have experienced an NDE will likely be relieved to learn that others have had similar experiences that are being explored scientifically.

An NDE is both an existential crisis and an intense learning experience. People are transformed by the conscious experience of a dimension where time and distance play no role, where past and future can be glimpsed, where they feel complete and healed, and where they can experience unlimited wisdom and unconditional love. These transformations are primarily fueled by the insight that love and compassion for oneself, others, and nature are important conditions of life. Following an NDE, people realize that everything and everybody are connected, that every thought has an impact on oneself and others, and that our consciousness survives physical death. People realize that death is not the end.

People with near-death experiences have been my greatest teachers. My many conversations with them and my in-depth study of the potential significance of an NDE have changed my views on the meaning of life and death. There is much to learn from the insights acquired through an NDE. We do not need our own near-death experience to gain new insights into life and death.

The acceptance of new scientific ideas in general and ideas about endless consciousness in particular requires us to have an open mind and to abandon dogma. And of course this extends beyond science to include all topical issues in contemporary Western society. As we open our minds to universal questions about life, death, and consciousness, our view of humanity may undergo a profound transformation. I sincerely hope that this book can make a positive contribution to this process.

A detailed report of an NDE and its impact on life can be found in chapter 1. Following a brief historical overview of the first scientific NDE studies, chapter 2 features a comprehensive account of the twelve universal NDE elements, illustrated with striking quotes. In chapter 3 I discuss the positive life changes people report after an NDE during a cardiac arrest that lasted only a few minutes. The many problems of coming to terms with the experience are also dealt with in this chapter. Regrettably, people with an NDE are still too often dismissed as dreamers, fantasists, attention seekers, or confused patients. Chapter 4 focuses on near-death experiences in children because it seems highly unlikely that their experiences could be the result of any outside influence. Young children recall the same NDE elements as adults and are also noticeably different from their contemporaries after an NDE. In chapter 5 I cite historical writings from Europe and Asia to show that experiences of an enhanced and endless consciousness and the idea of consciousness after physical death are not new but feature prominently in these writings.

All existing scientific explanations for an NDE are reviewed in chapter 6. A satisfactory theory that explains all the different aspects of the NDE must consider both the various circumstances under which an NDE can be experienced and the distinct elements that constitute an NDE. In chapter 7 I focus on the Dutch NDE study among 344 cardiac arrest survivors and compare its results and conclusions with those of comparable studies from the United States and the United Kingdom.[14] These four prospective studies all concluded that the reported NDE elements were experienced during the period of cardiac arrest, that is, during the complete loss of blood flow to the brain. How was this possible? Chapter 8 contains a detailed description of what happens in the brain in the event of acute lack of oxygen precipitated by the loss of the heartbeat and blood pressure. Complementing this, chapter 9 looks more closely at normal brain function and the limitations of our current scientific ideas about the relationship between the brain and consciousness.

As an interlude between the preceding, predominantly descriptive chapters and the subsequent, more analytical chapters, chapter 10 features a comprehensive report of two NDEs that a woman named Monique Hennequin underwent several years ago.

Chapter 11 explains the concepts and insights from quantum physics that may contribute to a better understanding of consciousness. In chapter 12 I draw on a theoretical overview to consider the relationship between the brain and consciousness and put forward some ideas for a possible scientific explanation. New insights into DNA's potential role in the continuous changes to our bodies are discussed in chapter 13. It is possible that DNA acts as the interface between nonlocal consciousness and the body and plays a role in the coordination of cells, cell systems, organs, and the organism as a whole. Chapter 14 focuses on the different aspects of nonlocal or endless consciousness, many of which have been demonstrated by empirical scientific research.

Some of the implications of NDE and nonlocal consciousness in relation to ethical, medical, and social issues in our predominantly materialist Western society are reconsidered in chapter 15. In the epilogue the concept of nonlocal consciousness and its consequences for science, health care, and our image of humankind is summarized. Finally, in the appendix I stress that knowledge about near-death experiences can be of great practical significance to health care practitioners and to dying people and their families. Everybody ought to be aware of the extraordinary experiences that may occur during a period of clinical death or coma, on a deathbed, or after death.

Chapter One

A Near-Death Experience and Its Impact on Life

Here is a test to find whether your mission on earth is finished: if you're alive, it isn't.
—RICHARD BACH

I want to begin this book with a report that is typical of a near-death experience (NDE) and of the difficult process of coming to terms with the experience afterward. This NDE was precipitated by serious complications during the delivery of a child.

On September 23, 1978, I get my first contractions. At that point I am nine months pregnant with, as we later learn, our second daughter. My entire pregnancy has been a textbook case. After some time my husband and I join the midwife and go to the hospital. I'm wheeled into the delivery room. The midwife regularly listens to the child's heartbeat through the large wooden horn [a natural stethoscope]. The waters are broken. The delivery room becomes extremely quiet. People are rushing around and talking to one another in soft yet urgent voices. When I ask what's happening, neither I nor my husband receives a reply. The contractions stop, but I'm feeling fine. Meanwhile the gynecologist has joined us, along with some more nurses. We have no idea what's happening. I'm told to start pushing. "But I have no contractions!" This doesn't seem to matter. There's a rattling of tongs, scissors, trays, and tissues. My husband passes out and is pulled out of the delivery room and left in the corridor.

Suddenly I realize that I'm looking down at a woman lying on a bed with her legs in supports. I see the nurses and doctors panicking, I see a lot of blood on the bed and on the floor, I see large hands pressing down hard on the woman's belly, and then I see the woman giving birth to a child. The child is immediately taken to another room. The nurses look dejected. Everybody is waiting. My head is knocked back hard when the pillow is suddenly pulled away. Once again, I witness a great commotion. Swift as an arrow, I fly through a dark tunnel. I'm engulfed by an overwhelming feeling of peace and bliss. I feel intensely satisfied, happy, calm, and peaceful. I hear wonderful music. I see beautiful colors and gorgeous flowers in all colors of the rainbow in a large meadow. At the far end is a beautiful, clear, warm light. This is where I must go. I see a figure in a light garment. This figure is waiting for me and reaches out her hand. It feels like a warm and loving welcome. Hand in hand, we move toward the beautiful and warm light. Then she lets go of my hand and turns around. I feel something pulling me back. I notice a nurse slapping me hard on my cheeks and calling my name.

After some time I realize where I am and I know that my child isn't well. Our daughter is no longer alive. This return hurts so much! I long to go back to—indeed, where to? The world goes on turning.

My NDE was caused by blood loss during the delivery. Initially, this blood loss went largely unnoticed by the nurses. Everybody must have been too focused on the birth of the child. They only intervened at the last moment by pulling the pillow from under my head, giving me blood and . . . I didn't see any more. By then I had reached the heavenly paradise.

When I returned from this beautiful world, this amazing experience, my reception here in this world was cold, frosty, and above all loveless. The nurse with whom I tried to share my beautiful experience dismissed it by saying that I would soon receive medication to help me sleep and then it would all be over. All over? I didn't want that. I didn't want it to be over at all. I wanted to go back. The gynecologist told me that I was still young and that I could have plenty more children; I should move on and look forward to the future.

I stopped telling my story. It was difficult enough to find words for my experience because how could words express what I had experienced? But what else could I do? Who could I talk to? What was the matter with me? Had I gone crazy?

The only person I could tell my story to, over and over again, was my husband. He listened and asked questions even though he didn't understand what had happened to me, what this experience meant or what it was called, and whether I was the only person with such an experience. I was, and still am today, delighted that he was such a good listener. My NDE didn't jeopardize our relationship. And I know now that this is very precious indeed. Speaking of unconditional love! But I did feel like I was the only person who had experienced such a thing. Nobody asked me anything; nobody was interested. To be fair, my situation made it harder because how do you react when you expect a birth announcement and you receive a death notice instead? For many people that's hard enough, even without having to listen to an experience like mine.

During that time I lived like an automaton. Although I looked after my husband and our eldest daughter and I walked the dog, my mind was elsewhere. My mind was on my experience. How could I reconnect with it? Where could I hear such beautiful music, see such lovely colors, find such gorgeous flowers, see such a dazzling light, experience so much unconditional love? And was I crazy for thinking these things? What was the matter with me?

In my undergraduate dissertation I proposed the following key recommendation for health care practitioners: "I would have been immensely grateful for only 1 percent of all the advice now found in books and articles on NDE!" In 1978 support was clearly not of the same high standard as it is today, but apart from regular nurses, the gynecologist, and the midwife, I didn't see anybody. The family doctor never came to see me, not even after a couple of weeks. He never got in touch with me. Did he assume I was doing okay? I didn't go and see him either; after all, what could I have told him? I had come to the conclusion that my experience was abnormal and that it was better to keep silent. My checkups at the gynecologist revealed no irregularities. On a mechanical level I was still functioning fine, and that was all that mattered. No further questions were asked.

So I kept silent.

I spent years dedicated to a silent search. When I finally found a book in the library that mentioned an NDE, I could scarcely imagine that I'd had such an experience. Surely not? I had stopped believing myself. Only very, very gradually did I find the courage and the strength to believe myself, to trust my experience, and to start accepting and integrating it into my life. It wasn't easy. Over the years I had developed a fairly successful survival strategy, or rather a flight strategy. I fled from my feelings, and I fled from myself. I had taken on more and more work. I had also thrown myself into sports—running, of all things. How symbolic! I was running away from myself and from my NDE. Initially this worked out well, also in the eyes of the world: I often found myself clutching flowers on the winners' podium. But these weren't the flowers I was looking for. I struggled to accept the opinions of others, of colleagues. My inner conflicts—between what I knew and what I felt—intensified. Everything became increasingly difficult.

Then my body intervened. From being overworked and stressed, from what I felt was a burnout, I slipped into a depression. I received treatment from a psychologist who worked in the homeopathic tradition. There's no such thing as coincidence. He was the first health care practitioner who listened to my story, to my experience. He believed it and even considered it normal! But this was more than twenty years after my NDE! He told me to sketch the experience or write it down, to actively engage with it. With his help I made a fascinating journey into my inner self. Everything was accepted and considered normal. Now I realize that I'm not crazy but that my NDE has changed me. This is why my fear of death has disappeared completely, in marked contrast with the years prior to my NDE, years in which I struggled with death and with the fear of death. This is why I have difficulties with the concept of time. These days I constantly lose track of time, whereas before I lived by the clock. Material things aren't important to me. The only thing that matters to me is unconditional love. And this is what I've always had with my husband. Yet recently I read in a study that unconditional love between human beings is an illusion. And they refuse to believe me! This is why I feel like an outsider sometimes. This is why I'm always, especially during vacations, on the lookout for landscapes, for colors and flowers that I've seen but can't find again. This is why I can't stand arguments—I want to go back to those peaceful surroundings. As a matter of fact, I'm incapable of arguing.

Having made my inner journey to where I am now, I'm glad that I had my NDE. I accept it as a beautiful experience, as something that gives me peace and allows me to be myself—a self that includes my experience. I can now enjoy life, with my experience. Integrating my NDE can only make this world a better place. It was only when I began to accept and integrate my NDE that I was able to take some pleasure in life again. My thoughts and feelings are relevant after all; they are neither strange nor crazy. I need them to cut through the chaos to carve out my own identity amid the masses. Of course I still face the task of raising awareness of NDE among people and especially health care practitioners. Having carried out a small-scale study among family doctors in my hometown, I was disappointed to learn that many of them still don't know what to do when somebody has an NDE.

But the most important thing for me now is that I can be who I am, with my experiences. I am who I am, no more, but certainly no less either! And that's good.

E. M.

Chapter Two

What Is a Near-Death Experience?

It is worth dying to find out what life is.
—T. S. ELIOT

Throughout history, in all times and cultures, people have been known to remember an extraordinary experience after a life-threatening crisis.[1] This is now known as a near-death experience. In my definition, a near-death experience (NDE) is the (reported) recollection of all the impressions gained during a special state of consciousness, which includes some specific elements such as witnessing a tunnel, a light, a panoramic life review, deceased persons, or one's own resuscitation. This special state of consciousness can occur during a cardiac arrest, that is, during a period of clinical death, but also in the course of a serious illness or without any apparent medical indication. The experience nearly always brings about fundamental and lasting changes in people's attitude to life and a loss of the fear of death. Because the experience is highly subjective and lacks any frame of reference, other factors, such as individual, cultural, and religious perceptions, determine the way it is described and interpreted. A child will use different words than an adult while a Christian description or interpretation will differ from that of a Buddhist or atheist.

To my mind, the American NDE researcher and professor of psychiatry and neurobehavioral sciences Bruce Greyson has formulated another good definition: "Near-death experiences are profound psychological events with transcendental and mystical elements, typically occurring to individuals close to death or in situations of intense physical or emotional danger."[2]

Professor Janice Holden, recent chair of the International Association of Near-Death Studies (IANDS), writes, "Near-death experiences are the reported memories of extreme psychological experiences with frequent "paranormal," transcendental, and mystical elements, which occur during a special state of consciousness arising during a period of real or imminent physical, psychological, emotional, or spiritual death, and these experiences are followed by common aftereffects."[3]

Circumstances That May Prompt an NDE Report

In the past these experiences were often known under different names, such as visions or mystical, religious, or enlightenment experiences.[4] In antiquity they were referred to as journeys to the underworld. The term *near-death experience* is confusing because the experiences are reported not just by people on the brink of death but also by those who are not in any physical or psychological danger. I will come back to this in more detail later. Although the experiences may occur under a range of different circumstances, they are most frequently reported after a period in which brain function is seriously impaired, such as in cardiac arrest. Other comparable clinical circumstances include brain damage and coma after a serious traffic accident or brain hemorrhage, unconsciousness through shock (low blood pressure) caused by severe blood loss during or after a delivery, or following complications during surgery. Near-drowning is a well-known NDE trigger in children. Other causes include asphyxiation and serious but not immediately life-threatening illnesses with high fever. These experiences are also reported during isolation, depression, or without any clear medical indication during walks in the countryside or during meditation.

Identical experiences, called *fear-death experiences,* are caused by acute fear of death and are reported after a seemingly inevitable death, such as a traffic accident or mountain-climbing accident. Similar experiences occur during the process of dying, when they are known as *deathbed visions* or *nearing-death awareness*.

The fact that an NDE can occur under very different circumstances, such as during a coma caused by a cardiac arrest, under general anesthesia, but also without any clear medical indication, is important for the study of its potential causes. The frequently cited argument that an NDE is caused by oxygen deficiency in the brain obviously does not apply to people who experience one during depression or isolation. The experience of a very lucid consciousness at a moment when all brain function has ceased raises important questions about the relationship between consciousness and the brain.

The Incidence of Near-Death Experiences

Better chances of survival due to improved resuscitation techniques and treatment options in recent years have prompted a rise in NDE reports. Many thousands of people are known to have had a near-death experience, but estimates put the number of people who have had one in the past fifty years at more than 25 million worldwide. Fairly recent studies in the United States and Germany suggest that approximately 4.2 percent of the population has reported an NDE. Based on these data, and the assumption that the Dutch population is comparable to its American and German counterparts, it is possible to deduce a similar percentage for the Netherlands and other countries. In other words, according to this research, about 600,000 people in the Netherlands, 2 million people in the United Kingdom, and more than 9 million people in the United States have had an NDE.[5]

Given this large number, the experience must be quite common during life-threatening clinical circumstances. But underreporting of NDEs is thought to be widespread in hospitals. A possible explanation for this is that the phenomenon confounds current medical knowledge. Seeing NDEs as implausible and inexplicable, most doctors and other health care practitioners are not open to them and rarely come across direct accounts of them. Research has also shown that most patients remain silent about their near-death experience because nobody believes them when they first try to talk about it.[6] This was confirmed during an NDE conference at an American university hospital in 1994 attended by some three hundred people. After a few presentations on NDE and somebody's personal story, a man got up and said, "I've worked as a cardiologist for twenty-five years now, and I've never come across such absurd stories in my practice. I think this is all complete nonsense; I don't believe a word of it." Whereupon another man stood up and said, "I'm one of your patients. A couple of years ago I survived a cardiac arrest and had an NDE, and you would be the last person I'd ever tell."

Their medical training makes it hard for doctors to accept NDEs, even when they themselves or a colleague experience one. After a cardiac arrest and emergency bypass surgery, a family doctor wrote to me, "I always thought that these things didn't really exist." Another family doctor, who suffered a cardiac arrest following complications during coronary angioplasty (the inflation of a small balloon in a coronary artery) and who tried to discuss it with colleagues later, wrote to me, "I encountered nothing but skepticism among cardiologists, although they always remained polite. Several years later I had a checkup with an internist, who also ignored my story. I was unable to share this experience with colleagues and other people."

Moody's NDE Classification

In 1975 psychiatrist Raymond Moody wrote his first book on near-death experience. In it he described twelve different NDE elements but emphasized that most people experience only a few. It is rare for all elements to be reported. Each NDE is unique and is experienced as a coherent episode rather than a series of clearly distinct elements. The order of the reported elements can also vary a little. At this point I should mention a comparative study that looked at whether there is any difference between NDE accounts recorded before and after 1975. The question was whether the publicity that followed the publication of Moody's book had any influence on the contents of an NDE. However, the study showed that all the NDE elements he mentioned were reported just as frequently before as after 1975; the sole exception was the tunnel experience, which was reported a little less often in the past. The NDE and its effect on patients appear to be essentially the same worldwide except for some culture-specific differences in content and interpretation. One intercultural study shows that certain NDE elements, such as the panoramic life review and tunnel experience, are reported less frequently by the indigenous peoples of North America, Australia, and the islands in the South Pacific.[7]

Moody described twelve possible NDE elements and the order in which they are usually reported. These elements will be discussed at length later in this chapter.

1. The ineffability of the experience
2. A feeling of peace and quiet; pain is gone
3. The awareness of being dead, sometimes followed by a noise
4. An out-of-body experience (OBE); from a position outside and above their bodies, people witness their own resuscitation or operation
5. A dark space, experienced by only 15 percent of people as frightening; people are pulled toward a small pinpoint of light in this dark space, which they describe as:
   • A tunnel experience; they are drawn rapidly toward the light
   • A frightening NDE; approximately 1 to 2 percent of people linger in this dark space and experience their NDE as frightening (also known as a hell experience)
6. The perception of an unearthly environment, a dazzling landscape with beautiful colors, gorgeous flowers, and sometimes also music
7. Meeting and communicating with deceased persons, mostly relatives
8. Seeing a brilliant light or a being of light; experiencing complete acceptance and unconditional love and gaining access to a deep knowledge and wisdom
9. The panoramic life review, or review of life from birth: people see their entire life flash before them; there appears to be no time or distance, everything happens at once, and people can talk for days about a life review that lasted only a few minutes
10. The preview or flash forward: people have the impression that they

are witness to part of the life that is yet to come; again, there is no time or distance

11. The perception of a border: people are aware that if they cross this border or limit they will never be able to return to their body

12. The conscious return to the body, accompanied by great disappointment at having something so beautiful taken away

Other NDE Classifications

After Moody identified his twelve NDE elements in 1975, Kenneth Ring, a former professor of psychology and NDE researcher, confirmed them in his own study in 1980 but came up with a new classification, distinguishing five NDE phases. Two years later Michael Sabom, a cardiologist, identified three main types while in 1983 Bruce Greyson developed a classification featuring four components.[8] I would like to look more closely at these different classifications because each researcher made a substantial contribution to the debate by highlighting different aspects of a near-death experience. For someone with an NDE, however, the experience will always be unique and each classification utterly superfluous.

Ring's Five NDE Phases

Generally speaking, the early elements in Moody's classification are reported most often while the later elements occur with steadily decreasing frequency. Ring, however, believed that it was more illuminating to think of the experience as happening in phases, but he stressed that his study's findings are remarkably similar to Moody's results.

Ring starts with the affective phase, which includes *feelings of absolute peace, calm, surrender, and bliss,* with the end of pain. This phase is nearly always experienced as positive and occurs in 60 percent of all cases. For Ring the second phase is *leaving the body* and is reported in 37 percent of all cases. Whereas some people's experience is limited to the sensation of no longer having a body and not feeling any more pain or restrictions, others can actually see their lifeless body and their surroundings from a position outside and above their body.

They can clearly see and hear what is being said. They feel detached from their body and experience themselves as complete and transparent. In the third phase (23 percent), people *arrive in a dark, usually peaceful environment.* Some linger in this phase while others hurtle *through a tunnel toward a nonblinding, exceptionally bright light,* which radiates unconditional love and acceptance. This fourth phase is described by 16 percent. The fifth and final phase (10 percent) consists of *entering another, unearthly dimension of unbelievable beauty,* where people hear beautiful music and occasionally meet deceased friends and family. This is also where the life review and flash forward can take place. People have the greatest difficulty leaving this environment to return to the body.

Sabom's Three NDE Categories

In response to Moody's book, Michael Sabom was challenged by some friends to explain why he, a cardiologist, had never heard of an NDE. Extremely skeptical, he began to ask people who had been resuscitated whether they had any recollection of their period of unconsciousness. To his surprise, it was only a matter of weeks before he heard the story of an astonishing NDE. It inspired him to embark on a study looking mainly at heart patients. The study, in which he identifies three NDE categories, was published in 1982.

Sabom's first category is the *autoscopic* or out-of-body NDE, in which patients experience a separation of mind and body. They may be able to perceive their own resuscitation and surroundings, usually from an elevated position near the ceiling. This category was reported in 53 percent of all cases in his study. Communication with living persons proves to be impossible even though they can be seen and heard quite clearly. Patients move through mental force; as soon as they want to see or hear something and they think about it, they are taken there. In their nonphysical "body" people move straight through other people and walls. The moment of reentry into the physical body usually coincides with the moment of successful resuscitation, which patients sometimes observe. Astonished by the level of detail that patients know about their own resuscitation, hospital staff often respond with disbelief. The detailed resuscitation accounts led Sabom to conclude that the out-of-body experiences with verifiable perception could have occurred only during the cardiac arrest. In order to check whether the descriptions of resuscitations were merely reconstructions based on familiar images from medical TV programs, Sabom also interviewed twenty-five heart patients who had not reported an NDE after their resuscitation. Their descriptions contained at least one aspect that did not correspond with reality, unlike those of NDE patients with an out-of-body experience. At times the latter even recalled very specific and atypical details they could not have known.

The second category covers the description of objects and events that transcend the earthly dimension (*transcendental* characteristics), described by 54 percent of the NDErs as a stay in a dark space or void, accompanied by feelings of peace. This was usually followed by an encounter with a brilliant and nonblinding light at the end of a tunnel, after which people found themselves in an unearthly or heavenly environment. Here approximately half of the NDErs engaged in nonverbal communication with deceased friends or relatives or with a "spiritual entity." Their communication usually centered on the decision to stay or go back to earth, that is, return to the physical body left behind. There were also some reports of a life review or the perception of a border.

The third category included patients who reported a combination of the first and second categories in an uninterrupted sequence of the various elements. This was recorded in approximately 20 percent of the NDEs.

The term *autoscopic,* as used by Sabom, is actually incorrect for an out-of-body experience. In the event of an autoscopy, a patient (usually with psychiatric symptoms) observes a kind of double of the self from the vantage point of his or her own physical body. In the event of an out-of-body experience, however, people see their body, including verifiable details, from a position outside and above the lifeless body.

Greyson's Four NDE Components

From a total of eighty characteristics, Bruce Greyson reduced the constituent parts of the NDE to sixteen elements, grouped into four clusters: cognitive, affective, paranormal, and transcendental.[9] The *cognitive* component includes losing one's sense of time, having accelerated thoughts, the life review and preview, and the possibility of universal knowledge. The *affective* component covers feelings of peace, joy, cosmic unity, and the perception of or inclusion in a brilliant and nonblinding light. *Paranormal* refers to all those phenomena that defy explanation with our normal physical laws and universally accepted concepts. The paranormal component may include a hyperacute auditory and visual sense, the conscious experience of remote events, premonitions and prophetic visions, and an out-of-body experience. *Transcendental* literally means "climbing or going beyond." The transcendental component involves traveling to an unearthly realm, meeting or sensing the presence of a mystical being, seeing and communicating with deceased persons or religious figures, and reaching a border. The affective and transcendental components were reported most frequently, the paranormal and cognitive components less often.

Retrospective Versus Prospective Scientific Studies

There are two scientific approaches to the study of empirical data. *Retrospective* studies feature interviews with people who come forward in response to advertisements, articles, readings, or radio and TV broadcasts. These people present themselves randomly and voluntarily, which means that an NDE study can include people whose experience dates back ten or twenty years and whose medical and other circumstances can no longer be ascertained. We do not know why people volunteer or not. Presumably many people are afraid or unwilling to come forward or else they are unaware of the study. The results of retrospective studies are therefore less reliable. The four best-known and most important NDE researchers, Moody, Ring, Sabom, and Greyson, all based their NDE classification on findings from retrospective studies, although Sabom's study was in part prospective.

In a *prospective* study, researchers approach all consecutive patients with a predefined diagnosis within days of their coma or cardiac arrest and ask them whether they have any memories of the period of unconsciousness. This allows all medical and other data to be carefully recorded and lends a prospective study much greater scientific value. A prospective study is possible only among patients with an objective and life-threatening medical indication. As mentioned, in the past nearly all NDE studies were retrospective, but in recent years some prospective NDE studies among cardiac arrest survivors have been published.

The Depth of an Experience

Definitions are important scientific tools for describing and judging the depth and complexity of an experience. If research subjects report memories of their period of unconsciousness, their NDEs are coded with the help of an index based on the number of reported NDE elements. The more elements are reported, the deeper the NDE and the higher the total score. Some elements are given a higher score than others.

In order to determine the depth of an NDE, Kenneth Ring developed the WCEI score, the Weighted Core Experience Index.[10] In his retrospective study a score of 0 to 6 was deemed too low for the experience to merit the label NDE. An experience scoring between 7 and 9 was called a moderately deep NDE, while a score between 10 and a maximum of 29 was labeled a deep to very deep NDE.

Bruce Greyson adjusted the WCEI scoring system because as well as measuring the depth of the NDE, he wanted to create the possibility of eliminating certain NDE-like (false positive) elements in retrospective studies. This new Greyson Scale provides a better overall picture, is easier to use, and makes it possible to distinguish between NDEs and experiences resulting from brain damage, from other stress responses, or from an altered state of mind caused, for example, by the use of drugs. Greyson uses a scale of 0 to 32, in which a score of 7 or higher marks the cutoff point for genuine NDEs in retrospective studies. The WCEI is best for determining the depth of an NDE while the Greyson Scale is useful for screening a population to identify NDEs.[11]

In both scoring systems, experiences with a score of 6 or lower in retrospective studies are not seen as real NDEs. I am convinced, however, that in the more recent prospective studies, in which all patients are monitored from the moment they regain consciousness or wake from their coma, each reported memory of the period of unconsciousness, even an experience with just a single element (that is, with an extremely low score), merits the label NDE. I say so because in the Dutch study all people with a low score—with a so-called superficial NDE—displayed in later interviews the classic personality changes associated with an NDE, which we will look at in more detail later.

The Twelve NDE Elements, with Some Striking Examples

Moody's classification is a useful starting point for discussing the various aspects of an NDE. I understand full well that this is an artificial division and that while these elements may be distinct, they cannot be isolated because the NDE constitutes a continuous experience. It is a practical division, however, because each of the twelve elements raises different questions about a possible scientific explanation for an NDE. How is it possible for people to observe their own resuscitation from a position above their lifeless body? How can they have clear thoughts and retain their memories without a physical body? How is it possible for them to meet and recognize deceased relatives? How is it possible to experience a life review or a preview in mere minutes, as if time and distance do not exist in this other, unearthly realm?

The various elements will be discussed individually in the order in which they are usually experienced. All elements will be illustrated with typical examples. In some cases I include several quotations to better illuminate the various aspects of the element in question. Most of the quotations are taken from experiences that were shared with me in person or in writing by people I met in the course of my research. I reproduce the experiences anonymously and have made them as unidentifiable as possible. Translations follow as close as possible people's original words, even where this results in somewhat unwieldy language.

1. Ineffability

What happens in a life-threatening situation is often totally unfamiliar and indescribable and lies outside our normal sphere of experience. It is not surprising, therefore, that people run into difficulties when they try to put their experience into words.

"I was there. I was on the other side." For a long time that was all I could say. I still get tears in my eyes thinking about the experience. Too much! It's simply too much for human words. The other dimension, I call it now, where there's no distinction between good and evil, and time and place don't exist. And an immense, intense pure love compared to which love in our human dimension pales into insignificance, a mere shadow of what it could be. It exposes the lie we live in in our dimension. Our words, which are so limited, can't describe it. Everything I saw was suffused with an indescribable love. The knowledge and the messages going through me were so clear and pure. And I knew where I was: where there's no distinction between life and death. The frustration at not being able to put it into human words is immense.

I regret that words can't do my experience justice. I must admit that human language is woefully inadequate for conveying the full extent, the depth, and the other dimension I've seen. In fact, no pen can describe what I went through.

2. A Feeling of Peace and Quiet; the Pain Has Gone

For many people, the overwhelming feelings of peace, joy, and bliss constitute the first and best-remembered element of their experience. The intense pain that usually follows a traffic accident or a heart attack is suddenly completely gone.

And the pain, especially the pressure on my lungs, was gone. The atmosphere made me feel totally relaxed. I'd never felt this happy before.

3. The Awareness of Being Dead

It is often confusing to hear bystanders or doctors declare you dead at a moment when you feel extremely alive and whole. If a sound is heard at this point it is usually a buzzing or whistling sound, sometimes a loud click or a soft murmur.

> The weird thing is that I wasn't at all surprised or anything. I simply thought: Hey, I'm dead now. So this is what we call death.

4. An Out-of-Body Experience

During an out-of-body experience people have verifiable perceptions from a position outside and above their lifeless body. Patients feel as if they have taken off their body like an old coat, and they are astounded that despite discarding it they have retained their identity, with the faculty of sight, with emotions, and with an extremely lucid consciousness.

The out-of-body experience begins with a patient's sensation that his or her consciousness is leaving the physical body but continues to function unchanged. Sometimes this is accompanied by fear, followed by a (futile) attempt to return to the body, but patients often feel liberated and are amazed at the sight of the lifeless or seriously damaged body. The most common vantage point is from the ceiling, and because of this unusual position some people initially fail to recognize their body. People experience their new weightless body as a spiritual or nonphysical body that can penetrate solid structures such as walls and doors. It is impossible to communicate with or touch others who are present. To their utter amazement, people go unnoticed even though they can hear and see everything. The range of vision can extend to three hundred sixty degrees, with simultaneous detailed and bird's-eye views. Blind people too have the faculty of sight while deaf people know exactly what has been said. While this is happening, people discover that all it takes to be near someone is to think of that person.

This out-of-body experience is of scientific importance because doctors, nursing staff, and relatives can check and corroborate the reported perceptions and the moment when they were supposed to have taken place. "In a recent review of 93 reports of potentially verifiable out-of-body perceptions (or 'apparently nonphysical veridical perceptions') during NDE it has been found that 43 percent had been corroborated to the investigator by an independent informant, an additional 43 percent had been reported by the experiencer to have been corroborated by an independent informant who was no longer available to be inteviewed by the investigator, and only 14 percent relied solely on the experiencer's report. Of these out-of-body perceptions, 92 percent were completely accurate, 6 percent contained some error, and only 1 percent was completely erroneous. And even among those cases corroborated to the investigaor by an independent informant, 88 percent were completely accurate, 10 percent contained some error, and only 3 percent were completely erroneous."[12] This proves that an out-of-body experience cannot be a hallucination, which is a sensory perception that is perceived as real by the hallucinating person but that does not correspond with reality. Just like a psychotic episode or the effects of hallucinogenic drugs, a hallucination is not rooted in objective reality. It is neither a delusion, an incorrect interpretation of an actual perception, nor an illusion, an apparent reality or a false sense of reality. This raises the question whether the out-of-body experience may be a form of extrasensory perception.

The scientific importance of out-of-body experiences prompts me to include quite a few, very diverse examples. First up is the account of an out-of-body experience as told to me by a nurse at a coronary care unit. The account, taken from our article in *The Lancet*, was verified by us, and we asked the nurse to write it down as objectively as possible.[13]

During the night shift the ambulance crew brings in a forty-four-year-old cyanotic [purplish-blue skin discoloration], comatose man. About an hour earlier he had been found in a public park by passers-by, who had initiated heart massage. After admission to the coronary care unit, he receives artificial respiration with a balloon and a mask as well as heart massage and defibrillation. When I want to change the respiration method, when I want to intubate the patient, the patient turns out to have dentures in his mouth. Before intubating him, I remove the upper set of dentures and put it on the crash cart. Meanwhile we continue extensive resuscitation. After approximately ninety minutes, the patient has suffi-

cient heart rhythm and blood pressure, but he's still ventilated and intubated, and he remains comatose. In this state he is transferred to the intensive care unit for further respiration. After more than a week in coma the patient returns to the coronary care unit, and I see him when I distribute the medication. As soon as he sees me he says, "Oh, yes, but you, you know where my dentures are." I'm flabbergasted. Then he tells me, "Yes, you were there when they brought me into the hospital, and you took the dentures out of my mouth and put them on that cart; it had all these bottles on it, and there was a sliding drawer underneath, and you put my teeth there." I was all the more amazed because I remembered this happening when the man was in a deep coma and undergoing resuscitation. After further questioning, it turned out that the patient had seen himself lying in bed and that he had watched from above how nursing staff and doctors had been busy resuscitating him. He was also able to give an accurate and detailed description of the small room where he had been resuscitated and of the appearance of those present. While watching this scene, he had been terrified that we were going to stop resuscitating and that he would die. And it's true that we had been extremely negative about the patient's prognosis due to his very poor condition when admitted. The patient tells me that he had been making desperate but unsuccessful attempts at letting us know that he was still alive and that we should continue resuscitating. He's deeply impressed by his experience and says he's no longer afraid of death.

Here is the account of a patient who had an NDE with out-of-body experience caused by complications during surgery:

No, I'd never heard of near-death experiences, and I'd never had any interest in paranormal phenomena or anything of that nature. What happened was that I suddenly became aware of hovering over the foot of the operating table and watching the activity down below around the body of a human being. Soon it dawned on me that this was my own body. So I was hovering over it, above the lamp, which I could see through. I also heard everything that was said:

"Hurry up, you bloody bastard" was one of the things I remember them shouting. And even weirder: I didn't just hear them talk, but I could also read the minds of everybody in the room, or so it seemed to me. It was all quite close, I later learned, because it took four and a half minutes to get my heart, which had stopped, going again. As a rule, oxygen deprivation causes brain damage after three or three and a half minutes. I also heard the doctor say that he thought I was dead. Later he confirmed saying this, and he was astonished to learn that I'd heard it. I also told them that they should mind their language during surgery.

Next up is the account of psychologist Carl G. Jung of his out-of-body experience during his heart attack in 1944. His description of the earth from a great height is remarkable because it is quite consistent with what images from outer space taught us only some forty years ago—decades after Jung's experience.

It seemed to me that I was high up in space. Far below I saw the globe of the earth, bathed in a gloriously blue light. I saw the deep blue sea and the continents. Far below my feet lay Ceylon, and in the distance ahead of me the subcontinent of India. My field of vision did not include the whole earth, but its global shape was plainly distinguishable and its outlines shone with a silvery gleam through that wonderful blue light. In many places the globe seemed colored, or spotted dark green like oxydized silver. Far away to the left lay a broad expanse—the reddish-yellow desert of Arabia; it was as though the silver of the earth had there assumed a reddish-gold hue. Then came the Red Sea, and far, far back—as if in the upper left of a map—I could just make out a bit of the Mediterranean. My gaze was directed chiefly toward that. Everything else appeared indistinct. I could also see the snow-covered Himalayas, but in that direction it was foggy or cloudy. I did not look to the right at all. I knew that I was on the point of departing from the earth.

Later I discovered how high in space one would have to be to have so extensive a view—approximately a thousand miles! The

*sight of the earth from this height was the most glorious thing I
had ever seen.*[14]

This is the story of a woman in a deep coma who was about to be
taken off the ventilator because she had been declared brain-dead by
her treating neurologist. She had no measurable brain activity.

While she was thought to be in a deep coma without any apparent
brain activity, her specialist and husband were having a conversation by her bedside. The specialist predicted that his patient would
be a "vegetable" for the rest of her life and asked the husband to
consider taking her off the equipment that was keeping her alive.
The husband was still hopeful of a recovery, so she was kept on
the ventilator. Several months later the woman woke up, despite
the somber prognosis. It emerged that she had been able to hear
throughout most of her coma and had overheard the conversation
between her doctor and husband about passive euthanasia! She
said how awful this had been and that while she had been trying to
shout that she was still there, that she wanted to live, be with her
husband and children, they were discussing her possible demise.

The account of somebody who is color-blind:

I saw the most dazzling colors, which was all the more surprising
because I'm color-blind. I can distinguish the primary colors, but
pastels all look the same to me. But suddenly I could see them, all
kinds of different shades. Don't ask me to name them because I
lack the necessary experience for that.

Next up is the account of Vicki, a woman who was born blind.
She was born extremely premature in 1951, after a pregnancy of only
twenty-two weeks, and immediately placed in a very primitive incubator and administered 100 percent oxygen. Such a high concentration of
oxygen damages the development of the eyeball and optic nerve, which
doctors were not aware of in the early days of the incubator. Thousands of premature babies who survived such early incubators went

completely blind as a result. Vicki suffered complete atrophy (wither-ing) of the eyeball and optic nerve. The visual cortex, the part of the occipital lobe of the brain that processes light stimuli into images, also fails to develop properly when it receives no light stimuli from the non-functioning eyes and optic nerves.

Vicki's near-death experience is described in Kenneth Ring and S. Cooper's book, and she was also interviewed at length in the BBC documentary *The Day I Died*. In 1973, when Vicki was twenty-two, she was hurled out of her car in a traffic accident. A basal skull fracture and severe concussion left her in a coma, and she had fractured neck and back vertebrae and a broken leg. She caught a brief glimpse of the car wreck from above (as a blind woman she could see and recog-nize the smashed Volkswagen van), and later in the emergency room, where she had been taken by ambulance, she was able to see from a position above her body. In the room where she saw a body on a metal gurney, she also spotted two people and could hear them talking and expressing their concern. It was only when she recognized her wedding ring, which of course she knew only by touch, that she realized that it was her own body. And after she had gone up "through the ceiling," she saw the roof of the hospital and trees.

I've never seen anything, no light, no shadows, no nothing. A lot of people ask me if I see black. No, I don't see black. I don't see anything at all. And in my dreams I don't see any visual impressions. It's just taste, touch, sound, and smell. But no visual impressions of anything.

The next thing I recall I was in Harborview Medical Center and looking down at everything that was happening. And it was fright-ening because I'm not accustomed to see things visually, because I never had before! And initially it was pretty scary! And then I finally recognized my wedding ring and my hair. And I thought: is this my body down there? And am I dead or what? They kept say-ing, "We can't bring her back, we can't bring her back!" And they were trying to frantically work on this thing that I discovered was my body and I felt very detached from it and sort of "so what?" And I was thinking, what are these people getting so upset about?

Then I thought, I'm out of here, I can't get these people to listen to me. As soon as I thought that I went up through the ceiling as if it were nothing. And it was wonderful to be out there and be free, not worry about bumping into anything, and I knew where I was going. And I heard this sound of wind chimes that was the most incredible sound that I can describe—it was from the very lowest to the very highest tones. As I was approaching this area, there were trees and there were birds and quite a few people, but they were all, like, made out of light, and I could see them, and it was incredible, really beautiful, and I was overwhelmed by that experience because I couldn't really imagine what light was like. It's still . . . a very emotional thing when I talk about this . . . because there was a point at which . . . at which I could bring forth any knowledge I wanted to have.[15]

Vicki goes on to explain that in this other world she was welcomed by some acquaintances. As Ring and Cooper point out:

There are five of them. Debby and Diane were Vicki's blind schoolmates, who had died years before, at ages eleven and six, respectively. In life, they had both been profoundly retarded as well as blind, but here they appeared bright and beautiful, healthy and vitally alive. They were no longer children, but, as Vicki phrased it, "in their prime." In addition, Vicki reports seeing two of her childhood caretakers, a couple named Mr. and Mrs. Zilk, both of whom had also previously died. Finally, there was Vicki's grandmother— who had essentially raised Vicki and who had died just two years before this incident. Her grandmother, however, who was further back than the others, was reaching out to hug Vicki.[16]

Vicki's experience concludes with a forced reentry into her body:

And then I was sent back and then I went back into my body and it was excruciatingly painful and very heavy and I remember feeling very sick.[17]

The fact that somebody who has been blind from birth as a result of an atrophied eyeball and optic nerve and who has an undeveloped visual cerebral cortex can nonetheless see people and surroundings raises significant questions. How is it possible that this woman can see, from a position outside and above the body, at a moment when she is in a coma caused by brain damage sustained in a serious traffic accident? She has never been able to see. Besides, she perceives things from a position outside her body. How does she do this? What is responsible for this? How can she be aware of her perceptions during her coma? This is impossible according to current medical knowledge. The stories of Vicki and of other blind people with an NDE are forcing scientists to consider new ideas about the relationship between consciousness and the brain. Vicki's reported observations could not have been the product of sensory perception or of a functioning (visual) cerebral cortex, nor could they have been a figment of the imagination given their verifiable aspects.

5. A Dark Space

People feel like they are pulled rather abruptly into a dark space, which they describe as an enclosed space, a void, or a well. Approximately 15 percent of people experience their stay in this dark space as frightening.

> And then everything went dark, but to my mind I didn't lose consciousness, because my memories are as vivid as ever. . . . As I peered into the dark, the color changed from black to deep blue, not dark, but an intense cobalt blue that leaves you speechless. . . .

> Soon I found myself in a dark space, a kind of tunnel, which didn't seem to end. I couldn't go back, but plowing through it seemed an equally dreadful prospect. Would I ever get out? Or would I suffocate somewhere halfway? You see, there was very little space in this tunnel; it was really tight. After I had spent a long time—more than terrifying—squeezing through this tunnel, a glimmer of light appeared at the end, and after a real struggle I stood, or found myself, in this absolute light, which seemed to envelop me.

The Tunnel Experience

A little pinpoint of light appears in this dark space, and people are often pulled toward it at an incredible speed. They describe it as a tunnel experience.

People move through this dark, occasionally multicolored or spiral-shaped narrow space, sometimes accompanied by visible or invisible beings or by music. They approach the light, which slowly intensifies to become an exceptionally bright but nonblinding light. Eventually people are wholly enveloped by this light and feel completely absorbed by it. This process is coupled with an indescribable feeling of bliss, a sense of unconditional love and acceptance. The journey through the tunnel appears to be a passing from our physical world to another dimension where time and distance no longer play a role. This sensation of moving through a tunnel toward the light has become almost synonymous with near-death experience.

I felt that I was sliding deep down into another state of consciousness. That's to say, my consciousness traveled while my body remained motionless on the bed. I could see my body, but I couldn't feel it. I was being sucked away, as it were. I entered an extremely dark, long, and spiral-shaped tunnel, which struck me as frightening though not unfamiliar at first. I soared through this spiral-shaped funnel, and the further or the higher I got, the lighter it became. The intensity of the light changed to a deep purple/violet. "Above" me I saw an extremely bright, radiant white light. I whirled, floated, toward it.

I felt that I was letting go of my body and rising up. Through the roof. Over the hospital. Everything became smaller, and I began to accelerate. Everything around me was dark, with the exception of several stars that hurled toward me and I noticed their different colors. I had no time to look at anything because I was moving so fast. Things slowed down when I saw that I'd ended up in a kind of hourglass and that I was being "sucked" toward the opening. Then I realized that I wasn't alone because a flow of translucent beings

Jeroen Bosch, Visions of the Afterlife:
Earthly Paradise and Ascension to the Empireum.

was heading the same way as me and another flow was moving in
the opposite direction. When I thought about reincarnation, it later
dawned on me that it could well be this flow. Once I got through
the opening everything began to change. First off, I got this feeling.
It was so emotional that I can't possibly describe it. I was overcome
with a feeling of peace that I'd never known on earth. . . . An overwhelming feeling of love came over me, not the earthly feeling I
was quite familiar with, but something I can't describe. Above me

I saw a bright light, and on my way there I heard beautiful music and I saw colors I'd never seen before. As well as the feelings I just described, I had the impression that this was a different dimension altogether. And if anything was missing it was our earthly conception of time! I had what you might call an enhanced vision of this other dimension. On my way up, toward the all-encompassing light, I saw numerous other "beings" who were also heading there. When I had nearly arrived at this light, a kind of membrane in front of it stopped me from going any further. . . .

All of a sudden I knew that I was dead. This realization struck me as odd. I hovered about twenty feet above my body, which was still on the operating table. I was surrounded by doctors who were talking to one another, but I didn't hear their voices. I also saw my husband waiting on a bench in a darkish room somewhere in the hospital. He was nervous. He was rolling a cigarette. From one moment to the next I found myself flying through a tunnel. It was extremely long, and I flew through it head-first. The tunnel was virtually horizontal, but at a slight upward angle. It was about 10 feet in diameter. I heard a whizzing sound, like wind blowing past my ear, and in the distance I saw a bright light, which I was being sucked toward, but which still seemed a long way off. And all this time I felt scared, powerless, and lonely, because nobody knew that I was aware that I was dead. I wanted to either return or not be aware of my death. But I clearly had no choice in the matter. . . . The light that I was now approaching was of a kind that I'd never seen before and that differs from any other kind such as sunlight. It was white and extremely bright, and yet you could easily look at it.

A Frightening NDE

Perhaps 1 to 2 percent of people with a near-death experience linger in a frightening dark space, unable to escape. To their horror, they sometimes find themselves pulled even deeper into the profound darkness. The NDE ends in this scary atmosphere, from where people reenter their body. The experience is devoid of positive emotions and later stirs profound feelings of guilt. In fact, such a terrifying NDE usually produces long-lasting emotional trauma. Not surprisingly, it is also known as a "hell experience." The exact number of people who experience such a frightening NDE is unknown because they often keep quiet out of shame and guilt. Yet if people can accept and make sense of this negative experience, they too eventually exhibit positive change.[18] Evans Bush's study has shown that people who have a frightening experience are not necessarily bad people. One possible explanation is that everybody has some negative character traits and that during a frightening NDE one or more such negative aspects are magnified for later analysis. A person is not defined by such a negative character trait although this does not make the experience any less intense.

> Suddenly I stopped in this dark tunnel and began to fall at enormous speed, faster and faster and faster. Like I was literally hurled down, head-first, into this black hole. It was pitch-dark; I couldn't see a thing. And as I was falling, I began to hear screams, shrieks, heartrending, dreadful, terrible laughter, and the most disgusting stench you can imagine, and then the blackness changed to fire. . . . And there were all kinds of ghastly looking and terrifying creatures, some worse than others, who were snatching at me. . . . I begged for God's mercy. . . . And suddenly I was woken up by the voices of female ER doctors who had resuscitated me. . . .

Next is the account of a visit to the netherworld during the comprehensive NDE of George Ritchie, who nearly died as a twenty-year-old medical student. The account shows a remarkable similarity with the description of hell in Dante's *The Divine Comedy*.

> *We were moving again. We had left the Navy base with its circumference of seedy streets and bars, and were now standing, in this dimension where travel seemed to take no time at all, on the edge of a wide, flat plain. . . . I could see no living man or woman. The plain was crowded, even jammed with hordes of ghostly discar-*

nate beings. . . . All of these thousands of people were apparently no more substantial than I myself. And they were the most frustrated, the angriest, the most completely miserable beings I had ever laid eyes on.

"Lord Jesus!" I cried. "Where are we?". . .

Everywhere people were locked in what looked like fights to the death, writhing, punching, gouging. . . .

Although they appeared to be literally on top of each other, it was as though each man was boxing the air; at last I realized that of course, having no substance, they could not actually touch one another. . . .

If I suspected before that I was seeing hell, now I was sure of it. Up to this moment the misery I had watched consisted in being chained to a physical world of which we were no longer part. Now I saw that there were other kinds of chains. . . . These creatures seemed to be locked into habits of mind and emotion, into hatred, lust, destructive thought patterns.

Even more hideous than the bites and kicks they exchanged were the sexual abuses many were performing in feverish pantomime. . . . Whatever anyone thought, however fleetingly or unwillingly, was instantly apparent to all around him, more completely than words could have expressed it. . . .

And the thoughts most frequently communicated had to do with the superior knowledge, or abilities, or background of the thinker. . . .

What was it going to be like, I thought with sudden panic, to live forever where my most private thoughts were not private at all? No disguising them, no covering them up, no way to pretend I was anything but what I actually was. How unbearable. . . .

Perhaps in the course of eons or of seconds, each creature here had sought out the company of others as pride-and-hate-filled as himself, until together they formed this society of the damned. . . .

I didn't know. All I clearly saw was that not one of these bickering beings on the plain had been abandoned. They were being attended, watched over, ministered to. And the equally observable fact was that not one of them knew it.[19]

6. The Perception of an Unearthly Environment

People often find themselves in a dazzling landscape with gorgeous colors, remarkable flowers, and sometimes also incredibly beautiful music. Some see cities and splendid buildings.

> What I saw was too beautiful for words: I was looking at a magnificent landscape full of flowers and plants that I couldn't actually name. It all looked hundreds of miles away. And yet I could see everything in detail—even without glasses, although in real life I have bad eyesight. It was both far away and close. Exceptionally beautiful. The best way to describe it would be: a heavenly sight.
>
> I arrived in a royal realm, or at least that's what it smelled like. The atmosphere, insofar as you could call it that, was divine, a flowery, sweet-smelling environment, which was completely three-dimensional and about a thousand times more beautiful than my favorite holiday destination in spring.

7. Meeting and Communicating with Deceased Persons

Hamlet: My father! Methinks I see my father.
Horatio: Where, my lord?
Hamlet: In my mind's eye, Horatio.
—SHAKESPEARE, *Hamlet*

Some people are aware of the presence of deceased friends or family members whom they also clearly recognize. Sometimes these people look healthy again, even though the prevailing memory of them is as very sick and weak in the period before they died. If they died at a very young age, they may look like young adults now. Some NDErs see individuals whom they have never met before or of whose death they could not have been aware. They feel a strong connection with the thoughts and feelings of people who have died in the past.

> During my NDE following a cardiac arrest, I saw both my dead grandmother and a man who looked at me lovingly but whom I

didn't know. Over ten years later my mother confided on her death-bed that I'd been born from an extramarital affair; my biological father was a Jewish man who'd been deported and killed in World War II. My mother showed me a photograph. The unfamiliar man I'd seen more than ten years earlier during my NDE turned out to be my biological father.

At the age of sixteen I had a serious motorcycle accident. I was in a coma for nearly three weeks. During that coma I had an extremely powerful experience . . . and then I came to a kind of iron fence. Behind it stood Mr. Van der G., the father of my parents' best friend. He told me that I couldn't go any further. I had to go back because my time hadn't come yet. . . . When I told my parents after waking up, they said to me that Mr. Van der G. had died and been buried during my coma. I couldn't have known that he was dead.

Suddenly I recognized all these relatives. They were all around thirty-five years old, including the little brother I'd never known, because he had died during the war when he was two years old, before I was born. He had grown a lot. My parents were there too, and they smiled at me, just like the others.

8. The Perception of a Brilliant Light or a Being of Light

I have been in that Heaven that knows his light most, and have seen things, which whoever descends from there has neither power, nor knowledge, to relate.
—DANTE, *The Divine Comedy: Paradiso*

The light is described as an extremely bright, nonblinding light that permeates everything. People are ineluctably drawn to this light and are usually completely enveloped by it. Sometimes this light is experienced as a being, and some religious people identify it as Jesus, an angel, or a being of light. A person's religious background is a significant determining factor in the naming of this being of light. People always report direct communication with this being, as if it reads their mind and responds through the mind. While enveloped by this light, people experience total acceptance and unconditional love and have access to a deep knowledge and wisdom.

The most profound questions are answered before they are even asked.

In the distance I saw a light that I had never seen on earth. So pure, so intense, so perfect. I knew it was a being I had to go to. I don't know how this happened. I didn't have to think, I knew everything. I had no mobility problems anymore. I had no body anymore. This dead weight had gone. . . . I passed through everything. At once I realized: there's no time or space here. We're always in the present here. This gave me a great sense of peace. I felt it as I experienced the Light. It's the pinnacle of everything there is. Of energy, of love especially, of warmth, of beauty.

I was immersed in a feeling of total love. It was crystal clear to me why I'd had cancer. Why I had come into this world in the first place. What role each of my family members played in my life, where we all were within the grand scheme of things, and in general what life is all about. The clarity and insight I had in that state are simply indescribable. Words seem to diminish the experience—I was in a place where I understood that there's so much more than we can fathom in our three-dimensional world. I realized that this was a great gift and that I was always surrounded by loving spiritual beings of light.

That very same moment, in a split second, I gained access to a wealth of knowledge, a complete knowing and understanding. All knowledge. Universal knowledge. I understood the origins of the cosmos, how the universe works, and why people do what they do. Their positive actions, but also why they hurt one another, deliberately or not. Wars and natural disasters, everything has a purpose, a reason. It all makes sense. I understood the past, the present, and the future. I saw evolution. Everything and everyone evolves and develops together. I saw and understood—without any judgment—the connection, the coherence, the logical and sometimes major consequences of every single act. I mean at every level and down to the smallest detail. . . . The way all kinds of mechanical, electrical, and electronic equipment, gadgets, and engines work. Everything. I knew and understood all about mathematics, electronics, physics, DNA, atoms, quantum mechanics, and quantum physics. . . . I also saw where evolution is headed, what its ultimate goal is. I realized that this grand scheme not only includes me, but everything and everybody, every human being, every soul, every animal, every cell, the earth and every other planet, the universe, the cosmos, the Light. Everything is connected and everything is one. "I see!" I thought happily. "I get it. It's all so simple. So obvious. It all makes sense. . . ." No, I wasn't allowed to bring back the knowledge itself. Why, I don't know. . . . Perhaps we're not supposed to have such universal knowledge in the here and now, in our physical form? Perhaps we're here to learn? Perhaps there's another reason?

9. The Panoramic Life Review

The life review is usually experienced in the presence of the light or a being of light. During a panoramic life review, people experience not just their every action or word but also every thought from their past life, and they realize that everything is an energy that affects both themselves and others. All of life, from birth up until the present moment, can be relived as a spectator and as an actor. This makes it much more than a speeded-up film. People know their own and others' past thoughts and feelings because they have a connection with the memories and emotions of others. During a life review people experience the effects of their thoughts, words, and actions on other people when they originally occurred, and they also get a sense of whether love has been shared or withheld. Although this can be extremely confrontational, nobody feels judged: people understand how they lived their life and how this affected others. They realize that every single thought, word, or action has a lasting effect on themselves and others. They speak of a cosmic law in which everything they do to another person will ultimately have an effect on them as well, and this applies to both love and affection and violence and aggression.

A person's whole life comes up for instantaneous review; time and distance appear to be nonexistent. People are instantly taken to whatever they focus their attention on. People can talk for hours or even days about their life review, even though the cardiac arrest lasted only a couple of minutes. Everything appears to exist and be open to experience at once. Everything and everybody appear to be connected in an eternal present.

My whole life so far appeared to be placed before me in a kind of panoramic, three-dimensional review, and each event seemed to be accompanied by an awareness of good and evil or by an insight into its cause and effect. Throughout, I not only saw everything from my own point of view, but I also knew the thoughts of everybody who'd been involved in these events, as if their thoughts were lodged inside me. It meant that I saw not only what I had done or thought but even how this had affected others, as if I was seeing with all-knowing eyes. And so even your thoughts are apparently not wiped out. And throughout, the review stressed the importance of love. I can't say how long this life review and insight into life lasted; it may have been quite long because it covered every single subject, but at the same time it felt like a split second because I saw everything at once. It seemed as if time and distance didn't exist. I was everywhere at once, and sometimes my attention was focused on something and then I was there too.

To start with, I was shown images of my two previous lives. The first time I died was during a campaign in England during the Roman era. I was the chief of a cohort escorting a number of female prisoners to the coast, and along the way we were attacked by the natives of those parts. I also experienced my death in World War I. I was in a fighter plane in what must have been 1917 and became caught up in a battle with a German aircraft. I was shot down and crashed between the lines. I say 1917 because with the help of pictures taken that year I managed to identify the type of aircraft used by the English air force at the time. Why I also saw excerpts from these two lives, I don't know. And I have no way of verifying any of this. What I remember much more vividly are images from my latest or, if you like, current life. First I witnessed my own birth. I was brought into this world by our family doctor, unlike my brothers and sisters, who were delivered by the midwife. The doctor held me in his arms and said the following memorable words to my mother: "This is a special child. He will either be a great genius or a great scoundrel." I turned out to be neither of the two. I saw my first few steps. I saw how my unreasonable behavior hurt my mother. I saw myself playing with the neighbors' dog, Bello. It was a watchdog and watched over the farm. It was huge and obeyed only its masters. Strangely enough, I wasn't scared of this dog at all, and it put up with everything I did. Every now and then I even crawled into its kennel for a nap. I wouldn't even let Farmer Mast, its owner, anywhere near to try and get me out. I was on equally good terms with Bles, the big horse. When she was in the field and I slipped under the barbed-wire fence, she would come galloping and rear up right in front of me. Then she'd often drop to the ground, and I'd crawl between her legs and onto her belly. The spectacle used to terrify bystanders. I also saw my school years pass before my eyes, and the teachers I used to torment. The war years were quite prominent. I saw some people I'd known in the camp and whom I had, on occasion, robbed of the little food they had in a bid to survive. But some of my good deeds also flashed past. I saw the Indonesian girl whom I'd lived with for four years. I relived the intense love, but also that one time when I hurt her badly and I thought that she hadn't realized. Looking back, I saw a lot of situations in which, as a serviceman, I'd been rather ruthless. But my NDE also brought back some memories of incidents in which, in defiance of orders, I'd let mercy take precedence over law. Among them were quite a few incidents I had long since erased from my memory. For instance, there were things in my life that I'd come to accept as bad but that were now suddenly deemed good. The same applied to things that I'd always considered to be good and that were now branded wrong.

The next episode that I subjected to a thorough examination was the period of September 1944, the Battle of Arnhem. What struck me was that, despite the short time span, so many people passed before my mind's eye. I saw many whom I had taken to the hospital or who died in my arms. Many of them assured me that they'd welcome me as soon as I arrived on the "other" side. To my surprise, I only saw a single German. It was a German soldier who had fought an English soldier, and the two had wounded each other so badly that they both died, one after the other. He gave me his iron cross, which, astonishingly, I've managed to hold on to my entire life. He gave it to me because I let him have some drags on his English adversary's last cigarette. They both died within a short time of each other. The fact that I gave him a smoke was labeled a good deed, which I don't understand because I did it on the orders, or at the request, of the Englishman. I would have preferred to see the German stew in his own blood. What I mean to say by this is that up there they judge by different standards than down here.

10. The Preview or Flash Forward

People feel like they can see part of the life that is yet to come. At this stage of the experience too there appears to be no time or distance. The reports of the verifiable future events inevitably raise questions about free will and the extent to which people can determine their own future.

And in a flash I saw the rest of my life. I could see a large part of my future life: taking care of my children; my wife's illness; everything that would happen to me, both in and out of the workplace. I could see it all. I foresaw my wife's death and my mother's passing. One day I wrote down all the things I saw back then; over the years I've been able to tick them all off. For instance, I saw my wife on her deathbed, wrapped in a white shawl, just like the one she was given by a friend of hers shortly before she died. . . .

11. The Perception of a Border

People see a thick fog, a wall, a valley, a river, a bridge, or a gate and are aware that once they cross this border they will not be able to return to their bodies and resume their lives. At this stage there may be some communication with a deceased relative or with a being of light. People hear that they are not welcome because their time has not yet come. They must return to their body because they have a purpose in life. This purpose may be to care for a newborn baby, a child, or a relative. Sometimes people cannot quite remember why they were sent back. They either feel that the decision to return was theirs or that it was imposed by others.

First up is the account of a boy who was born deaf and almost drowned at the age of ten:

Then I reached a border. Even at the age of ten I needed no further explanation. I simply understood that I'd never be able to return if I crossed this border. But some of my ancestors were on the other side, and they caught my attention because they were communicating through a kind of telepathy. I was born profoundly deaf. All my relatives can hear, and they always communicate with me through sign language. Now I had direct communication with about twenty ancestors via some kind of telepathy. An overwhelming experience. . . .

He showed me a gate behind which I saw the same landscape. But now, with this gate in front, it suddenly looked extremely familiar. I came to the startling conclusion: I've been here before. It felt like a homecoming after an arduous journey. A state that led to complete peace of mind, a peace of mind I hadn't known for a long time. For me this was the highlight of the experience. Without a word the figure encouraged me to decide whether I wanted to remain in this state or whether I wanted to return to earthly life. I could either enter the gate or return to the lifeless body, which I immediately sensed below me. I had the impression that entry through this gate meant definitive physical death. Aware that this was my chance to go back knowing that this state of being is a reality that feels more real than what we call reality and thinking of my young wife and our three small children, I opted to return. . . .

12. The Conscious Return to the Body

The return to the body is usually quite abrupt. Sometimes people feel a great force sucking them back through the tunnel. Some people describe how they were pushed back into their body via the head, after seeing a nurse or doctor place the resuscitation equipment on their body. The conscious return to the body is an extremely unpleasant experience for most people. Back in the sick, damaged, and aching body, they are upset at having been denied something so beautiful. Some patients react with indignation, disappointment, or rebellion as soon as they regain consciousness after resuscitation or wake from a coma. Their attempts at talking to doctors, nursing staff, or family about the powerful experience often come to nothing, which only adds to the disappointment. In fact, some people remain silent on the subject for fifty years or more.

> When I came to in my body it was dreadful, so dreadful. . . . The experience had been so beautiful that I didn't want to come back. I had wanted to stay there . . . and yet I came back. From that moment it was a real struggle to live my life inside my body, with all the limitations I experienced at the time. . . . But later I realized that this experience was in fact a blessing, for now I know that the mind and body are separate and that there's life after death.

> Before I get a chance to turn around and dive into that heavenly light, I notice a slender hand on my back, from my right shoulder down to my waist. This large hand pushes me very firmly yet lovingly back into my body. For a moment I feel like I'm doing a couple of somersaults in the air. Then I realize that I've landed back in my body. Back to the pain and to the doctor's deafening

screams and slaps. I'm furious, incredibly furious! I don't know if I actually uttered all the insults that came to mind. . . . I think I did, because I felt a sense of relief afterward. I've never felt a fury like this rage. . . .

Empathetic NDE

An empathetic NDE is an experience based on empathy, the capacity to understand and identify with another person's emotions. This type of NDE is not rooted in one's own physical or psychological problems but is caused by the strong emotions felt upon the death of a dearly loved person. People share, as it were, in the death experience of another person who is dying or has just died. The content of this empathetic NDE is identical to that of a classic near-death experience.

I was in a relationship with Anne when she suddenly died in a serious traffic accident. Her son, who'd just turned seven, sustained severe head trauma. His brain virtually spilled out of his skull—it looked like a smashed watermelon—and it took him about five days to make the transition. He was the eldest grandson of a couple with nine children. Some sixty relatives had gathered around his hospital bed, and since I'd only been his mother's boyfriend, I was standing somewhere at the back by the window. The moment he died, when his EEG flatlined, I "saw" that his mother came to collect him. You must bear in mind that she'd died five days earlier. There was this incredibly beautiful reunion. And at one point they reached out for me and included me in their embrace. This was an indescribable, ecstatic reunion. Part of me left my body and accompanied them to the light. I know this must sound very strange indeed, but I was fully conscious and with Anne and her son as they went to the light, just as I was fully conscious and in the room where all the relatives were incredibly sad because their nephew and grandson had just died. And I joined them, we were heading toward the light, but at a certain point it was clear that I had to return, so I fell back. I simply fell back into my body. It was such an overwhelming experience, I glowed with happiness, but then I suddenly realized that I had a big smile on my face amid all these people who'd just lost a child dear to them. I quickly covered my face with my hands because I didn't want to be disrespectful toward all these mourning and crying people in the room. And I never said a word about the experience. Talking about it seemed completely inappropriate at the time, and besides I didn't have the words to describe what had happened to me. I used to think that I knew what was what. But my worldview underwent a radical transformation.

I was eighteen when my uncle in New York got cancer in his one remaining lung. I traveled to New York, and we talked a lot. When I came back my mother, his sister, immediately set off for the United States. She stayed with him day and night. He was absolutely terrified. Afraid there would be nothing. He was alone when he died. He wanted it that way. Now for my experience, which took place a couple of hours after his death. I was asleep and "dreaming." My uncle took me through a man-sized tunnel that opened out onto a beautiful landscape. A green meadow by a hill. A tree. A lake. All illuminated by a silvery sun. "I'm here now," he said. And he looked extremely happy.

Man, if thy spirit rise above Time and Space, each moment canst thou be in eternity.
—ANGELUS SILESIUS (JOHANNES SCHEFFER)

In conclusion, each of the NDE reports reproduced here is extraordinary and emotional. When people agree to share their experience, the sincerity is always palpable. But there is also a degree of reticence because NDErs are aware that their indescribable experience is difficult to put into words. They understand full well that other people, who have never had such an experience, struggle to believe or comprehend an NDE. Those who have had a near-death experience themselves have had the greatest difficulties understanding and coming to terms with their overwhelming experience. And scientists who have never spoken to anybody with an NDE and who are usually of the opinion that the experience lacks any scientific basis, find it hard to take an NDE report at face value. Scientists continue to dismiss people with an NDE as dreamers, fantasists, attention seekers, or confused patients.

I am of the opinion that people who have had a near-death experience and who are capable of putting their experience into words can teach us a great deal about the relationship between human consciousness and the brain. Finding an explanation for the cause and content of the near-death experience is a major scientific challenge.

Chapter Three

Changed by a Near-Death Experience

Our ideas about death define how we live our lives.
—DAG HAMMARSKJÖLD

A near-death experience elicits skepticism and critical questions from many. Is there any scientific explanation for comprehensive and permanent life changes after a two-minute cardiac arrest? Should the NDE be regarded as an existential crisis? An NDE is an overwhelming confrontation with the boundless dimensions of our consciousness. As long as people have not had an NDE themselves, they remain ignorant of the impact and profound consequences of such an experience.

It felt as if I'd become another person, but with the same identity.

People usually say that their NDE has transformed their views on what really matters in life, and that they have lost their fear of death.

It's possible to be physically dead while your mind lives on. Only one thing matters: your attitude toward other people. I think about everything now. Why? Everything continues and nothing continues. I feel very peaceful now; I'm no longer afraid of death. I now accept life as it is.

People often talk of attaching greater value and meaning to life and less importance to material things such as an expensive car, a big house, and a job with status or power. The near-death experience turns out to be a life-insight experience. Or as one person who experienced an NDE put it: NDE really stands for New Discernment through Experience. The newfound insight pertains to what matters in everyday life: acceptance of and unconditional love for oneself (including acceptance of one's dark side), others, and nature. It also pertains to insight into connectedness: everything and everybody is connected. Because of this sense of connectedness, some people describe the NDE as an experience of unity. They speak of a "cosmic law" in which everything they do to another person will ultimately have an effect on them too, and this applies to both love and affection and violence and aggression.

But is everything always rosy after an NDE? Are there any hitches in the process of coming to terms with the experience? How common are changed insights after an NDE? And what happens to these changed insights over time: do they intensify or wear off? Does everybody display the same patterns of change, or are there any cultural differences? In this chapter we look at changes that NDErs experience as a result of their NDE. We ask if these changes are mostly positive or if they actually cause any problems, and how the reactions of one's friends and family affect the ability of the person who had an NDE (NDEr) to integrate the experience.

The Consequences of an NDE

Irrespective of the immediate cause of a near-death experience, its survivors display permanent and fundamental change in their outlook on life, religious beliefs, values, and behavior. The depth of the experience and, above all, the panoramic life review and the encounter with the light appear to contribute to the intensity and inevitability of these changes.[1] Popular literature tends to concentrate on the positive nature of these changes, yet many people struggle to accept and integrate their newfound insight, especially when faced with negative reactions from family, friends, and health care practitioners.

My body, my life, and the whole world suddenly felt like a prison.

The reported changes are probably triggered by the conscious experience of a dimension where time and distance play no role, where past and future can be glimpsed, where people feel complete and whole, and where infinite wisdom and unconditional love can be experienced. After an NDE, the insight no longer rests on faith but on certainty.

It had such a profound effect on the rest of my life: the timelessness that I experienced; the knowledge that my consciousness will survive outside my body. It was enough to destabilize my life.

Some people feel nostalgic about their NDE because of the unforgettable feelings of peace, acceptance, and love they encountered during the experience. In fact, the life changes derive in no small measure from the new insight that love and consideration for oneself, others, and nature are paramount and that death is not the end of everything. The NDE teaches people that life goes on after physical death.

The experience changed everything for me: there's something after death, and it's good. Death is merely a release from the body.

To illustrate this, here are some extracts from an interview with a patient who had an NDE during his cardiac arrest eight years earlier:

I'm no longer afraid of death because I'll never forget what happened to me there. Now I'm certain that life goes on. Over the years I've undergone a number of changes. I feel a strong connection with nature. The garden now plays an important role in my life. I've become much more emotional. I've acquired a great sense of justice. I've become more patient and peaceful. I can see things in perspective now. My aggression is a thing of the past. I feel a strong inner urge to never lie again. I'd rather keep silent than tell a little white lie. I do struggle with deadlines: things must get done within a certain time. But I can handle it reasonably well. Before I never knew anything about spirituality; I wasn't interested. But now I've become aware of prophetic powers with which I can help others. I have a sixth sense. Over the years I've learned to live with it. It has simply become a part of my life. I really rely on my instincts these days. The moment I start thinking, everything goes haywire. But my hypersensitivity also causes a lot of problems, as it creates tension between rejection by those who don't understand and curiosity from those very same people. I've learned to listen to my body. I enjoy life immensely. Now I'm aware of things that I never noticed before in my experience. I believe that people have stopped living from the heart. And I prefer to do everything as simply as possible. I'm so glad and grateful that I can discuss it with my wife.

Studies of Life Changes After a Near-Death Experience

Over the past twenty-five years a great many books have been published with personal testimonies of near-death experiences and the often intense changes that follow. The most systematic (retrospective) studies of such changes have been published by Kenneth Ring, Margot Grey, P. M. H. Atwater, Cherie Sutherland, Melvin Morse, Peter and Elisabeth Fenwick, Kenneth Ring with Evelyn Elsaesser-Valarino, and Anja Opdebeeck.[2] These books show great similarities between the reported processes of change irrespective of age, cultural or religious background, or the medical cause of the NDE. However, the people who were interviewed for these (retrospective) studies were approached through advertisements, lectures, and the Internet, and thus they raise the aforementioned problem of selection. Who did or did not volunteer for an interview? Are we hearing only from people who struggle most to come to terms with the experience? Or those who struggle least? Unfortunately, we will never know.

The major failing of these interesting and edifying books filled with remarkable anecdotal evidence is that while they provide a clear picture of the various aspects of the process of change, they offer no reliable figures on the incidence of the changes and the exact times when they occur. Another limitation of these studies is that they fail to indicate the passage of time between the NDE and the interview. The sole exception here is Sutherland's study.[3] The passage of time between the NDE and the interview is significant in determining the degree to which the changes have been accepted and integrated. The longer this interval, the more positive changes are reported. In other words, the processes of change vary a lot depending on whether people are interviewed one year or twenty-five years after their NDE, and this complicates the interpretation of the data in these studies. Not all the changes listed in them occur, and the changes that do occur do not always occur at the same rate.

Cherie Sutherland carried out a somewhat more systematic study by examining changes in religious beliefs, spiritual interests, and aspects of heightened intuition before and after the NDE and comparing these data with the general population. Her study included fifty people with an average (young) age of 31 (7–76), and an average (long) interval between the NDE and interview of 19 years (2–52). There are only a few other retrospective studies in which post-NDE changes have been compared with a control group of people who suffered a serious medical crisis without an NDE. Such comparison plays an important role in ascertaining whether the reported changes are caused by the NDE or by the serious medical crisis. In his study Ring also interviewed friends and family in order to put the reported changes into perspective.[4]

The Dutch study that my colleagues and I conducted (see chapter 7) is the only prospective, longitudinal study of post-NDE processes of change. It drew on a control group of cardiac arrest survivors without NDE who matched the patients with an NDE in terms of age and gender. Our research, featuring interviews with NDErs and a control group at two- and eight-year intervals, has the longest follow-up published to date. Wherever possible, the interviews included patients' partners, which enabled us to put the processes of change into perspective. (We will discuss the effect of time on processes of change later in this chapter.)

Factors That Influence the Process of Change

The transformations reported by people with a near-death experience differ widely both in terms of content and in terms of the rate of acceptance and integration.[5] Factors that play no role are age (except in very young children), gender, and level of education. But personality traits of the person prior to the NDE, such as an extroverted or introverted character or a positive or somber disposition, are important factors in this process. If dealing with unexpected situations was always a challenge, coming to terms with an NDE is that much harder. And people with a strict religious upbringing cope with the experience differently than people who reject any form of religion, as for example in the former East Germany.[6]

Cultural factors play a role too: in countries such as India, where spirituality, meditation, and reincarnation are widely accepted, the content of the NDE and its resulting changes appear to be more easily integrated than in the West. Within Western culture the experience is completely at odds with conventional wisdom. And it is no surprise that people who were raised with material values find it harder to accept the new insight that money and power are no longer essential to happiness. It is not so much the content that makes the NDE so difficult to come to terms with but the fact that Western culture and science do not really accommodate this kind of spiritual experience.

As mentioned, the depth of the NDE is not the only important factor in the process of change. Significant too are the medical circumstances under which the NDE occurred, such as a car accident with prolonged rehabilitation, a heart attack with feelings of restriction and anxiety, or a brain hemorrhage with permanent paralysis. Patients must come to terms with the NDE itself but also accept and deal with the consequences of the serious medical crisis that precipitated it, which sometimes involves extensive rehabilitation, wheelchair use, and other lingering symptoms.

Alongside the personal acceptance and integration of the NDE, the social support of family and friends is equally crucial.[7] An NDE can put a severe strain on relationships (more on this later). Integration also depends on the response of health care practitioners, on whether they can lend a respectful ear and the support needed for processes of change. The skeptical response of most health workers is a source of extreme frustration to NDErs. The integration process can be accelerated if the person affected learns that he or she is not the only one who has had such an overwhelming experience and that there is a name for it (NDE). Contact with other NDErs and reading books on NDE and its consequences for later life can be equally helpful. This form of affirmation helps people recognize their experience, reduce anxiety, and facilitate better and faster integration.

Integrating the Experience

Acceptance of the experience and the changed insights often depends on the response of the partner and family, friends, acquaintances, doctors, nurses and other health care practitioners, particularly during the first few months and years after the NDE. Yet many people with an NDE discover that others are incapable of listening without prejudice and criticism. Retreating into years of silence may be the only way to cope with the experience. People feel transformed while those around them remain the same. The process of accepting and integrating the NDE cannot begin until people feel capable of sharing their thoughts and feelings.[8] With immense perseverance, often aided by positive reactions from those around them, people learn to live according to their newfound insights into what matters in life.

The integration process lasts at least seven years or more because the effort provokes a lot of resistance in both NDErs and those around them. I have come across people who were unable to talk about their NDE and its consequences until more than fifty years after the event. For fear of rejection, they kept their NDE a lifelong secret. These people had found it extremely difficult, if not impossible, to live their lives according to their newfound insights. Despite its largely positive content, the NDE is traumatic in these cases because the process of coming to terms with the experience is so difficult and painful.[9]

Positive and Negative Aspects of Processes of Change

In the first few years after the NDE especially, people are prone to depressions, feelings of nostalgia, and loneliness. As a rule, the positive aspects come to the fore only once people are capable of accepting and integrating their NDE. But the process of integration cannot get under way properly until the experience can be shared. When someone first tries to disclose the NDE, the other person's reaction is absolutely crucial. If this initial reaction is negative or skeptical, the process of accepting and integrating the NDE typically presents far greater problems than if this initial reaction is positive, sympathetic, or neutral. Evidence has shown that positive responses facilitate and accelerate the integration process. In fact, without the possibility of communication, the process of coming to terms with the NDE often fails to get under way at all. Most NDErs feel an urgent need to talk about their experience. This need stems not only from a desire for affirmation, but above all from a desire for support. That said, of course there will always be some people who prefer to come to terms with their NDE in silence.

An Overview of the Various Changes

What follows is an overview of the various aspects of the post-NDE process of change. Many of the potential changes described here are not common to everybody. Similarly, many elements of the transformation are not experienced until many years after the NDE. This overview is based on the findings of a number of articles, the aforementioned eight books, and my conversations with the hundreds of people who shared their NDE and its consequences with me.[10]

Self-Acceptance and a Changed Self-Image

The experience of transpersonal aspects during the NDE changes people's sense of who they really are. *Transpersonal* refers to those aspects of someone's consciousness that transcend the personal or the ego. This experience can be accompanied by a heightened sense of self-worth. Thanks to their changed self-image, people become less dependent on the approval of others, better at dealing with stress, and more adventurous, and they also take greater risks. It changes people's attitude toward their body and alerts them to new ways of thinking. They are more likely to look at the bigger picture and are capable of forming more objective opinions, even at the risk of seeming aloof. And because they are more easily engrossed in things, they are less aware of their surroundings. Increased levels of curiosity coupled with a hunger for knowledge spark a particular interest in theological issues, philosophy, psychology, and natural sciences (especially quantum physics), although their education often fails to satisfy this need for deeper knowledge. They also develop a noticeably greater interest in physical and psychological processes and the possibility of (self-) healing.

Compassion for Others

After the NDE, relationships with others change noticeably, and people are now capable of greater compassion.

> It's so clear to me now that my NDE has transformed my emotions toward life and my emotional life. Everything I do now is aimed at reliving and spreading this feeling of Love.

People are more forgiving, more tolerant, and less critical of others. They are also more emotional. Appreciation of relationships increases; people spend more time with family, friends, and relatives, and they are more willing and able to share emotions with others. They are more compassionate and caring and set greater store by unconditional love. Yet relationships are also more likely to run into problems. Sometimes there is more, sometimes less interest in sexual intercourse. Some people have trouble communicating with others because they struggle to find the right words. A greater sense of justice is coupled with the urge to tell the truth and say what is on one's mind. Any trace of past aggression is usually gone. It is replaced by the need to help and support others, which usually leads to a career change in favor of the care professions, such as nursing, care of terminal patients, or voluntary work with elderly people or low-income families. NDErs are also more likely to donate to charities or to dedicate themselves to a social cause.

Appreciating Life

The near-death experience brings about remarkable changes in what people see as the true purpose in life.

Apparently, I still have a task to fulfill in this life.

People who have a near-death experience seem positive of a new goal or new mission in life. They also appreciate the little things in life, pay more attention to the here and now, and enjoy the moment. They are less likely, however, to allow themselves to be restricted by social convention. People are more confident of their ability to handle problems, more open to change, and less preoccupied with time and schedules. But even though they struggle with the concept of time, they do tend to honor appointments. They are better at putting things into perspective, they take an unbiased view of life, and they are quick to smile while at the same time more serious. Their increased respect for life also reveals itself in the greater appreciation of and interest in nature. They are now much more aware of seasonal change and like doors and windows to be open to admit fresh air. They also take more pleasure in classical or soothing music and are less tolerant of noise. NDErs attach less importance to status, money, and material possessions and distance themselves from the competitive elements in contemporary society.

No More Fear of Death and a Belief in Life After Death

The experience of feeling utterly unchanged as a person after temporarily leaving the sick or lifeless body teaches people that death is something entirely different than they previously thought.

Dead turned out to be not dead.

In most cases this realization greatly reduces the fear of death and bolsters people's belief in life after death.

I'm no longer afraid of death. I see the experience as a gift. Now I know there's more after death. I'm grateful. I feel that I have to talk about it to help others, to reassure them if they're afraid of death. I feel privileged.

For figures on the reduced fear of death and the increased belief in an afterlife, see the table "A Different View of Death After an NDE." The table lists people's attitudes prior to their NDE alongside the changes that occurred at least twenty years after the NDE. Those who believed in life after death before their NDE have become absolutely certain after their experience.

A Different View of Death After an NDE

	Before NDE (percent)	After NDE (percent)
No fear of death		
Sutherland[11]	16	98
Grey[12]	37	100
Opdebeeck[13]	55	100
Belief in life after death		
Sutherland	38	100
Grey	24	76
Ring[14]	(number?)	86
Opdebeeck	25	96
Musgrave[15]	22	92

The fear of death decreases even further over the years after the NDE (see the table "Life Changes After a Cardiac Arrest" on p. 68). The loss of the fear of death also changes people's outlook on life; some aspects of life become important while others become completely irrelevant. After an NDE, people only want to spend time and energy on things of lasting value. Almost all ephemeral and material things, such as a lot of money, a big house, or an expensive car, become less important. People also identify much less with their own body, believing it to be "merely" the physical, material aspect of somebody's personality. "I can live without my body, but apparently my body cannot live without me."

The effect of the loss of the fear of death can be illustrated with the story of an eighty-two-year-old patient with serious heart failure who had been admitted to the cardiac ward where I worked as a cardiologist. This man was terminally ill, as his shortness of breath no longer responded to medication. Normally doctors and nurses give a wide berth to a dying patient's room; having finished treatment, they can do nothing more. But this was a completely different story. More than ten years earlier, this patient had suffered a massive myocardial infarction with cardiac arrest for which he had been successfully resuscitated. During his cardiac arrest he had experienced an extremely deep NDE and had lost his fear of death. The man knew he was dying. Yet there he was in bed, looking radiant despite his difficult breathing. Beautiful classical music could be heard at all times, and from morning till night his room was filled with nursing staff, family, and fellow patients, all of whom he tried to help. He was attentive and in good spirits, and it was a real joy to be in his presence. And this is exactly how he went: good-natured, cheerful, and supportive of those around him. After he died he lay in bed with a big smile on his face, and people kept coming into his room to pay their last respects.

A Decline in Religious Affiliation Coupled with Greater Religious Sentiment

A near-death experience can sometimes engender profound religious feelings and give people the impression of a personal bond with God.

> I now have a much, much stronger bond with God. I see and feel him as the greatest force in my life. He entered my life unbidden, but I welcomed him.

But an NDE leads some people to believe that they are one of God's chosen. This sense of salvation can make them feel relatively invulnerable and extremely important and may result in a strong urge to spread word of the NDE as a deeply religious experience. Such proselytizing is often seen as intrusive and stirs a great deal of resistance. But generally speaking, people's religious sentiment increases after an NDE while their interest in organized religion declines sharply (see table "Changes in Religious Affiliation After an NDE").

> I have strong religious feelings now. I no longer "believe" in God; I'm absolutely certain. But it has nothing whatsoever to do with the church.

Changes in Religious Affiliation After an NDE

	Before NDE (percent)	After NDE (percent)	General Population (percent)
No religion	46	84	16
Church of England	24	4	28.3
Roman Catholic	12	8	25.6
Methodist	4	0	4.3
Presbyterian	2	0	7.2
Jewish	2	0	0.4
Baptist	2	0	2.1
Lutheran	2	0	1.3
Calvinist	2	0	NA
Brethren	4	2	NA
Buddhist	0	2	0.2

Average interval between NDE and interview: 19 years (Sutherland[16])

The table shows Australian data from 1990. Because of major international variation in church attendance, the percentage of nondenominational people referred to in studies depends largely on both year and country of research.

It is important to bear in mind that, irrespective of NDEs, the percentage of nondenominational people in the Netherlands increased sharply in the twentieth century. In 1900 only 2 percent of the Dutch people were not affiliated with any church, by 1960 their number had risen to 18 percent, and by 1999 it was up to 63 percent. More and more people believe that religiosity has nothing to do with church attendance. Figures from 2002 also show that 37 percent of nondenominational people in the Netherlands believe in life after death, 25 percent believe in heaven, 19 percent believe in the sense of praying, and 31 percent believe in religious miracles. In other words, religious beliefs can exist independently of religious affiliation.[17]

There is no comparable decline in religious affiliation in the United States: 78 percent of Americans today are still Christian, of which the majority (55 percent) are Protestant, and only 16.1 percent have no religion. In the United States 28 percent of people never attend any religious service, and 40 percent attend at least once a week. In the United Kingdom there is similarly a high percentage of people with a religious affiliation: 71.6 percent are Christian and 2.7 percent Muslim, but only 14 percent attend church at least once a week (statistic from 2005). From 1964 to 2005 religion combined with church attendance decreased in the United Kingdom from 74 to 31 percent while at the same time the percentage of nondenominational people increased from 3 to 38 percent.[18]

Greater Spirituality

Happiness does not depend on outward things, but on the way we see them.
—Tolstoy

A near-death experience can evoke the sense that the old self has died and that a new person has been born. The NDE and its subsequent changes are thus experienced as a spiritual death and rebirth. If religious affiliation declines, people also report an increase in religiosity and a greater interest in spirituality, meditation, prayer, and surrender. With their precious life restored, people view themselves as having a unique mission, and they are fueled by a heightened sense of spiritual purpose. They feel part of a meaningful universe and adopt a more philosophical attitude to life.

By contrast, cardiac arrest survivors without an NDE display a marked decline in interest in spirituality (see the table "Life Changes After a Cardiac Arrest" on p. 68).

Physical Changes

People not only experience psychological transformations, they also report important physical changes. The NDE can precipitate hypersensitivity (hyperesthesia) to bright lights, in particular, sunlight. In fact, the experience can trigger a heightened sensitivity to all sensory impressions, such as sound, taste, touch, and smell. Some people can no longer stand the physical proximity of their partner. There are also

Changes in Post-NDE Religiosity

	Before NDE (percent)	After NDE (percent)	General population (percent)
Value of organized religion	36	20	56
Church attendance	38	20	34
Prayer	48	74	56
Meditation	12	60	NA
Quest for spiritual values	20	88	NA
Guidance	32	86	44

Average interval between NDE and interview: 19 years (Sutherland[19])

frequent reports of synesthesia, a phenomenon in which sensory impressions influence one another. In these cases, areas of the brain that play a role in processing data from the various faculties are more closely connected than usual and exchange information. In other words, they sort of get their wires crossed. People affected in this way talk of seeing smells and hearing or smelling colors (for example, "smelling red"). The NDE can also result in a greater sensitivity to loud noises, a desire for silence, and a newfound preference for soothing music.

After my NDE I felt like a child learning how to walk. The world around me overwhelmed me. I couldn't find my place in the world. For months I couldn't bear light and noise, TV and radio, not even music, which I used to love. Sometimes I think I've got a new problem because I've become hypersensitive to too much noise.

Some people experience an increased sensitivity to alcohol or develop an allergy to conventional medicines, which then sparks an interest in complementary or alternative medicine. It is possible for metabolic rates and energy levels to change, and people to recover faster and look younger. There are frequent reports of electric phenomena: at emotionally charged moments, in particular, the body can emit an electromagnetic field that interferes with electrical equipment—lights go out, the computer crashes, the car starter fails, or the supermarket checkout scanner refuses service.

> Another strange thing was that after my NDE every piece of equipment I touched, such as lamps, dishwasher, kettle, the light in the cooker hood, it broke. I gave off energy everywhere.

Some people do not wear a watch because it stops as soon as they wear it on their wrist. Thinking that the watch is broken, they buy a new one, only to have the same thing happen. Some discover the ability to direct healing power at others, which enables them to help people with physical or psychological problems. There are also reported cases of inexplicable self-healing after an NDE.[20]

Enhanced Intuitive Sensitivity

Without really wanting to, many NDErs feel inundated with information from or via another dimension. This appears to affect between 84 percent (Ring's study) and 92 percent (Sutherland) of people, making enhanced intuitive sensitivity one of the most common but also least spontaneously reported consequences of an NDE (see table "Symptoms of Enhanced Intuitive Sensitivity"). All of a sudden these people have a very acute sense of the emotions of others. Heightened intuition can cause major problems. Clairvoyance, enhanced sensitivity, and precognition can feel extremely threatening.[21]

NDErs rarely volunteer information about their heightened intuition in interviews. Neither do they have the right words for it. Indeed, what could they say? That they have become paranormal or something? A researcher or other interested party will have to ask targeted questions and explain that a great many people experience enhanced sensitivity after their NDE. Most of us have had the odd sensation of thinking about somebody only to find that when the phone rings it is the person we were thinking of. Synchronicity, as the phenomenon is called, is quite common and refers to the not strictly causal, or seemingly accidental, concurrence of events.

Symptoms of Enhanced Intuitive Sensitivity

Experience	Before NDE (percent)	After NDE (percent)	General population
Clairvoyance	38	71	38
Telepathy	42	86	58
Precognition	49	86	NA
Déjà vu	73	85	NA
Enhanced intuition	54	92	NA
Dream awareness	44	79	42
Out-of-body experience	8	49	14
Spirits	22	65	27
Healing ability	8	65	NA
Perception of auras	13	47	5
Psychic phenomena	55	98	39 (Sweden)

Average interval between NDE and interview: 19 years (Sutherland[22])

However, in the case of post-NDE enhanced intuitive sensitivity, such synchronicity is much stronger.

When I felt a bit better again, I was so unbelievably paranormal, it was creepy. Before this incident I'd had now-and-then premonitions, but this was different. From my bed in the town of Utrecht, I could see who was in the room next door, and I knew what food the priest had ordered. It's easing off now, but I still understand people better, and yes, sometimes I can literally read and intuit people's minds.

People suddenly sense emotions and sadness in others, or they sense that somebody has a serious illness. For example, there was the case of a young father who saw a brain tumor in his eighteen-month-old daughter. Unable to explain why he suspected that she might have a tumor in her head, he did not visit his family doctor. Three months later, when she suffered her first epileptic seizure, she was referred to a neurosurgeon who decided to operate on her brain. Similarly, people

sometimes sense a person's imminent death. And more often than not
these premonitions are correct. Enhanced intuitive sensitivity leaves
people feeling extremely insecure; struggling to deal with the information that comes pouring in unbidden, they often withdraw into
themselves. They avoid busy places, such as supermarkets or public
transport. They do not dare to talk about this new and unwanted
quality for fear of being rejected or even declared insane. After all,
other people tend to become very nervous when they know that somebody "can look right through them."

I couldn't talk about it, or I would have been committed to an institution.

Another post-NDE phenomenon is the ability to see auras. Even
children sometimes spontaneously report seeing all kinds of beautiful colors around people. Equally, NDErs are often aware of invisible
energy fields with which they can help others. Indeed, some decide to
volunteer in the palliative or terminal care sector, not just to share the
insight that death is not the end, but also to help terminal patients by
bringing peace of mind and alleviating pain.

Psychological Problems After an NDE

Were I a composer of books, I would keepe a register,
commented of the divers deaths, which in teaching men to die,
should after teach them to live.
—MICHEL DE MONTAIGNE

The NDE remains a largely unfamiliar and misunderstood phenomenon, as society cannot readily accommodate this kind of spiritual
experience. The resulting tension can lead to psychological problems.
Talking about the NDE is usually impossible. Prejudice means that
when people do try to broach the subject, they are often misunderstood and even ignored or ridiculed. Sutherland's study shows that
when people tried to discuss the NDE, 50 percent of relatives and 25

percent of friends rejected the NDE, and 30 percent of nursing staff, 85 percent of doctors, and 50 percent of psychiatrists reacted negatively.[23]

Despite the positive memories of my NDE, it was also a very lonely period because of the lack of understanding (and the resulting fear and self-aggression) I felt from those around me.

The issues arising from the inability to share the NDE with others—that is, the negative interaction with others—are known as "interpersonal" problems.[24] They include a feeling of exclusivity or a sense of isolation from others without an NDE; a fear of being ridiculed or rejected by others; problems reconciling personality changes with the expectations of friends and family; an inability to communicate the meaning and impact of the NDE; difficulties maintaining old patterns that have lost their meaning since the NDE; trouble accepting the limitations and shortcomings of human relationships compared with the unconditional love experienced during the NDE; conflicts with relatives about the marked personality changes, seen as a "social death of the former personality"; and sometimes absurdly high expectations of relatives who learned about positive post-NDE changes from popular science writing, radio, or TV programs.[25]

Friends and family often struggle to accept personality changes, but they may also put the NDEr on a pedestal, expecting an all-forgiving personality to emerge, with the patience of a saint, miraculous healing powers, and the gift of prophecy. And if the NDEr fails to meet these unrealistic expectations, disappointment and rejection ensue.

I had to find my place in the world, but how? I felt so vulnerable. For years I couldn't stand myself. For a long time I couldn't enjoy anything and suffered the vicissitudes of life.

People are transformed by their NDE. And nobody feels this more acutely than a spouse. The NDEr is no longer the partner he or she originally married. It is on account of such relationship problems that

Nancy Bush cites a divorce rate of up to 75 percent.[26] However, a number of social factors also play a role in divorce: a loss of interest in money, "social death" caused by the loss of a former job, estrangement within the family, and an inability to function in our material and competitive society. For all these reasons, people with an NDE struggle to resume their former role in everyday life and find it difficult, and in some cases nearly impossible, to maintain human relationships, with all their earthly limitations. Because of the negative reactions from those around them, people may gravitate toward repression and denial, which make the integration process even harder. In fact, some NDErs begin to doubt the veracity of their experience and keep silent for a long time.

> I knew nothing of NDEs, I thought I was the only one, and I had the impression that others thought I was a bit loopy.

The problems that affect people primarily on the inside, known as intrapersonal or intrapsychic problems, include constant anger and depression over the forced return to life.[27] The difficult and painful acceptance of the return is called the reentry problem. Other problems may include reconciling the NDE with previous religious and philosophical beliefs or overidentifying with the experience and seeing oneself first and foremost as an NDEr. Some people doubt their own sanity, afraid that the NDE is a manifestation of mental instability. At times, the NDE and the ensuing personality changes seem scarcely credible. People feel different, occasionally even better and more privileged than people without an NDE, but keep their distance for fear of being ridiculed or rejected. It is very difficult for NDE survivors to explain to others how and why they have changed so much. What follows is a period of intense loneliness coupled with feelings of depression at the rejection of what they perceive to be the most impressive experience of their life. The awareness of being back in the sick body, with all its physical pain and limitations and sometimes permanent symptoms, causes frustration and keen nostalgia, whereas the NDE itself had been a beautiful, blissful experience. Physical and psychological adjustment to everyday life can hinder the integration of the new values and insights and lead to a posttraumatic stress disorder, and even (though seldom) to suicidal tendencies.[28]

Coming to terms with a frightening NDE is even harder because such an experience is often accompanied by strong feelings of guilt. Just like a positive NDE may leave a person feeling privileged, a hell experience may prompt revulsion and in many cases self-rejection. People have a tendency to fully identify with the negative experience rather than acknowledge that every experience, including a frightening NDE, merely represents one aspect of their personality. Talking to others about the frightening NDE is virtually impossible. The result is total silence, which generally precludes any chance of integration and actually increases the fear of death.

If the NDE was experienced during a (failed) suicide attempt, people usually refrain from making a second attempt because the experience taught them that they remain burdened with the problems they tried to escape. They come to realize that it is better to solve problems in this life because the other dimension offers scant opportunity for solving their earthly problems.[29]

The Effect of Time on Processes of Change in People with and Without an NDE

What is the role of time in the process of coming to terms with both the psychological problems and the positive changes? In 2004 the psychologist Igor Corbeau carried out a survey in the Netherlands, asking 84 people about their problems coming to terms with their NDE. The effect of time on easing the problems received ample attention, and as far as I know this study is the only published research on specific psychological problems and changes after an NDE. The participants in this carefully executed but retrospective study were asked about interpersonal and intrapsychic problems in the first year after their NDE, in subsequent years, and at the time of the survey. The NDE had occurred at a mean age of twenty-nine (as early as two and as late as sixty) while the average interval between the NDE and the survey was twenty-four years (from three months to seventy-seven years). People were asked to indicate whether they had experienced minor, moderate, or major problems in coming to terms with certain aspects of their NDE. During the first and subsequent years, more than half of the people surveyed experienced moderate to major problems in nine of the fourteen subjects addressed: urgent need to talk about the NDE; anger, sadness, or depression at being back; a feeling that the NDE was everything and normal life had lost its appeal; feeling rejected by others without an NDE; fear of being considered crazy by others; other people's difficulties accepting personality changes; inability to communicate the meaning of the NDE; absence of unconditional love in human relationships; and discomfort with enhanced intuitive sensitivity.[30]

By the time the survey was done, an average of twenty-four years after the NDE, most of the problems had been dealt with to some degree of satisfaction. The results also show that the higher the percentage of positive responses to their personality changes, the better the NDErs were capable of dealing with the problems. That said, at the time of the survey, more than half remained incapable of communicating effectively about their experience. The absence of unconditional love in human relationships also continued to be a problem for more than half of the respondents.

Until the Dutch study that my colleagues and I conducted on survivors of cardiac arrest, there had never been any prospective, longitudinal research into the effect of time on processes of change in cardiac arrest survivors with and without an NDE. The published studies were all retrospective and often failed to mention the period of time elapsed between the NDE and the interview. In our study patients filled in identical questionnaires based on Kenneth Ring's life-change scale two and eight years after the cardiac arrest. The table on p. 68 contains the answers to twenty-eight of the thirty-four questions asked. The answers to the remaining six questions revealed only minor change (up to 36 percent) after eight years.[31]

Over time, change increases in nearly all cases. After *two years* more than two-thirds (66 percent) of the people with an NDE cited an appreciation of ordinary things and a sense of social justice as the only positive changes. However, after *eight years* at least two-thirds of the NDErs had experienced positive change relating to: showing emotions;

wanting to help others; accepting others; showing love; compassion for others; empathy with others; involvement in family; interest in meaning of life; importance of nature and environment; appreciation of ordinary things; and sense of social justice (see table "Life Changes After a Cardiac Arrest"). In many cases the positive changes had gotten under way only in the course of eight years because the problems of coming to terms with the NDE gradually ease as communication improves.

What struck us was that in due course cardiac arrest survivors without an NDE also showed change, albeit much later and to a much lesser extent. After eight years, more than two-thirds (66 percent) indicated changes relating to: listening to others; empathy with others; interest in meaning of life; what matters in life; and meaning of faith.

After eight years there were still some big differences between cardiac arrest survivors with and without an NDE. At the eight-year follow-up, the occurrence of or increase in enhanced intuitive sensitivity was only reported by people with an NDE. Another noticeable difference was the importance of a higher standard of living, which usually decreased among people with an NDE and increased among those without the experience. I have already mentioned the sharp decline in interest in spirituality among cardiac arrest survivors without an NDE compared to the greater interest after an NDE. Church attendance, by contrast, was clearly down after an NDE and slightly up without one. One surprising finding was that after eight years people without an NDE were often better listeners than people with an NDE while people without the experience attached more importance to the opinion of others.

To recap: after eight years, people with an NDE scored significantly higher in the following areas: showing emotions; less interest in the opinion of others; accepting others; compassion for others; involvement in family; less appreciation of money and possessions; increase in the importance of nature and environment; less interest in a higher standard of living; appreciation of ordinary things; sense of social justice; inner meaning of life; decline in church attendance; increased interest in spirituality; less fear of death; less fear of dying; and increase in belief in life after death. These different levels of

Life Changes After a Cardiac Arrest

	Increase (percent) with NDE		Increase (percent) without NDE	
	At 2 years	At 8 years	At 2 years	At 8 years
Understanding oneself	58	63	8	58
Positive outlook on future	26	57	58	50
Understand purpose of life	52	63	25	50
Showing emotions	42	78	16	58
Wanting to help others	26	73	8	58
Opinion of others	−31	−21	8	41
Listening to others	47	52	8	75
Accepting others	42	78	16	41
Showing love	52	68	25	50
Compassion for others	47	73	41	50
Empathy with others	36	73	8	75
Involvement in family	47	78	33	58
Interest in meaning of life	52	89	33	66
Appreciation of money and possessions	−47	−42	−25	−25
Importance of nature and environment	47	84	33	58
Importance of higher standard of living	−25	−50	0	33
Appreciation of ordinary things	78	84	41	50
Sense of social justice	75	68	16	33
What matters in life	42	57	33	66
Meaning of faith	52	57	33	66
Sense inner meaning of life	52	57	25	25
Religious beliefs	36	47	16	25
Church attendance	−15	−42	8	25
Interest in spirituality	15	42	−8	−41
Interest in death	21	47	8	8
Fear of death	−47	−63	−16	−41
Fear of dying	−26	−47	−16	−25
Belief in life after death	36	42	16	16

The figures in the table reflect the percentage of people indicating a slight to strong increase in the extent to which they had changed. A negative percentage indicates a decrease.

change are a consequence of the NDE and not of surviving a cardiac arrest.

The percentages of change in the Dutch study are significantly lower compared to those in the retrospective studies by Ring and Sutherland, among others, which cite percentages of 80 to 100 percent for the same changes. The most likely explanation is the much longer passage of time between the NDE and the interview (typically more than twenty years) and the inclusion of NDErs who had come forward voluntarily, which resulted in much deeper experiences than those reported by the patients in the prospective Dutch study. In our study, many people with a deep or very deep NDE died shortly after their cardiac arrest so that they could not be included in the later interviews about change. Neither does our table reveal the number of people who had no fear of death or who already believed in an afterlife prior to their NDE.[32]

Spiritual growth is the sole purpose of our life here on earth.
—ELISABETH KÜBLER-ROSS

In summary, there is no such thing as a classic near-death experience or a classic way of dealing with it. The often difficult and painful process of coming to terms with an NDE and the resulting positive changes depend on the depth of the NDE, personality structure, cultural background, and, above all, social factors. The latter includes the sometimes positive but usually negative or skeptical response of friends, family, and health care practitioners, which often prevents communication about the NDE and thus significantly slows or halts the process of coming to terms with the experience. As a result, the integration process suffers a serious setback while psychological problems eclipse a positive and loving attitude to life. The process of change cannot get under way until people have shared their experience with others and feel that both they and their NDE are accepted. This in turn facilitates the integration of the subsequent changes. Advising people to write down their NDE might boost and perhaps accelerate the process of change. It may help them find the right words for the experience and perhaps also write to others about it.

Chapter Four

Near-Death Experiences in Childhood

What if you slept? And what if, in your sleep, you dreamed? And what if, in your dream, you went to heaven and there plucked a strange and beautiful flower? And what if, when you awoke, you had that flower in your hand? Ah, what then?

—SAMUEL TAYLOR COLERIDGE

Children who have a near-death experience remember the same typical elements as adults; but how is this possible when children have never heard of near-death experience or, in some cases, have not even learned how to read yet? When it comes to the veracity of NDE reports, some people continue to believe that NDErs are simply telling a story based on prior knowledge of the phenomenon or on religious expectations about the content of an NDE. But this does not apply to young and spontaneous children. It seems inconceivable that children without any prior knowledge could fabricate a story that is entirely consistent with the NDE reports of adults. Young and uninhibited, children will talk about what really happened to them. In this chapter we look at the near-death experiences of children because it is unlikely that children's NDE reports are the result of any outside influence.

When I was five years old I contracted meningitis and fell into a coma. "I died" and drifted in a safe and black void where I felt no fear and no pain. I felt at home in this place. . . . I saw a little girl of about ten years old. I sensed that she recognized me. We hugged and then she told me, "I'm your sister. I died a month after I was born. I was named after your grandmother. Our parents called me Rietje for short." She kissed me, and I felt her warmth and love. "You must go now," she said. . . . In a flash I was back in my body. I opened my eyes and saw the happy and relieved looks on my parents' faces. When I told them about my experience, they initially dismissed it as a dream. . . . I made a drawing of my angel sister who had welcomed me and repeated everything she'd told me. My parents were so shocked that they panicked. They got up and left the room. After a while they returned. They confirmed that they had indeed lost a daughter called Rietje. She had died of poisoning about a year before I was born. They had decided not to tell me and my brother until we were old enough to understand the meaning of life and death.

Scientific Research into NDE in Childhood

Scientific research shows that children can indeed have an NDE. Pediatrician Melvin Morse carried out the first systematic study of NDEs in childhood at Seattle Children's Hospital.[1] Over a ten-year period he interviewed 121 children who had been critically ill; 3 of them reported a hallucination, but none had actually had an NDE. He interviewed 37 children who had been given potentially mind-altering drugs, but none of these reported an NDE. But of the 12 children who survived a cardiac arrest or a coma, 8 (67 percent) reported an NDE. Morse's research showed that psychological stress brought on by hospital admission for a serious illness or the use of powerful drugs was not enough to precipitate an NDE in children. The fact that children experience NDEs only in genuinely life-threatening circumstances is at odds with the findings of research among adults; adults' fear of imminent death can sometimes trigger an NDE, a topic we will look at in more detail in a following chapter. Perhaps children do not experience fear of death because they are unfamiliar with the concept of death.

P. M. H. Atwater, a researcher who has had three NDEs herself, has carried out research into childhood NDE for many years.[2] She has spoken both to children who had an NDE and to adults who had one when they were young. Atwater writes that children can have an NDE at any age and that in the course of her research she came across extremely young children who, as soon as they could talk, told their parents about the experience or made a drawing. Children of up to three or four seldom have spontaneous memories of their experience. But there are exceptions: I have spoken to some adults who experienced an extensive NDE before the age of three and who were able to remember a great many details, even of their out-of-body experience. Children between the ages of three and six usually remember their NDE although children are really able to share their experience with others only from the age of twelve. With hindsight, people who had an NDE as a child, but no memory of it, realize that they have always been different from their peers:

> I had a relatively normal childhood and thought that every now and then all people "sensed" things and had dreams like me. These things were never talked about. You get used to everything, and I was busy doing the things young women do: studying, getting married, having children, teaching. But then I suddenly came up against something. . . .

People with an NDE at a very young age, but without any memory of it, sometimes have a second one later in life. During this second NDE they suddenly realize that they had one when they were little. They recognize aspects of their previous NDE even though the content of a second NDE is seldom identical to the first. The Dutch NDE study that my colleagues and I carried out produced striking evidence that cardiac arrest patients who had an NDE earlier in life were significantly more likely to have a second one than the other patients.

Circumstances That May Prompt an NDE in Childhood

The most common precipitants of NDEs in children are near-drowning and coma after head trauma, such as from a serious traffic accident.

Other circumstances include coma caused by diabetes or by an inflammation of the brain, cardiac arrest caused by life-threatening arrhythmia, imminent asphyxia caused by an asthma attack, diphtheria, muscular dystrophy, or electrocution. It also seems that the anesthesia that was given in the past for tonsillectomy was a fairly frequent cause in children for experiencing an NDE.

The Content of a Childhood NDE

The content of a child's NDE is similar in many respects to that of an adult although it usually contains fewer elements. Atwater found that for more than three-quarters of the many children she interviewed as part of her retrospective study, the experience begins on a positive note: a loving environment, a friendly voice, the encounter with a kind being or an angel, a sense of peace, and often an out-of-body experience coupled with perceptions of the body and the hospital surroundings, and traveling through a tunnel. About a fifth of the children witness a heavenly environment, while 3 percent report a frightening experience. Another possible element is the conscious return to the body, accompanied by feelings of disappointment because the experience had been so beautiful. During their NDE children are more likely to meet their deceased grandmother or grandfather than their parents. If an NDE were based purely on wishful thinking, one would expect children to meet living family members such as their father and mother. Children do encounter favorite pets that have died more frequently than do adults. At a very young age, children rarely experience a life review, but these are reported from the age of six onward. And finally, like adults, children find it extremely difficult to talk about their experience. Their attempts at sharing it with family and doctors often fall on deaf ears. Between thirty and fifty years may pass before they can discuss the experience and the sense of feeling different.[3]

Changes After a Childhood NDE

The research of both Atwater and Morse into the changes that may follow a childhood NDE suggests that children too undergo a number

The out-of-body experience of a six-year-old girl
during her near death experience.

of profound and typical changes that determine their outlook on life.[4] But a major difference is that adults, who had accumulated much life experience before their NDE, have to abandon old, accepted beliefs in order to integrate their new insights. Children, by contrast, have not yet been socialized into society's prevailing mores, and so their understanding of life cannot be said to have changed. They accept their insight into life and death as normal and neither realize nor understand that other children and adults do not share these insights. Parents and teachers sometimes think that the child is difficult because he or she freely challenges their norms and values: "That's not true, Mom!"

At a young age these children are not yet aware of being different from their peers. They accept death as a part of life. They experience the death of their dog or cat differently than do their siblings and friends, and they do not realize that their daily reality of an enhanced consciousness, with empathy for others through a heightened intuitive sensitivity, is not shared by other children. Children with an NDE listen beyond the spoken word; they understand why things are said in a certain way.

After a near-death experience children feel a fundamental sense of loss that they cannot put into words. The beauty and peace that they encountered during their experience are gone. They have a tendency to withdraw from their peers, and they often watch from the sidelines rather than mix with other children. They cannot bear piercing sounds or noise and tend to like peaceful, classical music from a young age. They have an unspoken need for a sense of security, comfort, understanding, warmth, and genuine interest and attention. At a young age they are sometimes found to communicate with invisible beings they call angels or friends. And although they suspect that their understanding surpasses that of their peers, they are incapable of talking about this. They do not want intellectual talk but prefer to be addressed at an emotional level. During religious instruction in school they can drive everybody to distraction with their never-ending questions. Mentally they are far too mature for their age, and they may run the risk of avoiding the playful behavior typical of most other children their age.[5]

By high school, about a third of these children develop alcohol- or drug-related problems, to which they may be much more sensitive than their peers. A short attention span, caused by a flood of (involuntary) impressions and resulting in disruptive or hyperactive behavior, sometimes earns them an ADHD diagnosis. At this age they can also become depressed or develop suicidal tendencies. This period is often characterized by repression rather than acceptance and integration.[6]

Children who have had an NDE are alert, astute, and often highly intelligent. They often go on to study philosophy, theology, or physics. Alternatively, they may opt for one of the creative professions, such as painting, photography, or music, to express the emotions they struggle to put into words. Keen to help others, some opt for a career in health care and become nurses, doctors, or social workers. They want to work in a field that allows them to draw on their heightened intuition.

It is up to parents, guardians, teachers, psychologists, and other health care practitioners to try and approach these children without prejudice. Information about NDE and its effects is absolutely crucial to better understand these children and help them grow up—to facilitate a process of integration instead of the often-subconscious process of repression.[7]

Spontaneous Out-of-Body Experiences (OBEs)

A report of an out-of-body experience at an early age does not necessarily mean that somebody had an NDE as a child. In young children especially, spontaneous out-of-body experiences also occur in situations that are not life-threatening. About 10 percent of the general population and perhaps up to 25 percent of children and adolescents have had a spontaneous sensation of being outside the body, usually on the threshold between waking and sleeping. These data have been extrapolated from a scientific survey among 475 Dutch psychology students in 1993. Of these students, 22 percent reported a spontaneous out-of-body experience while 7 percent reported two to five such experiences. A comparable study from the United States found spontaneous out-of-body experiences among 25 percent of students and among 14 percent of the rest of the population in the same city.[8]

There are reports of adults who, after a childhood NDE, had frequent spontaneous out-of-body experiences, both during the day and while sleeping. Their NDE not only created a sense of enhanced consciousness with heightened intuition, but also increased the incidence of spontaneous out-of-body episodes.

Sexual abuse and the threat of physical or mental abuse, which may trigger an out-of-body episode as a protection from pain and humiliation (so-called dissociation), are much more common among children than previously assumed. These episodes do not qualify as spontaneous out-of-body experiences. Child abuse often goes undetected, and social workers tend not to ask targeted questions about either abuse or dissociation with out-of-body experiences. Raising the subject with parents and children remains taboo. And children seldom volunteer to talk about their experiences. Because children not only cease to have physical pain during such an experience but can sometimes also see their body and what happens to it from a position outside and above their body, an out-of-body experience is more than just dissociation. Specialist literature defines dissociation as the escape from the frightening reality of a trauma or "the disruption of the normal integrated functions of identity, memory, or consciousness." It does not make explicit mention of the possibility of verifiable perception from above and outside the body. Kenneth Ring's research provides reasonable evidence of a link between out-of-body experiences caused by childhood trauma and the later occurrence of an NDE.[9]

An Example of an NDE in Childhood

People often wonder, with good reason, how someone can remember an NDE in detail after fifty years. How can she or he be sure that such an account is not a total fabrication? For a number of reasons this seems unlikely. First of all, when someone shares an NDE after many years, the emotions are unmistakable, as if the event happened only yesterday. In tears, this person tries to find the words to describe the experience for the first time. Second, in the Dutch study people used near-identical wording in the taped interviews after two and eight years. Their words were virtually the same as when they first talked about their experience a few days after their successful resuscitation (more about this in chapter 7). Partners and other family members also confirm that many years after the event, the NDE is related almost word for word the same. Finally, it is highly unlikely that somebody could fabricate such convincing evidence of the changes that generally follow an NDE, such as the loss of the fear of death, the newfound insights into what matters in life, and the enhanced intuitive sensitivity.

The following account of a childhood NDE was told to me in person. I asked the NDEr to write down the experience and the subsequent changes.

As a child I had a near-death experience. But I didn't realize it until nearly forty years later! At the age of fifteen, I had a beautiful, profound experience when I nearly drowned in the Nieuwe Waterweg, a canal between Schiedam and Rotterdam. My friends and I challenged each other to a race across the canal. But the current was much stronger than anticipated, and the water was freezing too. At one point I was completely exhausted, but mustering all my strength, I managed to reach a buoy.

While clinging to that buoy I felt that my situation was hope-less, and I was about to let go and drown. Suddenly I found myself in another world—a world full of wonderful, green-glowing hills, covered with the most beautiful flowers you could never find here on earth. The most gorgeous light bathed me in unimaginable tran-quillity. This was a world I never wanted to leave again.

How long I spent in that world I don't know, but suddenly I was back on earth. And the surprising thing was that I immediately knew what to do to escape my plight. As if someone helped me, my eyes were drawn to the frayed ends of a cable that tied a ship to the mooring buoy I was clinging to. I regained a bit of strength and reached the other side of the canal. This saved me from a cer-tain death by drowning. The assistance from the other side was palpable to me.

For a long time I was amazed by what had happened to me. I couldn't talk to a soul about it, not even at home, because five years earlier a younger brother of mine had drowned and I couldn't open old wounds by mentioning my experience. Strangely enough, I had a lot of paranormal feelings afterward. I could read people's minds, and I knew what they wanted. Luckily this wore off later, but the intuitive feelings lingered and grew stronger. And yet I had become a loner. I was unable to share many of my emotions, and I had be-come hypersensitive to other people's sadness. Events here on earth didn't leave me cold. I became a conduit for emotions that didn't really have anything to do with me. My loneliness increased. I kept clashing with colleagues, my partner, and society as a whole. I had become too different from other people.

Some forty years later, in 1992, I was watching television, and to my utter amazement I heard these people talking about the feelings I had and heard that theirs were triggered by a near-death experi-ence. After the broadcast I immediately called the number given in the program and shared my story for the first time. I related my story and was told that I'd had an NDE. I received the recognition I so craved. Since then I've undergone a radical transformation. I've read hundreds of books on the subject. I soaked it all up, as it were. . . . And later I had several more out-of-body experiences. . . .

Chapter Five

There Is Nothing New Under the Sun

No other insight is as useful to always bear in mind: It is not new, it is age-old.
—FREDERIK VAN EEDEN

Better chances of survival due to improved resuscitation techniques and treatment options have prompted a rise in NDE reports in the last thirty years. But a near-death experience appears to be a personal rediscovery of an age-old, cross-cultural, but seemingly forgotten knowledge. In the past these experiences were often known under different names, such as visions or mystical, religious, or enlightenment experiences. In antiquity they were referred to as journeys to the underworld. Throughout history there have been many different views on death, but across all times and cultures, people have been convinced that the human essence, usually known as the soul, lives on after the death of the body. This chapter features stories of life after death as found in Hinduism, Buddhism, Judaism, Christianity, and Islam alongside quotations from ancient Egypt, Greece, and the Roman Empire. The insights from these disparate historic texts show striking similarities despite the fact that they were probably rarely shared, either in person or in writing. The sources are diverse, and the nations and peoples that recorded these insights were isolated by great distances over land and sea as well as separated by temporal, cultural, and linguistic differences.

Nothing New

It is an age-old idea that the soul lives on after death, that our lives are judged (the life review during the NDE), and that afterward, depending on how we lived our lives, we may dwell either in blissful spheres or, as a punishment, in fearful environs. The oldest written example of this universal idea about death originates in ancient Egypt, where, according to the *Egyptian Book of the Dead* from the papyrus of Ani (around 1250 B.C.E.), people believed that after death the soul leaves the earth and sets off on a journey through the underworld, where a final challenge awaits the deceased before he can enter the hereafter: the judgment of Osiris. This involved the enumeration of a long list of sins, which the deceased denied committing. The truth of this denial was then established by weighing the heart against a feather. If the deceased spoke the truth, and the heart was not weighed down by sin, he or she would be ensured of blissful immortality within sight of the sun god. But a dreadful fate awaited those who failed the test of truth, as these sinners were condemned to eternal oblivion.[1]

This view is not confined to ancient Egypt but reappears across times and cultures. According to old books and manuscripts, the idea that the soul can be experienced independently of the body has been known for thousands of years.[2] In ancient India it was said, "Coming and going is all pure delusion; the soul never comes nor goes. Where is the place to which it shall go when all space is in the soul? When shall be the time for the entering and departing when all time is in the soul?" Thousands of years ago this belief in the mortality of the body and the immortality of the soul gave rise to the doctrine of preexistence and rebirth. Plato and other Greek philosophers held the same ideas about a mortal body and an immaterial and immortal soul. The ancient Greek philosophers were probably influenced by ideas from India, the Persian Empire, and Egypt.

But ideas about an immortal soul are not confined to Asia. They are prevalent among many other peoples, including most tribes in Africa, the Aboriginals in Australia, the indigenous peoples of the Americas, as well as the Vikings, Celts, and Romans. The Roman general Julius Caesar wrote in his *Gallic Wars,* "The druids wish to inculcate this as one of their leading tenets, that souls do not become extinct, but pass after death from one body to another, and they think that men by this tenet are in a great degree excited to valor, the fear of death being disregarded."[3]

And the Roman poet Ovid wrote in *Metamorphoses:*

> *Then, death, so call'd, is but old matter dress'd*
> *In some new figure, and a vary'd vest:*
> *Thus all things are but alter'd, nothing dies;*
> *And here, and there th'unbody'd spirit flies.*[4]

The idea that during life consciousness is not confined to the body and brain is not new either. I recently saw a drawing by Robert Fludd, a doctor and philosopher in seventeenth-century England.[5] He believed that our intellect, with all its mental processes, our memories and emotions, and our dreams and visions, is largely located outside of our brain (see figure). The drawing also clearly shows the supposed energy links with our physical body, and with the brain in particular, by way of the crown and the forehead.

Mystical Experiences as a Source of Insight into Death

Near-death experiences have long been a source of new insights into the possibility of life after physical death. Many books on the subject of an afterlife are based on the writer's own mystical or religious experiences. An example of this is *The Divine Comedy* by Dante Alighieri. Dante's magnum opus, more than fourteen thousand verses based on his own visions, tells of his seven-day journey through hell and purgatory to heaven. He is initially accompanied by the poet Virgil, but in heaven he consorts with his beloved Beatrice. He encounters the souls of dead people whom he recognizes and with whom he can communicate. In hell he meets people enduring eternal punishment for their intemperance (lust, gluttony, avarice, prodigality, wrath), for their violence (against others, oneself, God, and nature), and for their deceit (usury, fortune-telling, bribery, theft, false counsel, forgery, and betrayal of family, homeland, and God). In purgatory he meets dead people who

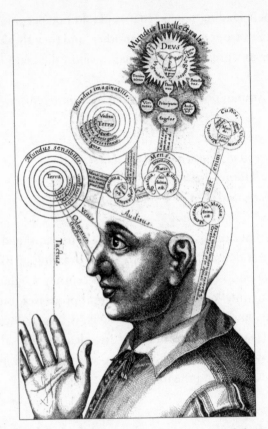

Robert Fludd, The Cabalistic Analysis of the Mind and the Senses. *1617.*

are there on account of pride, envy, wrath, lovelessness, avarice, gluttony, and lust. In paradise he encounters Beatrice, scores of saints, and ultimately God's eternal light. Dante wrote in the third part of *The Divine Comedy,* "Paradise (First Song)": "I have been in that Heaven that knows his light most, and have seen things, which whoever descends from there has neither power, nor knowledge, to relate."[6]

His descriptions echo what people try to express after an NDE:

My vision then was greater than our speech, which fails at such a sight, and memory fails at such an assault. I am like one, who sees in dream, and when the dream is gone an impression, set there, remains, but nothing else comes to mind again, since my vision almost entirely fails me, but the sweetness, born from it, still dis-

tils, inside my heart. . . . O Supreme Light, who lifts so far above mortal thought, lend to my mind again a little of what you seemed then. . . . Man becomes such in that Light, that to turn away to any other sight is beyond the bounds of possibility. Because the Good, which is the object of the will, is wholly concentrated there, and outside it, what is perfect within it, is defective.

In the eighteenth century the Swedish scientist Emanuel Swedenborg wrote *Awaken from Death* about the insights he derived from his many mystical experiences after the age of fifty-five.[7]

Between the late nineteenth century and the early twentieth century, several books were published in Europe on death and life after death based on people's own mystical experiences. The theosophist Annie Besant wrote *Death—and After?* while the anthroposophist Rudolf Steiner wrote *Death as Metamorphosis of Life*. Alice Bailey wrote *Death: The Great Adventure* while *Death Is an Illusion* describes the life's work of the nineteenth-century Danish mystic Martinus, who had his first mystical experience at the age of thirty.[8] All of these books draw on private mystical experiences to examine what happens after death. These texts, dating back more than a century, and the literature from antiquity and other times and cultures show a striking similarity with the content and consequences detailed in more recent NDE reports. In the past too, all fear of death was usually erased by an NDE and replaced by an inner conviction that consciousness survives physical death: "Dead turned out to be not dead."

Below are a few classic accounts, some dating back to the distant past, of what we now call a near-death experience. These reports underscore that a near-death experience is not just a contemporary phenomenon, made possible by modern resuscitation techniques. In the past these kinds of experiences were known under names that reflected the worldview or religious beliefs of the day. At the time, people spoke of enlightenment experiences, mystical experiences, religious experiences, or visions. In antiquity these kinds of experiences were described as a journey to the underworld or a sojourn with the gods, and in the early Middle Ages as a visit to paradise. These kinds of experiences were primarily ascribed to heroes, saints, and prophets, but sometimes also to normal people, as we will see in Plato's story about the soldier Er.

World Religions and Mystical Experiences

I quote here some reports with striking similarities between near-death experience and religious or mystical experiences from different world religions. The next sections feature stories of life after death as found in Hinduism, Buddhism, Judaism, Christianity, and Islam, but also includes stories about an immortal soul from ancient Greek philosophy.

Hinduism

Ancient India

The Upanishads are based on the Vedas, ancient Hindu stories that were passed down orally for thousands of years and documented around 800 B.C.E. In ancient Indian culture it was unthinkable that the human soul began with conception or birth. The ultimate aim of this belief was the realization that we are, in essence, immortal. This concept of immortality has no bearing on the physical body but rather involves the "Self." And the Self does not *become* immortal but is immortal already. As soon as a person becomes fully aware of this, the Self can unite with the Supreme (Brahman). Without this realization, self-regard evokes the illusion that we equal our mortal body. In this state, we remain caught in the cycle of birth and death.

Below are a few quotations from various texts from the Upanishads. The Katha Upanishad features an exchange between Naciketas, who offers himself as a sacrifice so that his poor father can keep his few worldly possessions, and Death, who tells him:

> The all-knowing Self was never born,
> Nor will it die. Beyond cause and effect,
> This Self is eternal and immutable.
> When the body dies, the Self does not die.
> If the slayer believes that he can kill

Or the slain believes that he can be killed,
Neither knows the truth. The eternal Self
Slays not, nor is ever slain.
Hidden in the heart of every creature
Exists the Self, subtler than the subtlest,
Greater than the greatest. They go beyond
All sorrow who extinguish their self-will
And behold the glory of the Self
Through the grace of the Lord of Love.
  . . .
The immature run after sense pleasures
And fall into the widespread net of death.
But the wise, knowing the Self as deathless,
Seek not the changeless in the world of change.
  . . .
The supreme Self is beyond name and form,
Beyond the senses, inexhaustible,
Without beginning, without end, beyond
Time, space, and causality, eternal,
Immutable. Those who realize the Self
Are forever free from the jaws of Death.
  . . .
When the ties that bind the Spirit to the body are unloosed
and the Spirit is set free, what remains then?
  . . .
What is here is also there; what is there,
Also here. Who sees multiplicity
But not the one indivisible Self
Must wander on and on from death to death.[9]

The Isha Upanishad features an almost literal description of end-less consciousness:

The Self seems to move, but is ever still.
He seems far away, but is ever near.
He is within all, and he transcends all.

Those who see all creatures in themselves
And themselves in all creatures know no fear.
Those who see all creatures in themselves
And themselves in all creatures know no grief.
How can the multiplicity of life
Delude the one who sees its unity?
The Self is everywhere. Bright is the Self,
Indivisible, untouched by sin, wise,
Immanent and transcendent. He it is
Who holds the cosmos together.

Contemporary India

The ancient wisdom is still alive in contemporary India, for instance in the work of the philosopher Swami Rama. He is known as the first yogi to submit to tests by Western scientists at the Menninger Institute in the United States. Swami Rama was found to be capable of controlling or altering automatic and unconscious bodily processes through willpower. For example, he could induce a seventeen-second arrhythmia of over three hundred beats per minute without losing consciousness; he could change his blood pressure and body temperature; he could manipulate the brain waves on his EEG into a pattern matching deep sleep; and he could perform feats of telekinesis (moving objects through mind power). Swami Rama writes:

> *It is not possible to understand what exists after death by intellectual arguments or discussions. The absolute Truth cannot be scientifically proven because it cannot be observed, verified, or demonstrated by sense perceptions. . . . That is why scientists cannot reach any concrete conclusions on the immortality of the soul and life hereafter, and nothing can convince them either. . . . The objective world is only one half of the universe. What we perceive with our senses is not a complete world. The other half, which includes the mind, thoughts, and emotions, cannot be explained by the sense perceptions of external objects. . . . The soul has not been created. It is essentially consciousness and is perfect. After*

the dissolution of the gross body, everything remains latent. The
soul survives. Our souls remain perfect and are not annihilated,
dissolved, or destroyed after death. . . . Life and death are only dif-
ferent names for the same fact—two sides of one coin. . . . Much
of the fear associated with death is the fear that death may be pain-
ful. The process of death itself is not painful; it merely changes
conditions. Lack of preparation and attachment are the cause of
the pain experienced at the time of death.[10]

According to this doctrine (Vedanta), pure consciousness has a
primary presence in the universe while our mind (our thoughts and
waking consciousness) is merely a spark or a reflection of this consciousness. The absolute or supreme consciousness is the source and
the foundation of the complete Self and of the entire universe.

Tibetan Buddhism

Buddhism originated in the fifth century B.C.E. in what was then
Hindu India, when Prince Siddhartha Gautama, the future Buddha,
was born. Buddhists believe in the cycle of death and rebirth. A person cannot die, but the body can. The terminal period can be either
extremely brief or last a considerable time, depending on the time the
consciousness needs to vacate the body. There is a continuity of the
spirit in a "spirit body," which is too subtle for the eye to see. Death is
like sleeping, and the bardo, the intermediate state immediately after
death, lasts a maximum of forty-nine days and is like a dream between
death and a new life. The soul begins its new life, according to the Tibetan Buddhists, usually on the fiftieth day after death.

The Tibetan Book of the Dead (Bardo Thödol) contains ancient
knowledge passed on orally from generation to generation. It may
have been disseminated in written form as early as the beginning of
our Christian era and was probably compiled in its current form in the
eighth century by Padmasambhava, the founder of Tibetan Buddhism.
In Tibet he is also known as Guru Rinpoche. The following quotation
from this book bears a great resemblance to an out-of-body experience:

When the consciousness-principle getteth outside the body, it sayeth to itself, "Am I dead, or am I not dead?" It cannot determine. It seeth its relatives and connexions as it had been used to seeing them before. It even heareth the wailings. . . . About this time the deceased can see that the share of food is being set aside, that the body is being stripped of its garments, that the place of the sleeping-rug is being swept; he can hear all the weeping and wailing of his friends and relatives, and, although he can see them and can hear them calling upon him, they cannot hear him calling upon them, so he goeth away displeased. At that time, sounds, lights, and rays—all three—are experienced.[11]

The ancient Tibetan texts that are recited to accompany the dying and the dead say:

Listen! When the expiration hath ceased, the vital-force will have sunk into the nerve-centre of Wisdom and the Knower will be experiencing the Clear Light of the natural condition. . . . At that time do not fear that bright, dazzling-yellow, transparent light, but know it to be Wisdom; in that state, keeping thy mind resigned, trust in it earnestly and humbly. If thou knowest it to be the radiance of thine own intellect—although thou exertest not thy humility and faith and prayer to it—the Divine Body and Light will merge into thee inseparably, and thou wilt obtain Buddhahood. . . . Be not fond of that dull bluish-yellow light from the human world. That is the path of thine accumulated propensities of violent egotism come to receive thee. . . . At that time fear not the glorious and transparent, radiant and dazzling green light, but know it to be Wisdom; and in that state allow thine intellect to rest in resignation.[12]

The dead spirit has to go through many more stages, but if the person fails to see everything for the illusion it is, the soul is headed toward rebirth:

Henceforth the body of the past life will become more and more dim and the body of the future life will become more and more clear. . . . Now the signs and characteristics of the place of birth will come. Enter upon the White Light of the devas, or upon the Yellow Light of human beings.

In his recent book *The Tibetan Book of Living and Dying,* Sogyal Rinpoche explains the ancient Tibetan wisdom of life and death and human transience in accessible terms. He frequently compares the ancient Tibetan insights with what Westerners know about the NDE and suggests that the former may help Westerners come to terms with death. His book is a good example of a modern interpretation of Tibetan Buddhism. He writes:

Isn't it about time now that the medical profession should understand that the search for the truth about life and death and the practice of healing are inseparable? . . . To learn how to die is to learn how to live; to learn how to live is to learn how to act not only in this life, but in the lives to come. To transform yourself truly and learn how to be reborn as a transformed being to help others is really to help the world in the most powerful way of all.[13]

Ancient Greek Philosophy

One of the greatest philosophers of all time is Plato (427–347 B.C.E.). He believed that the soul is imprisoned in the body and its sight restricted by the senses. Death to him was an awakening, a remembering of the eternal soul.

Plato attached more value to the immortal soul than to the mortal body. He was of the opinion that all human knowledge is memory, originating in a previous existence. It lies dormant in the soul as a memory and is awoken by a concrete perception. What we tend to regard as reality is, according to Plato, merely a shadow of true reality: the world of ideas. The realm of ideas, in Plato's philosophy, is a transcendent reality without space or time that is more real than the material world of concrete things.

In *Phaedo* Plato recounts what Socrates told his friends on the day he was to die by drinking from the poisoned cup: "And what is that which is termed death, but this very separation and release of the soul from the body?"

His friends observe, "But in what relates to the soul, men are apt to be incredulous; they fear that when she leaves the body her place may be nowhere, and that on the very day of death she may be destroyed and perish—immediately on her release from the body, issuing forth like smoke or air and vanishing away into nothingness."[14]

Socrates responds:

Like children, you are haunted with a fear that when the soul leaves the body, the wind may really blow her away and scatter her. . . . And is [death] anything but the separation of soul and body? . . . And being dead is the attainment of this separation; when the soul exists in herself, and is parted from the body and the body is parted from the soul—that is death. . . . Then the soul is more like to the unseen, and the body to the seen. . . . The soul is in the very likeness of the divine, and immortal, and intelligible, and uniform, and indissoluble, and unchangeable; and the body is in the very likeness of the human, and mortal, and unintelligible, and multiform, and dissoluble, and changeable. . . . If the immortal is also imperishable, the soul when attacked by death cannot perish. . . . That soul, I say, herself invisible, departs to the invisible world, to the divine and immortal and rational: thither arriving, she lives in bliss and is released from the error and folly of men. . . . And when the dead arrive at the place to which the genius of each severally conveys them, first of all they have sentence passed upon them, as they have lived well and piously, or not. . . . And these must be the souls, not of the good, but of the evil, who are compelled to wander about such places in payment of the penalty of their former evil way of life.

Medieval Jewish Mysticism

The Jewish tradition too teaches that death does not destroy the soul. Rather, death represents a transition from one level of consciousness to another, to a spiritual, disembodied consciousness.[15] The medieval

Midrash and Zohar are the best sources of Jewish views on the afterlife. A Hasidic legend tells the story of Rabbi Elimelech's encounter with his dead friend Chaim, as agreed before his passing; his friend tells him about life after death:

> At the moment of death I did not feel any pain. . . . After those who had washed and cleaned my body had done their duty, I wished to rise up and send them away. I could not do so and all seemed to me like a dream. After they had put me into the grave and covered it with the ground, and the people who had accompanied my body to the grave had gone home, I rose from my grave. . . . Suddenly a man as tall as from the earth to the sky appeared to me. . . . And he took me up from the earth and put me before the Heavenly Court. There they began to weigh and measure my record. . . . As I looked into Gehinnon I saw many people of my acquaintance and heard their outcries and painful sobs. . . . I was also able to see the great bliss of the Righteous in Gan Eden on the other side. [16]

The story bears a striking resemblance to a typical near-death experience. *The Jewish Book of Living and Dying* describes a great many Jewish mystical insights into death that are reminiscent of the contents of modern-day near-death experiences.[17]

In many ways these medieval Jewish views on human consciousness are similar to the aspects of consciousness that we will discuss later. The Kabbalah, for instance, teaches that the human essence, our consciousness or soul, is a complex phenomenon consisting of different layers. First, there is the individual essence, which survives physical death and is known as *nefesh* (literally, soul of the flesh). The next layer is called the integrating essence. This is part of the nefesh, though separate, and is called *ruach* (literally, spirit). It is the essence of awareness and forms a link with the next layer, the collective essence, or the soul of many individuals; it is akin to Jung's collective unconscious and is known as *neshama* (literally, breath). The next layer transcends the individual and collective aspects of consciousness to a nonindividual layer of consciousness known as *chaya* (literally, lifeforce). This life-force essence is the starting point for merging with the supreme, the ultimate consciousness. Here is the unity, omniscience, and love of the supreme consciousness, the divine or cosmic consciousness, *yechida* (literally, singularity or union with the transcendent).[18]

Christianity

The message spread by NDErs that love and acceptance of oneself and others is what matters most in life matches one of the central tenets of Christianity. Love and forgiveness are connected, just as Jesus preached: "You shall love your neighbor as yourself" (Mark 12:31; all biblical quotes from the NRSV). "In everything do to others as you would have them do to you; for this is the law and the prophets" (Matthew 7:12).

According to the gospel, Christ was capable of miracles and miraculous healings. He experienced visions and premonitions: he knew that he would be betrayed and put to death. In other words, he had what is now known as an enhanced intuitive sensitivity. Following Jesus's death on the cross, Christians believe he arose on the third day when his disciples recognized him in a new body. Within the Christian tradition a particular form of life after death is generally accepted, although the various Catholic and Protestant denominations all have their own, very different interpretations. At times, the chances of entering paradise seem to hinge on whether or not people have been baptized or confirmed rather than on their way of life; according to some Christian denominations, the possibility of eternal life is virtually nil for nonbelievers.

The Old and New Testaments feature many references to an imperishable soul and a material, perishable body. I have selected a few quotations on death and the soul's sojourn outside the body. Ecclesiastes 12:5–7 says about death, "Because all must go to their eternal home, and the mourners will go about the streets; before the silver cord is snapped, and the golden bowl is broken, . . . and the dust returns to the earth as it was, and the breath returns to God who gave it." About the dimension without time and distance, in which the whole past and future can be experienced in one's consciousness, Ecclesiastes 3:15 says, "That which is, already has been; that which is to be, already is; and God seeks out what has gone by."

Paul wrote in 2 Corinthians 5:6–8, "So we are always confident; even though we know that while we are at home in the body we are away from the Lord—for we walk by faith, not by sight. Yes, we do have confidence, and we would rather be away from the body and at home with the Lord." And in 2 Corinthians 12:2–4 Paul wrote, "I know a person in Christ who fourteen years ago was caught up to the third heaven—whether in the body or out of the body I do not know; God knows. And I know that such a person—whether in the body or out of the body I do not know; God knows—was caught up into Paradise and heard things that are not to be told, that no mortal is permitted to repeat."

For centuries the following quotes from the Bible have stirred the debate about the possibility of reincarnation. In John 3:6–7 it is written, "What is born of the flesh is flesh, and what is born of the Spirit is spirit. Do not be astonished that I said to you, 'You must be born from above.'" And in Matthew 11:13–14 we read that Jesus said, "For all the prophets and the law prophesied until John came; and if you are willing to accept it, he is Elijah who is to come." It was not until the Second Council of Constantinople in 553 C.E. that the Christian Church officially and definitively rejected the idea of rebirth.

The Christian tradition also features a great many mystics who claimed to have had direct contact with the divine during their ecstatic visions. In terms of content these visions are quite similar to near-death experiences. The best-known mystics are Francis of Assisi, Meister Eckhart, Teresa of Ávila, and John of the Cross. Some mystics even received the stigmata, displaying wounds to their hands, feet, and right side, something that is viewed by the Christian Church as a clear physical manifestation (materialization) of the Christ consciousness.

Islam

Muslims believe that the Qur'an is a direct revelation from Allah, which is why the scriptures are holy to them. There is no God but the One, and Muhammad (570–633 C.E.) is his prophet. Islam believes in the judgment at the end of time, when each person will be led before Allah for an individual trial. All will be accountable for their actions.

Surah 99:7–8 says, "Then shall anyone who has done an atom's weight of good, see it! And anyone who has done an atom's weight of evil, shall see it."[19] There are echoes here of the life review during an NDE. Islam too knows the concept of an eternal paradise for those who have lived a good and pure life, and also an everlasting hell for disbelievers and apostates is described, without any hope of redemption.

Similarities and Differences Between Various World Religions, and Parallels with NDE

For those readers who want to know more about the differences and similarities between the various world religions, and the parallels between these religions and the content and consequences of an NDE, several books can be recommended. In his book *Bijna dood ervaringen en wereldreligies: Getuigenis van universele waarheid* (Near-Death Experiences and World Religions: Testimony of Universal Truth), Bob Coppes focuses mainly on the parallels between these world religions and the content and consequences of an NDE. Of course this book is only one of many that describe the similarities between near-death experience and religious or mystical experiences within the various world religions, and especially between NDE and views on death within Christianity.[20]

Some Ancient NDE Reports

Reports of visions and mystical or religious experiences, often caused by life-threatening situations such as near-drowning, suffocation, exhaustion, or high fever, are prevalent across all times and cultures. Nowadays, we classify these cases as near-death experiences. The classical visions of three legendary people by the names of Er of Pamphylia, Aridaeus-Thespesius of Soli, and Timarchus of Chaeronea are relayed in the writings of Plato and Plutarch. I will quote here only from the vision of Er because it must be treated as the oldest report of an NDE. Readers who are interested in mystical experiences from ancient, medieval, and modern times may want to consult the book *Oth-*

erworld Journeys by Carol Zaleski as well a recent article about the three already mentioned ancient reports of near-death experiences.[21]

Plato: The Vision of Er

In *The Republic* Plato has Socrates narrate the myth or vision of Er. It is an extensive meditation on the destiny of the human soul after death and the way in which our next life on earth is determined. The quotation here has been slightly abridged but preserves the essence of Er's near-death experience. Socrates recounts:

> *I will tell you a tale . . . of a hero, Er the son of Armenius. . . . He was slain in battle, and ten days afterwards, when the bodies of the dead were taken up already in a state of corruption, his body was found unaffected by decay, and carried away home to be buried. And on the twelfth day, as he was lying on the funeral pyre, he returned to life and told them what he had seen in the other world.*
>
> *He said that when his soul left the body he went on a journey with a great company, and that they came to a mysterious place at which there were two openings in the earth; they were near together, and over against them were two other openings in the heaven above. In the intermediate space there were judges seated, who commanded the just, after they had given judgment on them and had bound their sentences in front of them, to ascend by the heavenly way on the right hand; and in like manner the unjust were bidden by them to descend by the lower way on the left hand; these also bore the symbols of their deeds, but fastened on their backs. He drew near, and they told him that he was to be the messenger who would carry the report of the other world to men, and they bade him hear and see all that was to be heard and seen in that place. Then he beheld and saw on one side the souls departing at either opening of heaven and earth when sentence had been given on them; and at the two other openings other souls, some ascending out of the earth dusty and worn with travel, some descending out of heaven clean and bright. And arriving ever and anon they*

*seemed to have come from a long journey . . . and those who knew
one another embraced and conversed, the souls which came from
earth curiously enquiring about the things above, and the souls
which came from heaven about the things beneath. And they told
one another of what had happened by the way, those from below
weeping and sorrowing at the remembrance of the things which
they had endured and seen in their journey beneath the earth (now
the journey lasted a thousand years), while those from above were
describing heavenly delights and visions of inconceivable beauty.*

*The story, Glaucon, would take too long to tell; but the sum
was this: He said that for every wrong which they had done to any
one they suffered tenfold; or once in a hundred years . . . and the
rewards of beneficence and justice and holiness were in the same
proportion. I need hardly repeat what he said concerning young
children dying almost as soon as they were born.*[22]

Er explains that the souls resume their journey after seven days and
in due course arrive at a place where three sirens sing of the past, present, and future. The souls present are told about their destiny and to
"behold a new cycle of life and mortality." Every soul is free to choose
his or her lot, and responsibility for the new life lies with the one who
chooses. As Er explains:

*And it was true of others who were similarly overtaken, that the
greater number of them came from heaven and therefore they had
never been schooled by trial, whereas the pilgrims who came from
earth, having themselves suffered and seen others suffer, were not
in a hurry to choose. And owing to this inexperience of theirs, and
also because the lot was a chance, many of the souls exchanged a
good destiny for an evil or an evil for a good. . . . Most curious,
he said, was the spectacle—sad and laughable and strange; for the
choice of the souls was in most cases based on their experience of
a previous life.*

Once everybody had chosen their next life, they went in the order
of their lots to the siren of the past, and via the siren of the present

to the siren of the future, and afterward they jointly marched to the plain of forgetfulness. It was evening when they set up camp by a river. They were all told to drink a certain quantity of water, and as soon as somebody drank they forgot everything. At midnight, after they had fallen asleep, it thundered and the earth shook. Suddenly they were driven upward in all different directions, heading toward birth, like shooting stars.

He himself was hindered from drinking the water. But in what manner or by what means he returned to the body he could not say; only, in the morning, awaking suddenly, he found himself lying on the pyre.

An Eighth-Century NDE Report

From the many fine early medieval examples of religious and mystical experiences, I would like to recount the story of Drythelm as chronicled in the eighth century by the Anglo-Saxon monk Bede. One evening Drythelm died of a serious illness but regained consciousness at sunrise the following morning; his grieving relatives were "terrified beyond measure" as he suddenly rose from his deathbed. Drythelm first told his detailed story to his wife and later to a monk, who in turn passed it on to Bede:

I was guided by a man of shining countenance and wearing bright robes. We went in silence in what appeared to me to be the direction of the rising of the sun at the summer solstice. As we walked we came to a very deep and broad valley of infinite length. It lay on our left and one side of it was exceedingly terrible with a raging fire, while the other was no less intolerable on account of the violent hail and icy snow which was drifting and blowing everywhere. . . . Since a countless multitude of misshapen spirits, far and wide, was being tortured in this alternation of misery as far as I could see, and without any interval of respite, I began to think that this might be hell, of whose intolerable torments I had often heard tell. But my guide who went before me answered my thoughts,

"Do not believe it," he said, "this is not hell as you think.". . .

As we entered this darkness, it quickly grew so thick that I could see nothing except the shape and the garment of my guide. . . . I saw, as the globes of fire now shot up and now fell back again ceaselessly into the bottom of the pit, that the tips of the flames as they ascended were full of human souls which, like sparks flying upward with the smoke, were now tossed on high and now, as the vaporous flames fell back, were sucked down into the depths. . . .

It was the one who had guided me before. . . . He began to lead me in the direction of the rising of the winter sun and quickly brought me out of the darkness into a serene and bright atmosphere. As he led me on in open light, I saw a very great wall in front of us. . . . When we had reached the wall we suddenly found ourselves on top of it. . . . There was a very broad and pleasant plain, full of such fragrance of growing flowers that the marvelous sweetness of the scent quickly dispelled the foul stench of the gloomy furnace which had hung around me. . . . In this meadow there were innumerable bands of men in white robes, and many companies of happy people sat around. . . . I began to think that this might perhaps be the kingdom of heaven of which I had often heard tell. But he answered my thought: "No.". . .

When we had passed through these abodes of the blessed spirits, I saw in front of us a much more gracious light than before, and amidst it I heard the sweetest sound of people singing. So wonderful was the fragrance which spread from this place that the scent which I had thought superlative before, when I savoured it, now seemed to me a very ordinary fragrance; and the wondrous light which shone over the flowery field, in comparison with the light which now appeared, seemed feeble and weak.[23]

Drythelm is then told to return to his body and promised that if he mends his ways he will be "among the joyous band of the blessed spirits" when he dies. When he regains consciousness he tells his wife, "Do not be afraid, for I have truly risen from the dead by which I was held fast, and have been permitted to live again among men; nevertheless, from now on I must live not according to my old habits, but in a much different manner." Drythelm accordingly distributed his possessions, retired to a Benedictine monastery, and lived a life of austerity, piety, fasting, and cold baths.

Bede was particularly impressed by the story's conclusion: "It is a greater miracle to convert a sinner than to raise up a dead man. . . . And it is an even greater miracle yet if the tale of a dead man's recovery and spiritual transformation change the hearts of its hearers."

The content of this detailed NDE and Drythelm's subsequent life changes are remarkably similar to Dante's *Divine Comedy* and to George Ritchie's account of his own NDE.[24]

A Nineteenth-Century NDE Report

In 1849 Admiral Francis Beaufort described the experience he had when, as a young man, he fell into Portsmouth harbor in the south of England and nearly drowned:

> *From the moment that all exertion had ceased . . . a calm feeling of the most perfect tranquility superseded the previous tumultuous sensations. . . . Though the senses were thus deadened, not so the mind; its activity seemed to be invigorated, in a ratio which defies all description. . . . The whole period of my existence seemed to be placed before me in a kind of panoramic review, and each act of it seemed to be accompanied by a consciousness of right or wrong. . . .*
>
> *My feelings while life was returning were the reverse in every point of those which have been described above. . . . Instead of being absolutely free from all bodily pain, as in my drowning state, I was now tortured by pains all over me.*[25]

Two Nineteenth-Century Medical NDE Reports

Nineteenth-century medical literature also contains reports of near-death experiences. In 1859 Dr. Brierre de Boismont described a woman who was delirious and who appeared to suddenly die. She was revived by those in attendance.

Instead, however, of thanking the persons who had taken such pains to restore her to life, she complained to them of their having recalled her soul from a condition of indescribable repose and happiness, such as it was not permitted to enjoy in this life. . . . She added that she had heard the sighs and lamentations of her father, and all that had been said with regard to her funeral.[26]

Frederic Myers published an article in 1892 about a doctor who appeared to have died of typhoid fever in 1889. His treating physician declared that "he was actually dead as fully as I ever supposed anyone dead." He had no perceptible pulse or heartbeat and was not breathing. Nonetheless, this patient and doctor had

a vivid and complex experience of seeming to leave his body and see it, as well as the actions of the people in the room. He went to a place of great beauty where he felt a presence and saw the face of an unidentified person who radiated great love. He also saw a dark cloud and dark pathway. He seemed to be given a choice of staying or returning, but when he chose to stay and tried to cross an apparent boundary, he was stopped from proceeding and then suddenly he found himself back in his body.[27]

Throughout the experience he had the sensation of being in a body that had "perfect health and strength," and he said that "memory, judgement, and imagination, the three great faculties of mind, were intact and active."

In conclusion, near-death experiences are by no means exclusive to the past thirty years, although there has of course been much more interest in the phenomenon since the publication of Raymond Moody's book *Life After Life*. It is astonishing to discover so many identical experiences throughout history. Across all religions and cultures, these experiences have greatly influenced views on death and the possibility of a life after this life. Of course local factors always prompt different interpretations. For example, for some communities, especially some indigenous peoples of Asia, Africa, and America, these experiences hint at the existence of a spirit world where ancestors dwell until their rebirth in another body.

Nowadays more and more questions are asked about how to explain the cause and content of these experiences that have been reported in all times. Many scientists assume that an NDE is caused by oxygen deficiency in the brain. Others believe that psychological reactions such as fear of death play a role, or that a combination of physical and psychological factors can offer an explanation. In the next chapter all existing scientific explanations for an NDE will be reviewed.

Chapter Six

Research into Near-Death Experiences

Skeptics and believers are all alike. At this moment scientists
and skeptics are the leading dogmatists. Advance in detail
is admitted: fundamental novelty is barred. This dogmatic
common sense is the death of philosophical adventure. The
universe is vast.

—ALFRED NORTH WHITEHEAD

In 1969 the psychiatrist Elisabeth Kübler-Ross managed to break the taboo in Western society against talking about death with her book *On Death and Dying*. In 1975 Raymond Moody greatly boosted interest in the topic with his descriptions of near-death experiences in his first and much-read book, *Life After Life*. His research into experiences of a lucid consciousness during a life-threatening situation and the possibility of a life after this life raised a lively debate and critical questions worldwide. Since then, a raft of fascinating but not always scientifically verifiable reports and books has been written on the subject.[1]

As interest in NDE proliferated, so did the question whether people who have this experience are special and whether an NDE occurs under exceptional circumstances. People began to speculate about a scientific explanation for these experiences. A frequently asked question was how people could have a lucid consciousness and memories during a spell of unconsciousness or coma, when there are objective indications of complete brain dysfunction. The prevailing scientific view that consciousness and memories are produced by the brain began to be called into question.

Most attempts at explaining near-death experience are based on anecdotal evidence of NDE and on retrospective studies involving a self-selected group of patients with a past NDE. This makes it impossible to ascertain the precise medical circumstances surrounding the NDE. Prospective research, meeting stricter scientific criteria and yielding better and more reliable results, is a much more recent trend.

This chapter starts with a brief overview of the history of scientific research into NDEs. This is followed by research data on the incidence of NDEs: who is affected, how often, and under what circumstances. Finally, this chapter reviews a number of theories with which scientists have tried to explain the incidence, the special character, and particular elements of NDEs.

The History of Scientific NDE Research

The first systematic study of near-death experience was written in 1892 by the Swiss geologist Albert von Sankt Gallen Heim. In the *Yearbook of the Swiss Alpine Club,* under the heading "The Experience of Dying from Falls," Heim described a total of thirty cases that were told to him in person; among them are the stories of Alpine mountaineers who survived near-fatal falls, including his own experience. He also included accounts of soldiers who sustained life-threatening injuries during war, of workers who fell off scaffolding, and of people who nearly drowned or died in accidents. He was the first to describe the experiences of people, such as soldiers, who actually nearly died of their serious injuries, and of those who merely expected to die, such as mountaineers. He concluded that people close to death or who were considered to be dead reported that "no grief was felt, nor was there paralyzing fright of the sort that can happen in instances of lesser danger. . . . There was no anxiety, no trace of despair, nor pain; but rather calm seriousness, profound acceptance, and a dominant mental quickness and sense of surety."[2]

MEDICAL RESEARCHER EXAMINING NEAR-DEATH EXPERIENCE

Between 1975 and 2005 a total of forty-two studies covering more than 2,500 patients with an NDE were published in scientific journals and in some monographs.[3] Many of these studies considered only a limited number of patients or drew on only letters instead of personal interviews. Yet despite the different designs, medical indications for the experience, and selection criteria for the interviewees, these studies produced remarkably similar results and conclusions about the content and consequences of an NDE.

Until 2000 most of the studies were retrospective, with participants coming forward in response to newspaper and magazine advertisements or appeals on radio and television. Retrospective studies thus lack a scientifically sound selection procedure. Because participation is random and voluntary (participants are self-selected), retrospective research can include people whose experience dates back ten to twenty years and whose medical and other circumstances can no longer be ascertained. Out-of-body experiences, for example, can no longer be verified.

It is not known why people do or do not volunteer. Those who are afraid or unwilling to come forward may have made earlier attempts at sharing their NDE with a doctor or relative but elicited negative reactions. The same is true for people who are unable to put their experience into words. It is equally likely that people who do not recognize their experience as an NDE will not come forward for research either. And finally, people may not be aware of the research project.

While these factors render the results of retrospective research less reliable, the studies have produced useful data on NDE, which will be presented below. Yet these results often fail to answer certain questions about the cause or content of an NDE or how a lucid consciousness can be experienced during a cardiac arrest.

In recent years a few prospective studies that address these problems have been initiated and published in medical journals, and in the next chapter I will look more closely at the content and conclusions of these studies. Prospective research is possible only in patients with a clearly diagnosed and critical medical condition. In a prospective study all participating patients are approached within a few days of their coma or cardiac arrest with the question whether they have any memory of their period of unconsciousness. All medical and other data can be carefully documented. The content and moment of any reported event during an out-of-body experience can also later be verified by doctors, nurses, and bystanders who were involved in the resuscitation procedure. A control group can be set up to establish if there are any differences compared to cardiac arrest or coma survivors without an NDE. All this gives a prospective study much greater scientific value.

The Incidence of NDEs

A representative Gallup poll in 1982 concluded that about 5 percent of the U.S. population may have had a near-death experience. A more recent survey in Germany, carried out in 1998, found a comparable figure of 4.2 percent.[4] The number of reported NDEs has probably risen in recent years because of growing media interest and also because professionals in the health, mental health, and social care sectors have become more interested in NDEs. And since coronary care units, intensive care units, and well-run emergency departments have been fitted with proper resuscitation equipment, patients have better chances of surviving a critical illness and experiencing an NDE. The fact that perhaps as many as one in ten people now report an NDE after a life-threatening medical crisis has made it easier to examine this phenomenon.

The different designs of the studies hinder efforts to determine the incidence of NDE in life-threatening circumstances. The more careful the design, the lower the percentage of reported NDEs. In retrospective studies the percentage of reported NDEs fluctuates greatly: between 14 and 80 percent or more of patients report an NDE. In studies among patients who were resuscitated at some point in the past, 27 to 43 percent of these (usually somewhat older) people report an NDE. Bruce Greyson's 1998 article offers a good overview of the figures. Three recent prospective studies among cardiac arrest survivors find figures ranging from 11 to 18 percent.[5]

The considerable differences between the percentages found in retrospective and prospective studies are due in part to the fact that the prospective studies ruled out self-selection of patients. But even the results of prospective studies cannot always be compared. For example, the Dutch study found that people who required several resuscitations after a cardiac arrest had a greater chance of experiencing an NDE. And people who had spent days or weeks in a coma after a complicated resuscitation and who required artificial respiration for a prolonged period of time were found to report significantly fewer NDEs. A good short-term memory appears to be essential for recalling the experience.[6]

The Effect of Age

All studies identify a link between age and the number of reported NDEs. The younger the patient, the greater the chance of an NDE report. Melvin Morse found a percentage of 67 percent in children. In his study, Kenneth Ring found that 48 percent of patients, with a mean age of thirty-seven, reported an NDE. Cardiologist Michael Sabom found that 43 percent of his patients, with a mean age of forty-nine, remembered an NDE. It must be mentioned here that Sabom's study included mostly cardiac arrest patients, which is comparable to prospective studies among heart patients with a higher mean age and fewer reported NDEs. In the retrospective NDE studies the life-threatening crisis was rarely caused by a cardiac arrest and the patients were younger. These retrospective studies found a higher percentage of NDEs at a mean age of twenty-two to thirty-two. Greyson and Ring also found more frequent NDEs, and more frequent deep NDEs, at a younger age. The Dutch study, in which the average age of patients was sixty-three, also found evidence of more frequent NDE reports below the age of sixty.[7]

Who Will Have an NDE?

People from all walks of life from all over the world have NDEs. The various studies found no link between the experience of an NDE and the following markers: social class, race, gender, standard of education, profession, hometown, or marital status. Prior knowledge of the existence and content of an NDE made no difference either. Researchers found no link between religion and the experience of an NDE. It did not matter whether people were Protestant, Catholic, Jewish, Muslim, Hindu, Buddhist, agnostic, or atheist. Neither did it matter whether or not people actually practiced their religion through regular church attendance. But because individual, cultural, and religious factors do play a role in coming to terms with the experience, adherents of the different religions describe and interpret the NDE differently.

Few intercultural NDE studies have been published. Brief mention should therefore be made of a prospective study from Japan, which looked at people who had been in a deep coma as a result of a cardiac arrest, an apnea, a brain hemorrhage, or a serious traffic accident. Upon waking from their coma, 37 percent of patients in this study reported an NDE with all the typical elements we know from Western studies. Further analysis found that in Japan too there were no differences between people with and without an NDE in terms of gender, profession, religion, or standard of education.[8]

Circumstances That May Prompt an NDE

As explained earlier, near-death experiences are reported under extremely diverse circumstances—not only in life-threatening situations, but also in situations without any physical or psychological danger.

Circumstances That May Prompt an NDE

A. *Brain function (seriously) impaired*

1. Cardiac arrest in patients suffering a myocardial infarction or serious arrhythmia
2. Coma caused by brain damage after a traffic accident or brain hemorrhage
3. Coma caused by near-drowning, especially in children
4. Coma caused by diabetes, asphyxia, or apnea
5. Coma caused by a failed suicide attempt or intoxication
6. Unconsciousness caused by shock (low blood pressure) as a result of:
   - Severe blood loss during or after a delivery or during surgery
   - An allergic reaction
   - A serious infection (sepsis)
7. Under general anesthesia, usually following complications from surgery
8. Electrocution (electric shock)

B. *Brain function unimpaired*

9. Serious but not immediately life-threatening illnesses with high fever
10. Isolation (such as shipwrecked people), extreme dehydration, or hypothermia
11. Depression or existential crisis
12. Meditation
13. Without a clear medical indication, such as a walk in nature
14. Similar experiences, so-called fear-death experiences, are reported after a seemingly inevitable death, such as almost having a traffic or mountaineering accident.

Although NDEs can occur under extremely diverse physiological and psychological circumstances, they are most frequently reported after severe impairment of brain function. But is there a reason why only a small number of patients, whether or not in a critical condition, report an NDE? Sabom, whose study looked at mostly heart patients, identified no differences that might explain whether or not an NDE occurred. The medical causes of the life-threatening situations, the methods of resuscitation, the estimated duration of unconsciousness, and the passage of time between the medical crisis and the interview all failed to provide an explanation. To our surprise, the Dutch study produced no statistical differences between patients with and without an NDE in terms of the duration of the cardiac arrest, the duration of the period of unconsciousness, or the medication administered. Psychological factors such as fear of death and prior knowledge of NDE, gender, standard of education, and religion played no role either. The study could not explain why some people do but most people do not experience a lucid and enhanced consciousness when there is no sign of brain function during a cardiac arrest.[9]

Tentative Conclusions About the Occurrence of NDEs

The numerous retrospective and few prospective studies provide conclusive evidence that NDEs can occur under diverse circumstances and not just in life-threatening situations. But no clear medical or psychological indicators have been found that explain why some people do but most people do not experience an NDE. Medication or demographic factors such as gender or standard of education play no role. The NDE occurs in all kinds of circumstances, in all ranks of society, in all sections of the population, in all religions, in all cultures, and in all times. Only a younger age seems to elicit more frequent NDE reports. The younger a person is, the greater the chance of an NDE.

The universal experience of a clear and enhanced consciousness during a period of deep unconsciousness, with lucid thoughts, emotions, and memories from earliest childhood and sometimes with perception from a position outside and above the lifeless body, raises fundamental questions. It resembles neither a dream nor the incoherent stories that are sometimes told upon waking from a coma with brain damage nor a hallucination. It bears no resemblance to the familiar side effects of medication or to memories of birth. But then what is it?

Theories About the Cause and Content of an NDE

Many scientists assume that an NDE is caused by oxygen deficiency in the brain. This used to be my own firm belief. Others believe that psychological reactions such as fear of death play a role or that a combination of physical and psychological factors offer an explanation. It is highly likely that the brain plays some role because certain NDE-like phenomena can be induced by stimulating a particular place in the brains of epileptics. The use of certain hallucinogenic drugs, such as LSD, also produces an altered state of consciousness. However, these actively induced altered states usually consist of fragmented experiences and memories and feature no life review or out-of-body experience. Besides, a drug-induced experience is seldom followed by reports of life changes.

Below, under the headings "Physiological Theories" and "Psychological Theories," I present a systematic review of most of the hypotheses that have been published in the years following Moody's first book. These two categories are complementary because the first covers certain physical functions (in this case brain function) and the second covers psychological functions. Each theory will be explained objectively and then appraised. Some NDE elements, specifically the lucid consciousness and verifiable perception during the loss or serious impairment of brain function, challenge the prevailing view of the relationship between consciousness and the brain, which sees consciousness as a product of brain function. This is why so many scientists struggle to understand near-death experience and why research into the subject can be seen as a threat to scientific dogma. Scientists do everything they can to explain the NDE with the help of existing theories and models and often end up giving a rather one-sided and simplified account of the NDE in an attempt to reconcile the comprehensive phenomenon with existing approaches. This has resulted in theories that can account for one or more aspects of the NDE but not for the complex phenomenon in its entirety. Other theories start from unverified and unverifiable assumptions or from speculation based on a few neurochemical studies of animal brains, which disqualifies them as a proper foundation for further debate.[10]

A theory that seeks to consider the special nature of an NDE and, where possible, explain the phenomenon within familiar frameworks will have to

- Acknowledge empirically proven elements of the NDE even when these do not conform to commonly accepted views;
- Establish a link between NDEs and the circumstances under which they occur in human beings; and
- Determine the special nature of NDEs on the basis of similarities and differences with phenomena that bear some resemblance to NDEs.

Our current knowledge appears to preclude an integrated theory of NDEs, forcing us to accept a multifaceted approach that aims to clarify individual aspects of the NDE. Bruce Greyson has produced a fine overview of the various biological aspects of an NDE while the psychologists Edward Kelly and Emily Williams Kelly provide a very well-documented historical overview of the different theories in the chapter on near-death experiences in their recent book, *Irreducible Mind*.[11]

Physiological Theories

Oxygen Deficiency

When a cardiac arrest disrupts the flow of blood to the brain or asphyxiation causes breathing to stop, the result is unconsciousness due to the *total* cessation of oxygen supply to the brain (anoxia). Breathing stops, all physical and brain-stem reflexes cease, and unless resuscitation is initiated within five to ten minutes, patients will die. However, in the case of oxygen *deficiency* in the brain (hypoxia), as seen in low blood pressure (shock), heart failure, or tightness of the chest, the result is not unconsciousness but confusion and agitation. Brain damage after waking from a coma is also associated with confusion, fear, agitation, memory defects, and muddled speech.

Nonetheless, the most common explanation for NDE is an extremely severe and life-threatening oxygen deficiency in the brain, resulting in a brief spell of abnormal brain activity followed by reduced activity and finally the loss of all brain activity. This results in the blockage of certain receptors in the brain and the release of endorphins, a kind of morphine produced by the body itself, causing hallucinations and a sense of peace and bliss.[12]

This theory seems inapplicable, however, because an NDE is actually accompanied by an enhanced and lucid consciousness with memories and because it can also be experienced under circumstances such as an imminent traffic accident or a depression, neither of which involves oxygen deficiency. Moreover, a hallucination is an observation that is not rooted in reality, which does not apply to descriptions of out-of-body experiences that are open to verification and corroboration by witnesses. In an out-of-body experience, patients during resuscitation have perceptions from a position outside and above their lifeless body, and doctors, nurses, and relatives can later verify the reported perceptions. They can also corroborate the precise moment the NDE with out-of-body experience occurred during the period of CPR. Besides, one would not expect hallucinations when the brain no longer functions because they require a functioning brain. Hallucinations will be discussed later on in this chapter.

Regarding the tunnel experienced by many NDErs, according to the psychologist (and consulting editor of the *Skeptical Inquirer*) Susan Blackmore, one possible explanation is oxygen deficiency in the (visual) cerebral cortex; others speculate that the tunnel experience is caused by the disruption of oxygen supply to the eye, gradually darkening one's range of vision and leaving only a short-lived pinprick of light in the middle that would be the tunnel.[13] However, a tunnel experience is accompanied by a sense of high speed, meeting deceased relatives, and sometimes by hearing beautiful music. Oxygen deficiency in the eye cannot explain this.

A study of fighter jet pilots is often cited as a possible explanatory model for NDE. Having been placed in a centrifuge, these pilots experienced momentary oxygen deficiency in the brain when the enormous increase in gravity caused their blood to drop to their feet. Fighter jet pilots can indeed lose consciousness and often experience seizures, like those seen in epilepsy, or tingling around the mouth and in the arms and legs as well as confusion upon waking. Sometimes they also experience elements that are reminiscent of an NDE, such as a kind of tunnel vision, a sensation of light, a peaceful sense of floating, or the observation of brief, fragmented images from the past.[14] They also see images of living persons but not of deceased people. There are no reports of a life review or out-of-body episodes. Life transformations, such as those often reported after an NDE, are not reported after such an event. In other words, these experiences are not identical to an NDE.

A similar kind of unconsciousness, sometimes accompanied by the experiences reported by pilots, occurs after fainting induced by hyperventilation (forcibly deep breathing) followed by a so-called Valsalva maneuver. The latter involves trying to push air from the body with the mouth and nose closed, which slows the heartbeat and lowers blood pressure and results in a short-lived oxygen deficiency in the brain. The effects of this type of faint have also been wrongly compared to an NDE.[15]

Carbon Dioxide Overload

Oxygen deficiency is accompanied by an increase in carbon dioxide in the body. This increased level of carbon dioxide in the blood has been cited as a possible cause of NDE. Over fifty years ago the Hungarian neurologist Ladislas Meduna, attempting a kind of treatment for his patients, asked people to breathe in carbon dioxide. Some experienced a sense of separation from the body, with occasional reports of a bright light, a tunnel, a sense of peace, or memory flashes. These images were quite rare, were usually extremely fragmented, and never involved a life review or an encounter with deceased persons. No process of life change followed.[16] In other words, inhaling carbon dioxide does not cause some of the characteristic NDE elements.

One practical problem is that during a frantic resuscitation it is difficult to measure these gases (oxygen and carbon dioxide) in the blood and impossible to measure them in the blood vessels in the brain. On the rare occasion when blood gases have been measured during resuscitation, it was usually only once the heartbeat and blood pressure had been stabilized, with the patient still unconscious on a ventilator and receiving extra oxygen.[17] In these cases the blood sample was taken from a vein or an artery in the arm or leg, and if the patient had an NDE, the level of oxygen saturation in the blood had been exceptionally high and the level of carbon dioxide extremely low.

Chemical Reactions in the Brain

Ketamine

Because low doses of ketamine, a drug formerly used as an anesthetic, can cause hallucinations, it has been postulated that this kind of substance is released in the brain during a period of stress or oxygen deficiency. Ketamine produces hallucinations because it blocks certain receptors (NMDA) in the brain. A small quantity of ketamine gives some people a sense of detachment from the body or tunnel experiences.[18] There are no known reports of an encounter with deceased persons or of a life review, nor have there been reports of positive changes. Ketamine usually causes such frightful and bizarre images, which are recognized as hallucinations, that research subjects prefer not to have the substance administered a second time. Because naturally occurring ketaminelike substances have never been found in the brain, this potential explanation must be abandoned. However, we cannot rule out that in some cases the blockade or malfunction of NMDA receptors may play a role in the experience of an NDE.

Endorphins

One of the first attempts at explaining an NDE was based on the fact that stress releases endorphins. These are morphines occurring naturally in the body in small quantities, which function as neurotransmitters. They are released in large quantities during stress. Endorphins can indeed get rid of pain and cause a sense of peace and well-being. However, the effects of endorphins usually last several hours whereas the absence of pain and the sense of peace during an NDE vanish immediately after regaining consciousness. Endorphins also fail to explain other elements of an NDE.

Psychedelics: DMT, LSD, Psilocybin, and Mescaline

The psychoactive substances DMT, psilocybin, and mescaline (but not LSD) are found in large quantities in nature, especially in plants in South America and Mexico but also in mushrooms ("magic" mushrooms).[19] They have been used for centuries in potions, powders, and inhalants to induce mind-expanding experiences. Dimethyltryptamine, or DMT, is effective only when it is injected into a vein or is being inhaled, but its effect is extremely short-lived because DMT is broken down very rapidly by the body.

All of these psychoactive substances are closely related to the neurotransmitter serotonin, which is found in large amounts in the body, and their chemical structure derives from tryptamine. These psychedelic substances have the same S2 receptor binding site in the brain as serotonin. DMT is produced in the pineal gland, or epiphysis cerebri. The pineal gland, which does not consist of brain tissue, is close to the emotional, visual, and auditory centers of the brain and transmits its substances directly to both the brain and the blood. The substances produced in this gland are responsible for regulating the body's water balance and sleep–wake rhythm and for developing the sexual glands until puberty. Perhaps they also play a role in dreams. The pineal gland also contains substances that can convert serotonin into DMT and substances capable of blocking the enzymatic breakdown of DMT. The latter also occur in plants, and because it greatly enhances the effect of DMT, the combination is used in ayahuasca in the Amazon.[20]

DMT is found not only in various parts of the brain but also in the lungs, the liver, the blood, and the eyes. DMT has an extremely short life span: it is quickly broken down by certain enzymes and is capable of passing the blood-brain barrier. This barrier in the blood vessels of the brain prevents certain substances from leaving the bloodstream, thus protecting the brain from their effects.

The production of DMT is stimulated by the adrenocorticotropic hormones cortisol, epinephrine (or adrenalin), and norepinephrine (or noradrenaline). These hormones are also found in the brain. During major psychological or physical stress, such as a traffic accident, a cardiac arrest, or acute pain, the body releases large quantities of these adrenocorticotropic hormones, which in turn activate a large amount of DMT. During the process of dying, large quantities of DMT are thought to be released by the death of cells in the pineal gland. During deep meditation, however, normal bodily function changes; serotonin levels and probably also DMT levels in the blood increase while cortisol and epinephrine (or adrenaline) levels drop.[21]

The experience induced by psychoactive substances is often surprisingly similar to a near-death experience, especially in the case of DMT although, depending on the dosage, confusing or frightening perceptions may also occur. These substance-induced experiences include the following elements: a sense of detachment from the body, out-of-body experiences, lucid and accelerated thought, an encounter with a being of light, a sense of unconditional love, being in an unearthly environment, access to a profound wisdom, and wordless communication with immaterial beings. Sometimes the characteristic post-NDE transformation, including the loss of the fear of death, is also reported after administration of DMT or LSD.[22]

It is a new and surprising hypothesis that DMT, which occurs naturally in the body, could play an important role in the experience of an enhanced consciousness during near-death experiences. Perhaps DMT, its release triggered or stimulated by events in our consciousness, lifts our body's natural inhibitions against experiencing an enhanced consciousness, as if it is able to block or disrupt the interface between consciousness and our body (and brain). Mention should be made here of the fact that zinc is essential for the synthesis of serotonin and related substances such as DMT. At a more advanced age, the body has lower levels of this metal, and, as mentioned earlier, NDE reports are less common at an older age.

Electrical Activity of the Brain

Epilepsy

An epileptic seizure is characterized by a kind of electrical storm, a short-circuit, which wipes out the electrical (and magnetic) activity in a certain area of the brain. As a result, normal activity of the brain cells (neurons) is blocked in that part of the cerebral cortex where the epileptic seizure originated. An epileptic seizure that originates in an area of the brain close to the temporal bone, the temporal lobes, may trigger muddled observations, mystical feelings, déjà vu experiences, a sense of detachment from the body, and olfactory (smell) or visual hallucinations. Sometimes these seizures are accompanied by unconsciousness or involuntary movements. After such an epileptic seizure of the temporal lobe, most patients have no memory of what happened to their bodies. They only remember what happened in their minds.

On the basis of these data some researchers have proposed a link between the near-death experience and either an increased activity or the cessation of all activity in the brain's temporal lobes. But studies with both superficial and deeper electrodes show that the symptoms of temporal lobe epilepsy are caused by underlying (limbic) structures and not by the cerebral cortex itself. A detailed study among epilepsy patients also shows that the characteristic elements of an NDE are rarely mentioned after an epileptic seizure of the temporal lobe.[23] Some elements are quite similar to an NDE, but déjà vu experiences are also frequently mentioned by healthy people. Needless to say, temporal lobe epilepsy cannot explain an NDE precipitated by fear, depression, or isolation.

Stimulation

Debate about the role of the cerebral cortex in extraordinary experiences in our consciousness has been intensified by studies in which epilepsy patients are subjected to electrical or magnetic stimulation of the cerebral cortex. We know that local electrical stimulation, which is usually applied during brain surgery, results in an inhibition or blockage rather than a stimulation of the affected part of the cerebral cortex. This happens because the stimulation, like an epileptic seizure, wipes out the brain cells' electromagnetic field. The effect depends on the duration and intensity of the electrical energy administered.

Some researchers claim that stimulation can trigger an out-of-body experience. Through local electrical stimulation of the temporal and parietal lobes during brain surgery for untreatable epilepsy, neurosurgeon Wilder Penfield occasionally managed to evoke memory flashes (never a life review); experiences of light, sound, or music; dreamlike experiences; and once an incipient out-of-body experience, during which a patient indicated, "Oh, God! I am leaving my body." Although he treated many hundreds of patients over the years, no real out-of-body experience with verifiable perception ever occurred and no transformation was ever reported. The effect of this stimulation was, in many respects, quite unlike an NDE.[24]

In 2002 neurologist Olaf Blanke described a female epilepsy patient who, after electrical stimulation (blockage), had an incomplete out-of-body experience with a distorted view of only her lower legs. The title of his article in *Nature* suggested that he had managed to locate the place in the brain where out-of-body experiences originate. The article received extensive press coverage and caused quite a (premature) stir. In an article in 2004 Blanke produced another possible neurological explanation for out-of-body experiences.[25] He described six patients, of whom three had an atypical and incomplete out-of-body experience—that is, without perception from the ceiling with verifiable elements of themselves or their surroundings—and four patients with an autoscopy, who saw their own double from the vantage point of their own body. In his article Blanke describes an out-of-body experience as an "illusion" caused by the temporary dysfunction or impairment of the temporal and/or parietal lobes. An illusion is an apparent reality or a false sense of reality whereas an out-of-body experience involves a verifiable perception—from a position outside and above the body—of a resuscitation, traffic accident, or operation and of the surroundings in which these took place. An observation with verifiable aspects is, by definition, not an illusion.

As far as we know, not one of the thousands of stimulated epilepsy patients around the world has ever reported a genuine out-of-body experience. The fact that in a single case, as described by Blanke, an abnormal bodily experience was reported does not warrant comparison between this stimulated or impaired area in the brain of an epilepsy patient and the brains of normal individuals. Generalizing this finding seems more than unjustified. Given the fact that none of Blanke's small number of patients ever showed damage or dysfunction in exactly the same area, we cannot cite the effect of stimulation of a certain area of an epilepsy patient's brain as evidence that this specific area actually causes the effect.

Similarly, transcranial magnetic stimulation (TMS), in which magnetic fields are aimed at certain parts of the brain, sometimes precipitates experiences in the mind caused by blockage (or stimulation) of electromagnetic fields in the brain. The neuropsychologist Michael Persinger has carried out many of these experiments, and he believes that the reported experiences resemble an NDE. However, closer inspection of his articles disproves this. The reported experiences, such as dreamlike, semimystical episodes with light or music, or the sense of somebody's presence, bear only a vague resemblance to the elements of an NDE. Suggestibility (that is, a placebo effect) appears to be the overriding factor in these reported experiences because Persinger also reports experiences in 33 percent of people without magnetic stimulation and because a double-blind control of his research in Sweden failed to corroborate his results.[26] The Swedish researchers had not informed the participants beforehand why, if, and when they would receive magnetic stimulation.

The EEG and Sleep Disorders as a Result of an NDE

As part of a recent study among people who had an NDE in the past, an EEG (measuring electrical activity in the brain) was made during sleep.[27] The rapid eye movement (REM) phase is the phase of sleep in which people dream. Patients with an NDE were found to have fewer periods of REM sleep than a control group without an NDE. The EEG also found anomalies in the left temporal lobe and symptoms of temporal lobe abnormalities, such as unusual visual, auditory, or olfactory experiences, but these are unlike the experiences that are reported during NDE (see also the section about epilepsy in this chapter). The NDErs also experienced a different pattern of sleep. However, the patients in this study were studied only *after* their NDE, which precludes a comparison with the EEG and sleep pattern *prior* to their experience.

Another study also found REM sleep pathologies after an NDE.[28] This study looked at the frequency of so-called REM intrusion. REM intrusion is accompanied by a sense of paralysis and confusing perceptions (hallucinations) at the onset of sleep. The content of these perceptions do not resemble an NDE. A higher percentage of these symptoms (42 percent) was found in a self-selected group of people who had had an NDE in the past than in a control group that had been recruited among hospital staff and that reported a much lower percentage of REM abnormalities (7 percent) than is common among the general public (20 to 30 percent). The study's conclusion that the brain disorders that underpin REM intrusion may also precipitate NDE is, at best, premature for several reasons. The study was poorly designed; nearly 60 percent did not report REM intrusion after an NDE; and patients were examined only after and not before their NDE.[29]

These studies therefore do not warrant any conclusions about either a neurological basis of an NDE or abnormal brain activity prior to an NDE. We can conclude only that compared to a control group without NDE, people with an NDE have a verifiably different sleep pattern, coupled with EEG anomalies in the temporal lobe. Perhaps the physical and psychological transformation that takes place after an NDE can shed a new light on the registered changes in electrical activity in the brain.

Reviewing the physiological approaches described above, we can conclude that most of them fail to offer an adequate explanation because they are based on:

- Physiological causes, such as anomalies or disorders in blood gases, in chemical or electrical brain activity, or at specific locations in the brain, for which there is no or insufficient evidence that they play a role in the origins of an NDE; and/or
- Effects that are not or not entirely consistent with the typical

NDE elements, especially the most striking and distinctive elements such as out-of-body experiences with verifiable perception, a panoramic life review, or an encounter with deceased persons.

By contrast, research into the effects of psychoactive substances such as DMT has found some striking similarities with NDE elements. It is a new and surprising hypothesis that DMT, which occurs naturally in the body, could play an important role in the experience of an enhanced consciousness during NDE. Perhaps DMT lifts the body's natural inhibitions against the experience of an enhanced consciousness.

Psychological Theories

Fear of Death

It is of course possible that in life-threatening situations people consciously or subconsciously evoke an experience to escape the fear of imminent death. In such a case, stress triggers a defense against (or a flight from) such imminent death. Another possibility may be stress-induced wishful thinking based on cultural and religious expectations. As mentioned earlier, elements of an NDE are sometimes experienced in life-threatening situations; this is known as a fear-death experience.

Expectations

An article from the 1930s describes experiences that were triggered by critical circumstances and that may have met certain expectations. These experiences were called "shock thoughts and fantasies" and were believed to be invoked as a defense mechanism against mortal danger. The fact that some NDE elements are phrased in religious or cultural terms could be seen as evidence of such expectations. Research has shown differences in the incidence and content of some NDE elements between people in the West and the native peoples of the Americas and Australia, while in India differences have even been found between people from the north and south of the country.[30]

But for many people the content of an NDE does not match their prior expectations of death. Their experiences are identical, irrespective of whether they believe that death is the end of everything or whether they believe in life after death. Children experience the same elements as adults. Prior knowledge of NDE does not affect the incidence or content of the experience, nor has its content changed since the publication of Moody's first book on the subject in 1975.[31]

Depersonalization

Depersonalization refers to the phenomenon of identity loss, coupled with a sense of detachment, alienation, and unreality. People are divorced from the world and their own identity and feel that life is unreal or like a dream. The condition is often accompanied by unpleasant emotions, fear, panic, or emptiness. Out-of-body experiences are never reported in such cases. Depersonalization, which is particularly common among young adult women, allows people to completely disconnect from their surroundings.

During an NDE, however, people retain their identity and experience an enhanced reality, a clear consciousness and lucid thought, and potential out-of-body episodes. An NDE is nearly always accompanied by feelings of peace and love, is reported by as many men as women, and occurs irrespective of age. Depersonalization also fails to explain NDEs among people who are not in a life-threatening situation.

Dissociation

The scientific literature defines dissociation as an escape from the frightening reality of a trauma by "the disruption of the usually integrated functions of identity, memory or consciousness." The definition makes no explicit mention of the possibility of verifiable perception from above and outside the body. This defense mechanism can kick in during physical or emotional abuse at a young age, such as sexual abuse or (the threat of) physical violence.

A study has shown that after a near-death experience some people are slightly more inclined toward dissociation, but never pathologically so.[32] And like depersonalization, dissociation cannot explain NDEs in people who experience them while not in a life-threatening situation.

Personality Factors

Could people with certain personality traits be more likely to have an NDE than others? The big problem with such questions is that systematic research prior to the NDE is impossible. Information about personality and character traits derives from retrospective research, which is usually carried out a long time after the NDE. However, these studies do tell us that post-NDE changes have consequences for someone's personality.

Generally speaking, NDEs occur in mentally stable people who function normally in everyday life and who, except in age, do not differ from control groups without an NDE.

Fantasies and Imagination

There has never been any evidence to suggest that before they had their experience NDErs were more inclined to fantasize. Nor does fantasizing explain why both children and adults across all ages and cultures have reported virtually identical NDEs. Similarly, the suggestion that an NDE is constructed on the basis of false memories or imagination can be refuted by the fact that people around the world report virtually identical NDEs.

The fact that some people are reluctant to share their profound and overwhelming NDE with others in the first few days after their experience and feel capable of disclosure only many years later does not necessarily mean that this experience is a false memory or fantasy. Evidence has shown that the content of an NDE, the words chosen to describe it, and the emotions it evoked remain essentially the same when, years later, people bring up their NDE again.

Fantasies During Out-of-Body Experiences

Some people continue to explain out-of-body experiences during an NDE as the product of fantasy and imagination. Those who are reluctant to accept a nonmaterialist explanation for the near-death phenomenon, especially, claim that reports of out-of-body experiences rest on fantasy and information about a resuscitation or operation shared by doctors and nurses afterward. In these critics' opinion, perceptions during unconsciousness or coma are by definition impossible.[33]

Psychologist Susan Blackmore maintains that the content of an out-of-body experience is simply a combination of memories, fantasies, lucky guesses, and expectations. She is not alone in this. These skeptics also assume that hearing remains intact despite the fact that the loss of brain function has rendered the patient unconscious. When heart patients without an NDE were asked to describe their resuscitation, they always made one or more essential errors, unlike patients who had an NDE during their resuscitation and who were able to recall surprising details of this procedure.[34]

Reports of out-of-body experiences are also said to be reconstructions of perceptions made shortly before losing consciousness or immediately after waking. However, research has shown that people who wake from a coma have no memories of the period just before or after their period of unconsciousness. And it is even harder to find a materialist explanation for perceptions at a considerable distance from the hospital or for verified perceptions by visually handicapped or blind people.[35]

Reports of out-of-body episodes can include verifiable facts that people could not have seen or heard with their normal senses and that doctors and nurses never mentioned afterward. These reported perceptions usually take place from a position outside and above the body and sometimes even from outside the room where the body lies. As mentioned, medical and nursing personnel were usually stunned by the level of detail patients knew about their resuscitation and almost always responded with surprise or disbelief.[36] The story of the dentures that were removed and stored during a resuscitation, which was published in *The Lancet* and told earlier, is inexplicable to most scientists because the patient knew details about his resuscitation and the appearance and actions of the doctors and nurses in attendance despite entering the hospital in a coma and being transferred to the intensive care unit for respiration while still comatose.

Out-of-body experiences are often difficult to corroborate if the NDEs took place many years ago. Additional prospective research is needed to verify out-of-body experiences shortly after resuscitation. But there are so many well-documented cases of people leaving their body, with a great many verifiable details, that it is virtually impossible to cast doubt upon them or to ascribe them to fantasy or imagination.

Memories or Fantasies During Surgery and Coma

Is it possible to experience consciousness during general anesthesia? Scientific research is divided on the question whether conscious memories of surgery under general anesthesia are possible. Memories of the surgical period are thought to be rare and usually conveyed in just a few simple words shortly after the operation. They tend to occur only after light sedation with a great deal of muscle relaxant.[37]

Registration of the electrical activity of the brain (EEG) in patients under general anesthesia has shown that unconsciousness during surgery is coupled with a clear change in the brain's electrical activity, with lower frequencies and less extreme discharges. This is consistent with the neurophysiological hypothesis that brain function is completely disrupted and impaired under general anesthesia. This picture is also confirmed by studies among patients under general anesthesia, in which neurological imaging techniques such as functional magnetic resonance imaging (fMRI) show a functional loss of nearly all the major brain functions. The same is true for the findings of fMRI research among patients in a vegetative state or some other form of coma. Although sounds and other stimuli continue to enter the brains of these patients, they no longer trigger the kinds of responses that are registered during conscious experiences under normal circumstances because the connections in the brain have been severed and information can no longer be exchanged.[38]

From a scientific point of view it therefore seems highly unlikely that under properly administered general anesthesia patients can have memories of the surgical period with lucid consciousness, with thoughts and emotions, and sometimes with perceptions from a position above the operating table. The same is true for patients in a coma.

And yet this exceptional situation exists. For some time now patients under general anesthesia have been known to perceive what a surgeon or anesthetist has said. Because current medical science does not yet widely accept the possibility of a near-death experience or an out-of-body experience, their later memories are in the medical literature referred to as "awareness" during surgery. The common assumption is that these patients were given insufficient anesthetic, but this is not always the case. Some patients are known to have awoken from general anesthesia feeling very agitated after overhearing somber comments about their prognosis during the operation. Under hypnosis some even brought up things that were said during surgery but of which they had no conscious memories.[39]

Awareness during general anesthesia is also experienced by patients with an NDE caused by complications during surgery (severe blood loss or a cardiac arrest), which leaves them with detailed memories of the operation, sometimes even with visual perception despite the fact that their eyes had been taped shut and their head blanketed with a drape. The many reports of near-death collected at the University of Virginia suggest that 23 percent of people experienced their NDE while under general anesthesia.[40] The fact that an NDE can be experienced during a coma, in which brain function is also seriously impaired, is an equally extraordinary phenomenon. When patients report memories of their operation or period of coma, we should not dismiss this as mere fantasy or question the amount of anesthetic administered or the coma diagnosis but instead seriously consider the possibility of an NDE.

Deceit

It has been alleged that NDErs tell deliberate lies to look interesting or to impress others. A personal meeting will quickly dispel such suspicions, not just because of what they say about the experience, but above all because of the obvious emotions and the struggle to find the right words when they share the experience. The fact that people often keep quiet for years for fear of rejection and that when they finally talk about their NDE they do so only reluctantly to a handful of friends also argues strongly against a deliberate lie to come across as interesting. Besides, it is tricky to fabricate a story about a different outlook on life and then actually show practical evidence of this change. Only a written and anonymous NDE account would make it harder to check whether and to what extent elements have been made up.

Memory of Birth

The description of a journey through a tunnel toward the light has prompted the suggestion that an NDE is a memory of birth or that it is an archetypical rebirth experience instead of a real memory. However, it is rare for an adult to be able to remember birth because the brain is believed to be insufficiently developed at birth. Besides, an NDE does not always involve a tunnel experience, and when it does it is as common among people who were delivered by natural birth as it is among those who came into this world via cesarean section.[41]

Hallucinations

Over the past thirty years there have been repeated claims that an NDE is a hallucination. A hallucination, however, is a sensory perception that is experienced as real by the hallucinating person but that is not rooted in reality. Hallucinations are unique and personal images with emotive elements, auditory impressions (sounds or voices), sensations of taste or smell, or visual images. Unlike an NDE, they contain no universal elements. A hallucination can feature vivid images and moving figures and can evoke a range of emotions, of which fear is usually the dominant one. A number of areas of the brain display an increased activity during hallucinations. But reports of a positive transformation are rare after such an experience. Hallucinations tend to be associated with psychiatric disorders such as schizophrenia and psychoses but also with migraines, (excessive) drug use, and alcohol withdrawal symptoms. By contrast, most people with an NDE are emotionally stable and did not use alcohol, medication, or drugs before their experience.

A hallucination is an observation without a basis in reality. The fact that an out-of-body experience during an NDE involves verifiable perceptions means that an NDE is, by definition, not a hallucination. The possibility of meeting and communicating with deceased people, of whose death one could not have been aware, also argues against hallucination. And NDErs who once hallucinated as a side effect of medication say that the contents of a hallucination and a near-death experience are extremely different.[42]

Dreams

Could the NDE be a kind of dream? Dreams usually occur during the REM phase of sleep, during which the brain displays a great deal of activity. An NDE, by contrast, is sometimes experienced at a moment when all brain activity has ceased. People with an NDE say that during their near-death episode they experienced a vivid reality, which was fundamentally different from anything they ever experienced in dreams. Dreams combine recent experiences with (subconscious) memories, and this can be coupled with strong emotions (nightmares).

Yet there are certain analogies between an NDE and a dream, such as the absence of time and distance: in a dream everything appears to be happening in a split second. Consciousness is so greatly enhanced during a dream that time and distance become irrelevant. Like NDEs, some dreams also feature visions of the future. In these prognostic dreams people dream certain events, such as a funeral, which come true years later and are then experienced as déjà vu. Similarly, people can meet deceased persons in lucid dreams, just as in NDEs. I believe we should take a closer look at the potential role of the aforementioned DMT in the origins and content of dreams.

As well as some similarities, there are clear differences between a dream and an NDE: in contrast to an NDE, a dream is usually forgotten after a while and is generally not followed by a transformation. And a dream usually has no fixed elements, unlike an NDE.

A summary review of the psychological approaches described above prompts the conclusion that most of them (expectations, depersonalization/dissociation, personality traits, hallucinations, fantasies, deceit) fail to explain the empirical features of an NDE. One psychological factor that may contribute to an explanation is that NDEs are precipitated not only by an actual life-threatening situation but also by a situation that is perceived—and feared—as such. A second lead for further research is provided by the similarities between NDEs and dreams (especially lucid ones); the role of DMT in both phenomena merits further attention.

Delusion Brought On by Medication

Another possibility is that certain types of medication, such as morphinelike substances or other strong painkillers that are administered to seriously or critically ill patients, could cause an NDE. Some believe that the near-death experience could be a delusion brought on by medication. But NDEs are frequently reported by people who do not use medication, so this assumption is incorrect.

Conversely, some medicines could have such a negative effect on memory that people are unable to remember their NDE. It remains unclear why only a small percentage of critically ill people report an NDE. People are kept in a coma while on a ventilator after an operation, a traffic accident, or a complicated resuscitation or they received general anesthesia during surgery. Is it possible that, as a recent article suggests, most of these patients have forgotten their NDE due to large doses of medication?[43]

This is not altogether likely because NDEs have been reported during surgery or coma, in which a heavily sedated patient was kept on a ventilator. The Dutch study included systematic research into the role of medication in NDE and statistically ruled out any effect.

In summary, a near-death experience is a special state of consciousness that arises during an impending or actual period of physical, psychological, or emotional death. Demographic, psychological, and physiological circumstances fail to explain why people do or do not experience an NDE.

Because an NDE can feature so many different elements, such as an out-of-body experience with verifiable perception, a tunnel experience, the experience of an unearthly environment, the sense of unconditional love in the presence of the light, the encounter with deceased persons, or a life review or preview, a range of different explanations have been put forward for each one of these elements. The various physiological and psychological factors set out in this chapter could all, to varying degrees, play a role but cannot explain the phenomenon in full.[44]

The theories on NDE set out above fail to explain the experience of an enhanced consciousness, with lucid thoughts, emotions, memories from earliest childhood, visions of the future, and the possibility of perception from a position outside and above the body. They also lack an adequate explanation for the fact that everything that is experienced during an NDE appears much more vivid and real than what happens during everyday waking consciousness. The fact that an NDE is accompanied by accelerated thought and access to greater than ever wisdom remains inexplicable. Current scientific knowledge also fails to explain how all these NDE elements can be experienced at a moment when, in many people, brain function has been seriously impaired. There appears to be an inverse relationship between the clarity of consciousness and the loss of brain function. There is no explanation for the fact that people across all ages and cultures have reported essentially similar experiences. Nor is there an answer to the question why some people have an NDE but most people cannot recall their period of unconsciousness after a life-threatening crisis. Interestingly, induced experiences are usually not completely identical to an NDE primarily because certain NDE elements are rarely if ever mentioned after drug use or brain stimulation but also because they are never followed by a process of change.

A satisfactory theory, one that explains the NDE in all its complexity, must consider both the range of different circumstances under which an NDE can be experienced and its various constituent elements. Perhaps such an all-embracing theory is indeed impossible and we should settle for a multifaceted approach that offers separate explanations for the separate aspects of an NDE. Oxygen deficiency appears to play a role at times, as do fear of death and perhaps also cerebral processes such as the blockage of certain receptors or the cessation of electrical activity in the temporal lobes. The role of DMT in triggering an NDE also merits further research.

Many fundamental questions remain unanswered. Most theories are based on anecdotal evidence and retrospective studies with self-selected patients and without accurate medical data. With the aim of meeting these objections and finding more definitive answers to the many unanswered questions, a group of researchers in the Netherlands launched a comprehensive and scientifically sound prospective study of the cause and content of near-death experience in 1988. The Dutch study, published in *The Lancet* in 2001, attracted worldwide interest. The next chapter considers it in detail.

Chapter Seven

The Dutch Study of Near-Death Experience

*If consciousness be a mere epiphenomenon . . . we shall of course
expect . . . that consciousness is exclusively linked with the
functional disintegration of central nervous elements, and varies
in its intensity with the rapidity or energy of that disintegration.
And ordinary experience, at least within physiological limits,
will support some view like this. Yet now and then we find a
case where vivid consciousness has existed during a state of
apparent coma . . . tranquilly and intelligently co-existing with
an almost complete abeyance of ordinary vital function. . . .
Until this new field has been more fully worked . . . we have
no right to make any absolute assertion as to the concomitant
cerebral processes on which consciousness depends.*
—F. W. H. MYERS

My scientific curiosity into the phenomenon of NDE was aroused when
I initiated a kind of pilot study in 1986. During a two-year period I
asked all the cardiac arrest survivors who attended my outpatient clinic
whether they had any recollection of their period of unconsciousness.
To my surprise, twelve out of the fifty patients (24 percent) reported an
NDE, often with extremely poignant details. But unfortunately I was
unable to explain how it is possible for people to have any memories
of this period of unconsciousness brought on by a cardiac arrest when,

according to prevailing scientific opinion, this should be impossible. In the previous chapter I reviewed all the existing explanatory models based on retrospective studies.

In order to corroborate or refute the existing theories on the cause and content of an NDE on the strength of more reliable data, we needed a well-designed scientific study. This is why in 1988 Ruud van Wees and Vincent Meijers, both psychologists specializing in NDE, joined me, a cardiologist, for a prospective study in the Netherlands. At that point, no large-scale prospective NDE studies had been undertaken anywhere in the world. Our study aimed to include all consecutive cardiac arrest survivors in the participating hospitals. In a prospective study such patients are asked, within a few days of their resuscitation, whether they have any memory of the period of their cardiac arrest, that is, of their unconsciousness. The patients' medical and other data are carefully recorded before, during, and after their resuscitation. In other words, this prospective study would include only patients with an objective life-threatening crisis. All of these patients would have died of their cardiac arrest had they not been resuscitated within five to ten minutes. This design also created a control group of cardiac arrest survivors without any memory of their period of unconsciousness.

The Organization

I started giving lectures to nurses and doctors at various hospitals in the hope of securing support to conduct the study of NDE in resuscitated patients at the coronary care units of these hospitals. We managed to include ten hospitals in our study, often thanks to the commitment of the nursing staff. The coronary care units of the four hospitals where I worked as a cardiologist at the time, which later merged to form the Rijnstate Hospital in Arnhem/Velp, took part in the study throughout the period from 1988 to 1992, as did the Antonius Hospital in Nieuwegein. Five smaller hospitals participated for a shorter period of time. We terminated a hospital's involvement when it emerged that because of the pressures of work not all consecutive revived cardiac patients were included in the study. The latter was particularly common when patients had not reported any recollection after their resuscitation. If

people without an NDE were excluded from the study, some of its results, for instance about the incidence of NDE after a cardiac arrest,
would be distorted. We had a contact for each hospital as well as a person on standby so that we were covered at all times. We also had someone who regularly visited the hospitals and monitored proceedings. We
applied for and received permission from the ethics committees of the
various hospitals. Patients were always asked if they wanted to participate; fortunately they all consented during their initial interview, probably because they were asked by a nurse or doctor at their own hospital.

The Mortality Rate of Cardiac Arrest Patients

For every one hundred successfully resuscitated patients we were able
to include in our study, at least two hundred people died of their cardiac
arrest in the same period. Few people realize just how many resuscitation attempts are made at a coronary care unit (CCU) every year and
that more than half of these patients do not survive their cardiac arrest.[1]

The Longitudinal Study

The longitudinal study of life changes was based on interviews after two and eight years with all NDE patients who were still alive, as
well as with a control group of postresuscitation patients who were
matched for age and sex but who had not had an NDE. The question
was whether the common life changes that are reported after an NDE
were the result of surviving a cardiac arrest or whether these changes
were caused by the NDE itself. This question had never been subject to systematic scientific research before. The two-year follow-up
interviews were coordinated by Ruud van Wees and Vincent Meijers
while the eight-year follow-up interviews were coordinated and conducted by life-span psychologist Ingrid Elfferich. All the work for our
prospective study, including the taped two- and eight-year follow-up
interviews, was carried out by nursing staff and university-educated
volunteers who had been briefed and trained by us. The study was
designed, planned, and coordinated by the Merkawah Foundation,
the Dutch branch of IANDS (International Association of Near-Death

Studies), and most of the volunteers were active members of this foundation. Throughout our ten-year study we did not receive any subsidies because research into near-death experience was not eligible for financial support from bodies such as the Dutch Heart Foundation.

The Design of the Study

We had a registration of the electrocardiogram (ECG) for all patients included in our study. An ECG displays the heart's electrical activity. In cardiac arrest patients this ECG always displays a lethal arrhythmia (ventricular fibrillation, or the chaotic flailing of the heart, resulting in a cardiac arrest that can be treated only by electric shock, or defibrillation) or an asystole (a flat line on the ECG). In the event of resuscitation outside the hospital, we were given the ECG done by ambulance staff.

For patients who were successfully resuscitated, we recorded their demographic data, including age, sex, standard of education, religion, prior knowledge of NDE, and whether or not they had experienced an NDE before. Patients were also asked whether they had been afraid prior to the cardiac arrest. We also carefully recorded all medical information: What was the duration of the actual cardiac arrest? What was the length of time of unconsciousness? How often did the patient require resuscitation? What was the exact nature of their cardiac arrhythmia? Was intubation (a tube inserted into the trachea for artificial respiration) needed because of a prolonged coma following a complicated resuscitation? Was the patient resuscitated inside or outside the hospital? Did the cardiac arrest occur during electrophysiological stimulation (EPS) during a heart catheterization, when patients are usually defibrillated through an electric shock to the chest within fifteen to thirty seconds? Was this the patient's first heart attack, or had the patient had a previous one? What medication, and in what dosage, did the patient receive before, during, and after resuscitation? (In the case of prolonged artificial respiration these are often extremely powerful drugs, which can keep the patient in a kind of coma.) We also recorded how many days after resuscitation the interview took place, whether the patient was lucid during the interview, and whether his or her short-term memory was functioning well.

The Initial Interview

During the initial interview, usually within five days of resuscitation, the patient was asked only a single, open question: "Do you have any recollection of the period of your cardiac arrest?" If the answer to this question was yes, an initial, unstructured interview was conducted and recorded, preferably by one of the study's principal researchers, although this was not always possible. One drawback of this method was that if the patient was on record as having thought that "he was going to die," this was coded as a possible NDE with the lowest score (score 1). However, two years later it emerged that some of the patients with this minimum score had not experienced an NDE. Likewise, a few patients who were listed as not having had an NDE did report an experience two years later. They had kept quiet about their NDE immediately after resuscitation, a common occurrence because people are scarcely able to grasp their extraordinary experience and remain silent for fear of being ridiculed or disbelieved.[2] I will come back to these findings when discussing the results of the longitudinal study.

A Hidden Sign, Visible Only During an Out-of-Body Experience

At one of the hospitals in Arnhem, the top cover of the surgical lamp in the resuscitation room was decorated with a hidden sign, invisible from a normal position. None of the attendant doctors or nurses were informed of this hidden sign so they would not influence patients. Even I never knew which sign (a cross, circle, or square, in red, yellow, or blue) had been applied by a colleague of mine. Unfortunately, no patients who were resuscitated in this room ever reported an out-of-body experience with perception. Because people are resuscitated everywhere—on the street, in the ambulance, in a CCU room, on the ward—we had estimated the chances of a hit to be relatively low. Still, one verified out-of-body experience would have been sufficient. Luckily, during our study a nurse told us about the case with the dentures, as described in an earlier chapter, although the resuscitation room in question featured no hidden sign.

The Design of the Longitudinal Study

The two-year and eight-year follow-up interviews were recorded on tape and transcribed. This allowed us to compare the content of the NDE with the experience as it had been reported to us in the hospital immediately after the cardiac arrest. Remarkably, after two and eight years patients related their NDE in almost the exact same wording, down to the very last detail. This is nearly impossible in the case of a dream or an invented story. The later interviews were accompanied by Kenneth Ring's life-change inventory, which all participants were asked to complete.[3] This inventory features thirty-four questions about self-image, compassion for others, material and social issues, religious and spiritual matters, and attitude toward death. To specify the level of change, patients were asked to answer these questions on a five-point scale. For the eight-year follow-up, the inventory was expanded with surveys on medical and psychological aspects drawn up by the Dutch Heart Foundation; this included a list of questions on coping with problems and a questionnaire on feelings of depression. These questionnaires were added for the purpose of qualitative analysis because after eight years very few people were still alive and the group under investigation had become quite small.

All the findings of the prospective study and the longitudinal study underwent statistical analysis to identify significant differences, with $P \leq 0.05$; P is the probability of getting a statistical significant difference, and $P \leq 0.05$ means that there is a 5 percent or less probability that the result is only by chance. The lower the P value, the more significantly different is the result.

Findings of the Prospective Study

The Dutch study was published in *The Lancet* in December 2001.[4] Within a four-year period, between 1988 and 1992, 344 consecutive patients who had undergone a total of 509 successful resuscitations were included in the study. In other words, all patients in our study had been clinically dead. Clinical death is defined as a period of unconsciousness caused by a lack of oxygen supply to the brain (anoxia) because of the arrest of circulation, breathing, or both following cardiac arrest in patients with an acute myocardial infarction. If no resuscitation is started, the brain cells suffer irreparable damage within five to ten minutes and the patient always dies. People who survived a resuscitation-with-complications outside the hospital were significantly younger, and only twelve patients survived a cardiac arrest that lasted more than ten minutes. Statistics show that only 10 percent of people who suffer cardiac arrest outside the hospital leave the hospital alive because they frequently undergo irreparable brain damage, which results in brain death and, ultimately, death.

Recorded Data of the 344 Patients in the Study

Number of resuscitations	509
Mean age	62 years
Number of men	73 percent
Number of women	27 percent
Religious	72 percent
Secondary education	66 percent
Prior knowledge of NDE	57 percent
Previous NDE	4 percent
Fear of death	2 percent
First myocardial infarction	86 percent
Resuscitation in hospital	234 patients (68 percent)
Duration of cardiac arrest <2 minutes	*190 patients (81 percent)*
Duration of unconsciousness <5 minutes	*187 patients (80 percent)*
Resuscitation outside hospital	110 patients (32 percent)
Duration of cardiac arrest >2 minutes	*88 patients (80 percent)*
Duration of unconsciousness >10 minutes	*62 patients (56 percent)*
Duration of unconsciousness exceeding 1 hour	104 patients
Artificial respiration in long-term coma	12 percent
Short-term memory defects	41 patients

Results of the Prospective Study

Percentage of NDEs

If patients reported memories of the period of unconsciousness, the experiences were coded according to the Weighted Core Experience Index (WCEI; see chapter 2 for more information). The higher the number of elements reported, the higher the score and the deeper the NDE. Our study found that 282 patients (82 percent) had no recollection of the period of their unconsciousness whereas 62 patients (18 percent of the 344 patients) did report an NDE. Of these 62 patients with memories, 21 patients had some recollection; having experienced only a few elements, they had a superficial NDE with a low score. A total of 18 patients had a moderately deep NDE, 17 patients reported a deep NDE, and 6 patients had a very deep NDE.

WCEI Classification of the 344 Patients (Depth of NDE)

	WCEI score	NUMBER
1. No memory	0	282 (82 percent)
2. Some recollection	1–5	21 (6 percent)
3. Moderately deep NDE	6–9	18 (5 percent)
4. Deep NDE	10–14	17 (5 percent)
5. Very deep NDE	15–19	6 (2 percent)

62 patients (18 percent) reported memories (NDE)

Compared to results from retrospective studies, the much lower percentage of NDEs here is conspicuous. It is a direct consequence of the study's prospective design and of the fact that we looked at a group of much older patients. Only 12 percent of patients had an experience with a score of 6 or more (a moderately deep, a deep, and a very deep NDE). This is known as a core experience because in retrospective studies this score of 6 marked the cutoff point for defining a remembered experience as an NDE. If we set the percentage of NDEs against the number of resuscitations, we are left with only 5 percent. Women, who had a higher mean age, reported significantly deeper NDEs, as did people who had been resuscitated outside the hospital and patients who had been extremely fearful prior to their cardiac arrest.

Another striking finding was that people with a deep NDE, and especially those with a very deep NDE, were significantly more likely (P≤0.0001) to die within thirty days of their cardiac arrest although medically they were no different from the other patients. I cannot offer an adequate explanation for this. However, it is possible that after a deep or very deep NDE people lost their fear of death to the extent that they were able to let go and leave their bodies. We know that people can exercise a certain control over the time of their death. For example, if a family is very sad to see mother or father die and observes a tearful twenty-four-hour deathbed vigil, this person will be unable to let go. The patient will usually die when the family is momentarily absent. And when somebody is close to death but waiting for a daughter to arrive from Australia, then this person will defy all medical prognoses and not die until the daughter has arrived several days later. Anybody working in a hospital or hospice has come across cases like this.

Identified NDE Elements

The table "Frequency of NDE Elements" shows which common elements of a near-death experience were reported in our study and how frequently they occurred.

Half of the patients with an NDE were aware of being dead and had positive emotions; 30 percent had a tunnel experience, observed a celestial landscape, or met with deceased persons; approximately a quarter had an out-of-body experience, communicated with "the light," or saw colors; 13 percent had a life review; and 8 percent perceived the presence of a border. In other words, all known NDE elements were reported in our study, with the exception of a frightening or negative NDE.

Frequency of NDE Elements in the 62 Patients

Awareness of being dead	31	(50 percent)
Positive emotions	35	(56 percent)
Out-of-body experience	15	(24 percent)
Moving through a tunnel	19	(31 percent)
Communicating with "the light"	14	(23 percent)
Perception of colors	14	(23 percent)
Perception of a "celestial landscape"	18	(29 percent)
Meeting with deceased friends and relatives	20	(32 percent)
Life review	8	(13 percent)
Presence of a border	5	(8 percent)

Factors That Did Not Influence the Occurrence of NDEs

Are there any reasons why some people have but most people do not have any recollection of the period of their unconsciousness? To answer this question, we compared the recorded data of the 62 patients with an NDE with the data of the 282 patients without one. To our big surprise, we identified no significant differences in the duration of the cardiac arrest, no differences in the duration of the period of unconsciousness, and no differences in whether or not seriously ill patients who remained in a coma for days or weeks after a complicated resuscitation needed intubation for artificial respiration. Neither did we find differences among the 30 patients who had a cardiac arrest during electrophysiological stimulation (EPS) in the catheterization laboratory and whose heart rhythms were always reestablished through defibrillation (electric shock) within fifteen to thirty seconds. So we failed to identify any differences between patients with a very long or a very brief cardiac arrest. The degree or severity of the oxygen deficiency in the brain (anoxia) appeared to be irrelevant. The administered medication played no role either. Most patients suffering a myocardial infarction receive morphine-style painkillers, while people who are put on a ventilator following a complicated resuscitation are given extremely high doses of sedatives. A psychological cause, such as the infrequently reported fear of death, did not affect the occurrence of an NDE, although it did affect the depth of the experience. Whether or not patients had heard or read anything about NDE in the past made no difference either. Any kind of religious belief, or its absence in non-believers and atheists, was irrelevant, and the same was true for the standard of education reached (see the table that follows).

Factors That Did Not Influence the Occurrence of NDE

1. Duration of cardiac arrest	N.S.
2. Duration of unconsciousness	N.S.
3. Intubation (complicated resuscitation)	N.S.
4. Induced cardiac arrest (EPS)	N.S.
5. Medication	N.S.
6. Fear of death	N.S.
7. Prior knowledge of NDE	N.S.
8. Religion	N.S.
9. Standard of education	N.S.

N.S. = not significant

Factors That Do Influence the Occurrence of NDE

Factors that do affect the frequency of an NDE are an age below 60 and a first myocardial infarction, in which case the patients were also younger than the mean age of 63. If patients required several resuscitations during their hospital stay, they were more likely to report an NDE. Remarkably, all patients who had experienced an NDE in the past reported them significantly more often in our study.

Factors That Influence the Occurrence of NDE

More frequent NDE:	
1. Age below 60	P = 0.012
2. First myocardial infarction (younger!)	P = 0.013
3. More than one resuscitation in hospital	P = 0.029
4. Previous NDE	P = 0.035
Less frequent NDE:	
Lasting memory defects	P = 0.011

P is the probability of a statistical significance
P≤0.05 denotes a significant difference

A complicated resuscitation can result in a long coma, and patients who have been unconscious on a ventilator for days or weeks are more likely to suffer short-term memory defects caused by permanent brain damage. The longer the coma, the greater the risk of these cognitive problems, which also occur after severe concussion or a stroke and which may wipe hours, days, and sometimes even weeks from a patient's memory.[5] These patients reported NDEs significantly less often, which suggests that a good memory is a prerequisite for remembering an NDE.

Conclusions of the Prospective Study

The possible causes of NDE outlined so far, that is, physiological or medical causes (such as anoxia), psychological (such as fear of death), or pharmacological (causes related to medication), could not be corroborated by this first large-scale prospective study of NDE.

We were particularly surprised to find that medical factors failed to explain the occurrence of an NDE. All the patients in our study had been clinically dead, and only a small percentage reported an enhanced consciousness with lucid thoughts, emotions, memories, and sometimes perceptions from a position outside and above their lifeless body during resuscitation. If this enhanced consciousness had a physiological cause, such as oxygen deficiency in the brain (anoxia), all patients in our study should have reported an NDE. They had all been unconscious because of their cardiac arrest, which resulted in a loss of blood pressure and the cessation of breathing and all body and brain-stem reflexes. The severity of the clinical picture, such as a lengthy coma after a complicated resuscitation, also failed to explain why patients did or did not report an NDE, except in the case of lingering memory defects. The psychological explanation is improbable because most patients experienced no fear of death preceding their cardiac arrest; its onset was so sudden that they failed to notice it. In most cases they were left without any memories of their resuscitation. This is borne out by Greyson's study, in which the subjective data of resuscitated patients show that most of them did not even realize they had had a cardiac arrest.[6] The situation is comparable to fainting. When people regain consciousness after fainting they have no idea what happened. We were also able to exclude a pharmacological explanation, as the medication administered had no effect on whether or not patients reported an NDE.

Findings of the Longitudinal Study

At the two-year follow-up, 19 of the 62 patients with an NDE had died and 6 declined to be interviewed, leaving 37 eligible for the second interview (see the table "Number of Patients Interviewed"). Of the 17 patients with a low score, 7 remained unchanged, 4 had the lowest score of 1 with only positive emotions, while in retrospect 6 patients did not have an NDE after all. After the initial interview these 6 people had been classified as a potential NDE with score of 1 because "they had had the feeling they were dying," but the in-depth interview two years later proved this false. Following the second interview these 6 people were grouped with the patients without an NDE. In order to put together a control group of cardiac arrest survivors without an NDE but matched for age and sex, we approached a total of 75 patients before we found 37 people prepared to participate in a second interview. This group turned out to contain four more patients with an NDE, two with a low score and two with a core experience. Following the second interview, these four patients were grouped with the people *with* an NDE. The first postresuscitation interview must have been too soon for them to be able or willing to talk about their NDE.[7]

Number of Patients Interviewed During the
1st, 2nd, and 3rd Phases of the Study

344 patients

	With NDE	Without NDE
1st phase (1988–1992)	62 patients (18 percent)	282 patients (82 percent)
2nd phase (1991–1993)	37 (−6+4) = 35 patients 19 died (31 percent) 6 declined	37 (+6−4) = 39 patients 38 died or declined
3rd phase (1997–1998)	23 patients 11 died 1 no interview	15 patients 20 died 4 no interview

At the two-year follow-up we were able to interview a total of 74 patients: 35 patients with an NDE affirmed by the second interview and 39 patients without an NDE. After eight years we approached the same patients for a third interview. Of the people with an NDE, 11 had died, while of the 24 patients who were still alive, 1 person was unable to participate because of memory defects. Ultimately we were able to conduct third interviews with 23 NDErs. Of the people without an NDE we could interview only 15. After eight years 20 patients had died and 4 were unable to participate for reasons including dementia. This means that we were able to compare the patterns of change after two and eight years in 23 patients with an NDE and in 15 patients without an NDE.

Results of the Longitudinal Study

As mentioned, the later interviews were conducted using a standardized inventory featuring 34 life-change questions.[8] Among the 74 patients who consented to be interviewed at the two-year follow-up, 13 of the 34 factors listed in the questionnaire were significantly different for those with and without an NDE (see the table "The Significant Differences Between People With and Without an NDE"). The second interviews showed a significant decrease in fear of death among people with an

NDE and a significant increase in belief in an afterlife. There were further significant differences between people with and without an NDE with respect to a number of social and religious factors such as showing emotions, accepting others, a more loving attitude to life, and more love and compassion for oneself and others. Other differences pertained to a greater involvement in family, a greater interest in spirituality and the meaning of life, and greater appreciation of ordinary things, coupled with less interest in money, possessions, and social norms ("keeping up appearances"). We compared these thirteen factors, which yielded such significant differences after two years, again after eight years for the same two groups of patients with and without an NDE.

We noticed that after eight years the people without an NDE were also undergoing unmistakable processes of change. Clear differences remained between the two groups although they were now a little less marked. We were also surprised to find that the processes of change that got under way after two years in people with an NDE clearly intensified after eight years. The same was true for the people without an NDE. Summing up, we found that, eight years after their cardiac arrest, all patients had changed in many respects: they had a greater interest in nature, the environment, and social justice; they showed more love and emotions; and they were more supportive and more involved in family life. Nonetheless, those who had experienced an NDE during their cardiac arrest remained significantly different. Specifically, they were less afraid of death and had a stronger belief in a life after death. They showed a greater interest in spirituality and questions of meaning as well as a greater acceptance of and love for oneself and others. Likewise, they had a greater appreciation of ordinary things whereas their interest in possessions and power had decreased. People without an NDE showed a marked decline in interest in spirituality.

By the time the later interviews were conducted, the NDE had become an experience that provided a fresh insight into everything that matters in life: compassion, unconditional love, and acceptance of oneself (including acceptance of one's negative qualities), others, and nature. Fear of death was usually gone. The interviews also revealed a sharp increase in heightened intuitive feelings after an NDE along

The Significant Differences Between People with and Without an NDE—Changes at Two and Eight Years

Life-change inventory (n = number of patients)	After two years		After eight years	
	NDE n = 23	No NDE n = 15	NDE n = 23	No NDE n = 15
Social attitude				
1. Showing emotions	+42	+16	+78	+58
2. Acceptance of others	+42	+16	+78	+41
3. More loving and empathetic	+52	+25	+68	+50
4. Understanding others	+36	+8	+73	+75
5. Involvement in family	+47	+33	+78	+58
Religious attitude				
6. Understand purpose of life	+52	+33	+57	+66
7. Sense inner meaning of life	+52	+25	+57	+25
8. Interest in spirituality	+15	−8	+42	−41
Attitude toward death				
9. Fear of death	−47	−16	−63	−41
10. Belief in life after death	+36	+16	+42	+16
Other				
11. Interest in meaning of life	+52	+33	+89	+66
12. Understanding oneself	+58	+8	+63	+58
13. Appreciation of ordinary things	+78	+41	+84	+50

The table shows the percentages of all positive (+1 and +2) and negative changes (−1 and −2) after two and eight years for the same patients with and without an NDE. For example, the possible answers to the question "Are you interested in spirituality?" were: strongly increased (+2), somewhat increased (+1), no change (0), somewhat decreased (−1), and strongly decreased (−2). After two years 15 percent of people with an NDE scored either +1 or +2; after eight years the percentage of NDErs with an interest in spirituality had increased to 42 percent. Meanwhile, 8 percent of the people without an NDE scored either −1 or −2 after two years, and this percentage of interest in spirituality fell further to −41 percent after eight years.

with a strong sense of connectedness with others and with nature. Or, as many of them put it, they had acquired "paranormal gifts." The sudden onset of this heightened intuition can be quite problematic; NDErs suddenly have a very acute sense of other people's emotions, which can be extremely intimidating. They also experience clairvoyance, premonitions, and visions. This intuitive sensitivity can be quite intense, with people recognizing feelings and sadness in others or sensing when somebody will die—which usually proves to be accurate. As one respondent put it, "It felt as if I'd become another person, but with the same identity." As mentioned earlier, the process of integrating and accepting an NDE may take many years because of its profound impact on people's values and outlook on life. Finally, the lifelong transformational effects of an experience that lasts only a few minutes was a striking and unexpected finding.

Commentary on the Dutch NDE Study

Thanks to *The Lancet*'s global press release, our Dutch prospective NDE study attracted a great deal of attention in December 2001. The study not only made it onto the front pages of all the major newspapers in Europe, the United States, Canada, and Australia, but was also front-page news in countries such as Russia, China, India, Sri Lanka, Japan, Brazil, and Argentina. Our group of researchers never anticipated this huge interest. For a couple of days I had to reschedule appointments at my outpatient clinic in order to satisfy all the requests for interviews with national and international newspapers, radio, and television. We received hundreds of e-mails with positive responses from NDErs who felt supported and recognized by this study. We also received feedback from doctors who had experienced an NDE themselves and who had never been able to discuss it with colleagues. The following is an extract from the commentary published by Dr. Pam Kircher, a family doctor and practicing physician at a hospice in the United States.

Van Lommel's article in The Lancet is a landmark that should be read by every doctor. It reports on the largest prospective study to

ask people about NDEs after a cardiac arrest. . . . Equally impor-
tant, the study followed people for eight years after their NDE. . . .
I would encourage [Vital Signs] readers to obtain the complete
Lancet article to share with your doctor. You will be doing your
family doctor a great favor. . . . I believe that the findings in Van
Lommel's study challenge hospitals to ask people about their
NDEs after a cardiac arrest.

Dr. Jeffrey Long, a physician practicing radiation oncology (use of radiation to treat cancer) and active in NDE research, wrote, "On behalf of all people with an NDE I congratulate Van Lommel for his excellent research."

But feedback from scientific and medical circles was mixed, and at my own hospital I received many positive as well as some more indifferent reactions while some colleagues never even mentioned the publication.

Since 2001 the Dutch study has been frequently cited in scientific articles and books (121 times), in science programs on radio and television, and in other publications. Our NDE study was the reason why Professor Janice Holden awarded me the Bruce Greyson Research Award on behalf of the International Association of Near-Death Studies in the United States in September 2005. And in September 2006 the president of India, Dr. A. P. J. Abdul Kalam, awarded me the Lifetime Achievement Award in New Delhi following a lecture I presented on our study at the World Congress on Clinical and Preventive Cardiology 2006.

To the best of my knowledge, no negative commentaries were ever published in any peer-reviewed scientific journals, with the exception of the mildly critical commentary in *The Lancet* itself. I did, however, receive some extremely critical comments in the Netherlands from Dr. C. Renckens, gynecologist and chair of the Dutch Association Against Quackery. As well as linking our study with "multiple personality disorder, chronic fatigue syndrome, fibromyalgia and alien abduction syndrome," he described me as "a failed prophet with the personality of a pre-morbid quack."

In Belgium I received some blunt comments from W. Betz, professor of family medicine in Brussels and a member of Skepp (the Belgian study group for the critical evaluation of pseudoscience and the paranormal). Betz's initial response to our study appeared in an article in Belgian newsmagazine *De Tijd* on 29 December 2001: "When scientists start spouting nonsense, the public must be warned." According to the magazine, he was "livid," criticized the study and me as "postmodern deception," "pseudoscience," "nonsense," and "a veritable cult." "Van Lommel belongs to a sect," he wrote, and he associated the research with "astral bodies, the paranormal, and graphology." Describing NDE as "a hallucination," he tried to refute the published out-of-body experience that includes the story of the dentures by hinting at a lack of integrity on the part of both the nurse who wrote the report and the authors of the article: "enthusiastic researchers, convinced of being in the right, are only too keen to 'help' the victim of an NDE retrieve his memories." Betz suggested that patients "can be talked into believing they had an NDE" even years after a cardiac arrest. In an interview in another publication, Belgian magazine *Humo*, he described our study as "complete nonsense" and claimed that "the publication lacks any kind of cohesion." He concluded by saying, "Imagine there were any truth to Van Lommel's claims . . . admit it, wouldn't that be most peculiar?"

Comparison with Prospective NDE Studies in the United States and the United Kingdom

One American and two British studies among cardiac arrest patients, with the same prospective design as our Dutch study, found near-identical percentages of NDE after a successful resuscitation.[9] None of these four studies, comprising a total of 562 patients, could produce a definitive scientific explanation for the phenomenon (see table "Four Prospective NDE Studies Among Cardiac Patients").

Four Prospective NDE Studies Among Cardiac Patients

Dutch study (2001): 344 patients
18 percent NDE: 12 percent NDE score of 6 or higher, 6 percent
score of 1–5

American study (2003): 116 patients
15.5 percent NDE: 9.5 percent NDE score of 6 or higher, 6 percent
score of 1–5

British study 1 (2001): 63 patients
11 percent NDE: 6.3 percent NDE score 6 of higher, 4.8 percent
score of 1–5

British study 2 (2006): 39 patients
23 percent NDE: 18 percent NDE score of 6 or higher, 5 percent
score of 1–5

Conclusions of the four studies comprising a total of 562 patients:
1. Evidence of the same percentage of NDE during cardiac arrest
2. No physiological or psychological explanation for an NDE
3. An NDE occurs during the cardiac arrest
4. Cardiac arrest involves a loss of all brain function

The American Study

As part of Bruce Greyson's prospective study in the United States,
a total of 1,595 patients were interviewed at the cardiac units of the
University of Virginia Hospital. It emerged that 5 percent of these
patients had experienced a previous NDE. Excluding cardiac arrest
diagnoses, only 1 percent of the heart patients reported an NDE. The
comparative study, however, looked at 116 cardiac arrest patients, of
whom 9.5 percent reported an NDE with a score of 6 or higher, and
6 percent reported an NDE with a low score. A total of 15.5 percent of the cardiac arrest survivors reported an NDE that satisfies
our more liberal criteria. This study also identified a younger mean
age of people with an NDE. The medical files were not systematically analyzed for physiological, psychological, and pharmacological

factors. Diagnoses such as "clinically dead," "close to death" or "no mortal danger" were not based on objective criteria but were made by the patients themselves. This was why so few people in the study were described as having been clinically dead—because most patients were unable to recollect their resuscitation. Similarly, the diagnoses "loss of consciousness," "diminished consciousness," and "normal consciousness" were made by the patients themselves. So unfortunately this study recorded mostly subjective and few objective medical data. In his conclusion, Greyson writes,

> *No one physiological or psychological model by itself explains all the common features of near-death experiences. . . . The paradoxical occurrence of heightened, lucid awareness and logical thought processes during a period of impaired cerebral perfusion [blood flow to the brain] raises particularly perplexing questions for our current understanding of consciousness and its relation to brain function. . . . A clear sensorium and complex perceptual processes during a period of apparent clinical death challenge the concept that consciousness is localized exclusively in the brain.[10]*

The First British Study

The British prospective study by Sam Parnia, an intensive care physician, and Peter Fenwick, a neuropsychiatrist, looked at 63 cardiac arrest survivors at Southampton General Hospital over a one-year period. Of these, 4 patients (6.3 percent) reported an NDE, and 3 patients (4.8 percent) had an experience with a low score, bringing the total to 11 percent according to our more liberal criteria. The only objective data to be recorded were arterial blood gases (oxygen and carbon dioxide) and the drugs administered. The number of patients in this study was too small for statistical analysis. Significantly, hidden signs were affixed near the ceilings of the patient rooms at the coronary care unit. But unfortunately, as in our study, none of the patients had an out-of-body experience with perception of one of these signs. According to the authors, the data suggest that the NDE arises during unconsciousness. "This is a surprising conclusion," in their view,

because when the brain is so dysfunctional that the patient is deeply
comatose, the cerebral structures which underpin subjective experi-
ence and memory must be severely impaired. Complex experiences
such as are reported in the NDE should not arise or be retained in
memory. Such patients would be expected to have no subjective
experience (as was the case in 88.8 percent of patients in this study)
... as those cerebral modules which generate conscious experience
and underpin memory are impaired by cerebral anoxia.[11]

Another frequently cited explanation might be that the experiences occur either during the early stages of unconsciousness or during the recovery of consciousness. However, Parnia and Fenwick claim that the verifiable elements of an out-of-body experience during unconsciousness, such as patients' reports of their resuscitation, render this extremely unlikely.

The Second British Study

Over a period of four years, Dr. Penny Sartori, a senior intensive care nurse, carried out an even smaller study of NDE. Only 1 percent of the 243 patients who survived their stay in intensive care in a Welsh hospital reported an NDE. However, her study focused on 39 cardiac arrest patients, of whom 18 percent reported an NDE and 5 percent only an out-of-body experience without any of the other NDE elements, bringing the total to 23 percent according to our more liberal criteria. Sartori notes that only two patients with a deep NDE reported their experience "spontaneously"; the other NDEs were reported during the purposive interviews. This may be a result of reluctance to discuss this extremely profound experience. Three patients with an NDE died soon after their cardiac arrest, which is another parallel with our study. Similarly, Sartori's study featured hidden signs, which were not noticed during an NDE. One patient, however, recounted an extremely detailed out-of-body experience, many aspects of which proved to be accurate upon inquiry. A control group of people who had been successfully resuscitated but who had not had an NDE made a great many fundamental mistakes when asked to describe their own resuscitation. The cardiologist Sabom reached a similar conclusion in his study.[12] The medication administered or the arterial blood gases (oxygen and carbon dioxide) measured in a few patients failed to explain why patients in Sartori's study did or did not experience an NDE. But this study was also too small in scope for statistical analysis. Sartori concludes,

> *The phenomenon remains unexplained when considered from the current scientific perspective of consciousness being a by-product of neurological processes. . . . The fact that clear, lucid experiences were reported during a time when the brain was devoid of activity . . . does not sit easily with current scientific belief.*[13]

Only the large-scale Dutch study allowed statistical analysis of the potential contributing factors to an NDE. The results failed to confirm the aforementioned physiological, psychological, and pharmacological explanations. Our study was also the first to include a longitudinal component with follow-up interviews after two and eight years, which allowed us to compare the processes of change in people with and without an NDE. We identified a distinct pattern of change in people with an NDE and found that integrating these changes into everyday life is a long and arduous process. Cardiac arrest patients without an NDE also underwent a gradual but in many ways different process of change.

On the strength of the four prospective studies among cardiac arrest survivors, we concluded that they experienced all the previously mentioned NDE elements during their cardiac arrest, during impaired blood flow to the brain. Nonetheless, the question how this is possible remains unanswered.

> *Although the content of consciousness depends in large measure on neuronal activity, awareness itself does not. . . . To me, it seems more and more reasonable to suggest that the mind may be a distinct and different essence.*
> —WILDER PENFIELD

Scientific research into the phenomenon of NDE highlights the limitations of our current medical and neurophysiological ideas about the various aspects of human consciousness and the link between consciousness, memories, and the brain. According to the prevailing paradigm, memories and consciousness are produced by large groups of neurons or neuronal networks. For lack of evidence for the usual explanations for the origins and content of an NDE, the commonly accepted but never proven concept that consciousness is localized in the brain should be questioned.

How can an extremely lucid consciousness be experienced outside the body when the brain has momentarily stopped functioning during a period of clinical death? What happens when blood supply to the brain ceases? And what do we really know about normal brain function? The next chapters will look at these important questions in more detail.

Chapter Eight

What Happens in the Brain When the Heart Suddenly Stops?

The task is, not so much to see what no one has yet seen, but to think what nobody has yet thought, about that which everybody sees.

—ERWIN SCHRÖDINGER

The four prospective NDE studies discussed in the previous chapter all reached one and the same conclusion: consciousness, with memories and occasional perception, can be experienced *during* a period of unconsciousness—that is, during a period when the brain shows no measurable activity and all brain functions, such as body reflexes, brain-stem reflexes, and respiration, have ceased. It appears that at such a moment a lucid consciousness can be experienced independently of the brain and body. This conclusion was reached on the basis of compelling evidence that the NDE occurs *during* the period of clinical death and not shortly before or after the cardiac arrest. It was the studies' prospective design that enabled this conclusion. If the cardiac arrest involved an NDE with clear perception of the patient's surroundings, its contents could be verified immediately after the report. The story of the lost dentures in chapter 2 is a good example of this.

The precise onset of an NDE is important because it rules out any conclusion other than that the NDE is experienced at a point in time when the brain shows no activity and all brain function has ceased. If the prevailing hypothesis, that consciousness is produced by the brain, were correct, there could be no sign of consciousness at the moment when the brain shows no activity. Indeed, this is reported in most cases of clinical death, coma, or brain death. But as the NDE studies have shown, there are exceptions to this rule. This finding all but forces us to reconsider the relationship between the brain and consciousness. After all, how can people experience an exceptionally lucid consciousness during a period of temporary loss of all measurable brain function?

The Paradox of a Lucid Consciousness During the Loss of Brain Function

As mentioned, the four prospective NDE studies reached remarkably similar conclusions. In our article in *The Lancet* we argued, "NDE pushes at the limits of medical ideas about the range of human consciousness, and the mind-brain relation."[1]

Bruce Greyson concluded,

The paradoxical occurrence of heightened, lucid awareness and logical thought processes during a period of impaired cerebral perfusion [blood flow] raises particularly perplexing questions for our current understanding of consciousness and its relation to brain function. As prior researchers have concluded, a clear sensorium and complex perceptual processes during a period of apparent clinical death challenge the concept that consciousness is localized exclusively in the brain.[2]

Sam Parnia and Peter Fenwick wrote in their conclusion:

The data suggests that in this cardiac arrest model, the NDE arises during unconsciousness. This is a surprising conclusion, because when the brain is so dysfunctional that the patient is deeply comatose, the cerebral structures which underpin subjective experi-

ence and memory must be severely impaired. Complex experiences such as are reported in the NDE should not arise or be retained in memory. Such patients would be expected to have no subjective experience, . . . as those cerebral modules which generate conscious experience and underpin memory are impaired by cerebral anoxia.[3]

Finally, Penny Sartori concluded,

The phenomenon remains unexplained when considered from the current scientific perspective of consciousness being a by-product of neurological processes. . . . The fact that clear, lucid experiences were reported during a time when the brain was devoid of activity . . . does not sit easily with current scientific belief.[4]

These NDE researchers start from the assumption that when a cardiac arrest disrupts the supply of blood to the brain, the brain becomes devoid of activity. Indeed, all brain function appears to be lost. To prove this hypothesis, we need conclusive evidence of this loss of function. This makes it imperative to ascertain what happens in the brain in the absence of a blood supply when the heart has stopped beating. The loss of blood pressure and breathing results in immediate unconsciousness and the loss of all body and brain-stem reflexes. Does it really mean that all brain function has ceased? Can this be measured? And has all electrical brain activity ceased as well, resulting in a flat EEG? Has there been any research in this area?

Measuring the Loss of Brain Activity During a Cardiac Arrest

Research in both humans and animals has shown that during an induced cardiac arrest the loss of function of both the cerebral cortex and the brain stem results in unconsciousness within seconds. All brain-stem reflexes are gone too: there is no cornea reflex (the blinking of the eye upon touch) and no gag reflex, and the dilated pupils do not react to light. The respiratory center near the brain stem has also stopped functioning, as evidenced by the suspension of breathing (apnea).[5]

Patients' blood flow to the brain stops altogether when a cardiac arrest is induced for threshold measurements during the implanting of internal cardiac defibrillators (ICDs). These ICDs are implanted in patients with recurring life-threatening arrhythmias that are not or not sufficiently responsive to medication. Blood flow in the brain can be measured very accurately in the middle cerebral artery with the use of ultrasound (Doppler ultrasonography). This test shows that blood flow stops completely at the onset of the cardiac arrest and is restored within seconds of an electric shock (defibrillation) reestablishing the heartbeat.[6]

Some studies in humans have also used electroencephalograms (EEG) to register the electrical activity of the cortex, and in animals also the electrical activity of the deeper structures of the brain have been measured. Results have shown that after a very short time the electrical activity in the cerebral cortex and in the deeper structures disappears completely.[7] The first symptoms of oxygen deficiency are recorded, on average, 6.5 seconds after the onset of the cardiac arrest. If the heartbeat is not immediately restored, the complete loss of all electrical activity in the cerebral cortex *always* results in a *flat EEG* after ten to twenty (a mean of fifteen) seconds.[8] In tests on animals, auditory evoked potentials, or measures of brain-stem viability, can no longer be induced, which means that the reaction caused in a normal functioning brain stem by sound stimulation is no longer produced.[9]

If the cardiac arrest lasts longer than thirty-seven seconds, the EEG does not normalize immediately. After a complicated resuscitation with persistent coma, it can take hours or days for the EEG to return to normal. Despite maintaining normal blood pressure in the period following resuscitation, this ultimately depends on the duration of the cardiac arrest.[10] The longer the cardiac arrest, the greater the brain damage, the longer the coma, and the longer the EEG remains flat or highly irregular.

Normalization of the EEG may actually create an overly positive impression of the recovery of the brain's metabolism. After the heart begins beating again and blood flow resumes, oxygen supply to the brain may be reduced for a long time. Following a cardiac arrest lasting longer than thirty-seven seconds, measurements of blood flow to the brain after the heartbeat is restored initially show an increase of blood flow (an overshoot) followed by a significant decrease of up to 50 percent less than normal blood flow as a result of the swelling of the brain (edema). The result is undersaturation of oxygen in the brain during this period of time.[11]

Many argue that the loss of blood flow and a flat EEG do not exclude some activity somewhere in the brain because an EEG primarily registers the electrical activity of the cerebral cortex. In my view this argument misses the point. The issue is not whether there is some immeasurable activity somewhere but whether there is any sign of those specific forms of brain activity that, according to current neuroscience, are considered essential to experiencing consciousness.[12] And there is no sign whatsoever of those specific forms of brain activity in the EEGs of cardiac arrest patients. A flatline EEG is also one of the major tools for diagnosing brain death, and in those cases the objection about not ruling out any brain activity is never mentioned. Besides, there are circumstances in which the EEG registers brain activity, yet no waking consciousness is experienced. This phenomenon occurs under general anesthesia, during which, depending on the medication used, the EEG shows clear changes but certainly not complete loss of brain activity. The same happens during deep dreamless sleep (non-REM sleep), when no consciousness is experienced despite demonstrable activity in the EEG. Later I will look in more detail at the brain structures that must actively cooperate to allow the experience of waking consciousness.

It is also highly unlikely that the out-of-body experience takes place immediately after regaining consciousness, as is sometimes claimed. The reason is that the time between the restoration of blood circulation after a successful resuscitation and the recovery of consciousness varies from five minutes to seventy-two hours, with a mean of six hours, which is much later than when the reported and verifiable perceptions during resuscitation must have taken place.[13]

Patients with a myocardial infarction who suffer a cardiac arrest at the coronary care unit are usually successfully resuscitated within one to two minutes; at a nursing ward, however, this takes at least two to five minutes. In the event of a cardiac arrest in the street (an out-of-hospital arrest) it takes, at best, five to ten minutes for a patient to be successfully resuscitated and usually longer, resulting in the death of nearly 90 percent of these patients. Only patients with an induced cardiac arrest, as part of electrophysiological studies or for threshold measurements during the implantation of ICDs, are successfully treated within fifteen to thirty seconds.

Needless to say, no EEG is carried out at the moment patients with a myocardial infarction suffer a cardiac arrest. Medical staff want to resuscitate the patient as quickly and effectively as possible. However, we know from the already-mentioned blood flow and EEG registrations that all cardiac arrest patients included in the prospective NDE studies have suffered a loss of blood flow and electrical brain activity. Their clinical picture also reflects the loss of all cerebral cortex and brainstem activity. In this state the brain can be compared to a computer that has been disconnected from its power supply, unplugged, and all its circuits disabled. Such a computer cannot function; in such a brain even so-called hallucinations are impossible. Nonetheless, during such temporary loss of all measurable brain function, a number of these patients experienced a period of exceptionally lucid consciousness.

What Happens in the Brain When the Heart Stops?

What exactly happens in the brain when the heart stops? The brain accounts for only 2 percent of overall body weight, but it uses 15 to 20 percent of the body's total energy supply, primarily for maintaining the membrane potential (the electric charge across a cell membrane) of the nerve cells, or neurons. Oxygen deficiency causes a functional loss of all cell systems and organs in the body. But some cells respond better to oxygen deficiency than others. Neurons respond badly because their sole source of energy is glucose. Unlike the muscle cells in our body, our brains do not store glucose in the form of glycogen as a ready supply of cell energy. The parts of the brain that are most susceptible to oxygen deficiency are the neurons in the cerebral cortex, the hippocampus, and the thalamus.[14] Oxygen deficiency reduces these structures, which form an important link between the brain stem and cerebral cortex, to utter chaos and wipes out their connections. Synapses are the junctions that enable communication between neurons, and when these synapses stop functioning cooperation is no longer possible. But research drawing on magnetic resonance imaging (MRI), for example, has shown that the joint and simultaneous activity of the cerebral cortex and brain stem, with their shared pathways (hippocampus and thalamus), is a prerequisite for conscious experience.

If the absence of blood flow to the brain prevents the supply of glucose and oxygen, a neuron's first symptom is the inability to maintain its membrane potential, resulting in the loss of neuronal function.[15] The acute loss of electrical and synaptic activity in neurons can be seen as the cell's inbuilt defense and energy-saving response (a pilot-light state). When these functions cease, the remaining energy sources can be deployed very briefly for the cell's survival. In the case of short-term oxygen deficiency, dysfunction can be temporary and recovery possible because the neurons will remain viable for a few more minutes.

The Difference Between Temporary and Permanent Dysfunction

Cardiologists draw on a comparable temporary loss of function of the cardiac muscle to check if somebody with chest pain after exertion (angina pectoris) shows signs of oxygen deficiency in a certain part of the heart muscle. I cite this example not only because I am a cardiologist, but also because this process is much easier to explain in the heart than in the brain. During an exercise test, which induces oxygen deficiency in the heart, ultrasound (echo) or nuclear imaging (SPECT scan) registers contraction of the heart muscle. As soon as oxygen deficiency sets in, the ECG changes and part of the heart muscle no longer contracts. This part of the heart muscle has stopped functioning normally because obstruction of the coronary artery has caused oxygen deficiency. As soon as the test is discontinued and the oxygen flow is restored, the heart muscle resumes normal function. The loss of function was temporary and reversible, and is known as "stunning" of the heart. A similar kind of stunning (pilot-light state) takes place in the neurons, but if the oxygen deficiency is too prolonged, the death of cells causes irreparable damage and the loss of function will be permanent and irreversible. In the heart this is known as a myocardial infarction. Permanent loss of all brain function as a result of a cardiac arrest is known as brain death because after five to ten minutes the neurons are irreversibly damaged by the disintegration of the cell membrane, which leads to an influx of calcium and the formation of so-called free radicals. The proteins in the neurons break down, and the cell dies.[16]

This difference is clearly visible in patients with a temporary or permanent loss of blood flow in part of the brain. If a blood vessel (artery) in the brain is blocked by a blood clot, part of the cerebral cortex will receive no more blood and therefore no more oxygen and glucose. The resulting loss of function of this part of the brain will cause one-sided paralysis of the body, partial blindness, or the loss of speech. If the blood clot dissolves within five to ten minutes, the loss of function is temporary and the paralysis and other symptoms will disappear. Such temporary dysfunction is known as a transient ischemic attack (TIA). However, if the blood clot continues to block the blood vessel, the neurons die, leading to permanent dysfunction of that part of the brain. Patients are left with permanent paralysis or other symptoms, and this is called a cerebral infarction. It is also known as a stroke or a cerebrovascular accident (CVA). The loss of brain function is no longer transient because the neurons have sustained permanent damage and die as a result of prolonged oxygen deprivation (anoxia).[17]

During a cardiac arrest the entire brain is deprived of oxygen, resulting in the loss of consciousness, reflexes, and respiration. This is known as clinical death. It is usually reversible—that is, temporary, if resuscitation is begun within five to ten minutes. But a long delay in resuscitation may result in the death of a great many brain cells and thus in brain death. Most patients will ultimately die. A study carried out at a coronary care unit showed that patients whose resuscitation was started within one minute had a 33 percent chance of survival compared to only 14 percent for those who were resuscitated more than a minute after the onset of unconsciousness.[18]

What Happens During Resuscitation?

During resuscitation, blood gases (oxygen and carbon dioxide) are sometimes measured to determine the severity of the oxygen deficiency in the blood. However, normal levels do not guarantee that enough blood, and thus enough oxygen, will reach the brain during resuscitation.

Research has shown that external heart massage cannot pump enough blood to the brain to restore brain function. Nobody has ever regained consciousness during external resuscitation of the heart. This always requires defibrillation (an electric shock). Once the cardiac rhythm is restored, blood pressure usually stabilizes. Blood pressure is typically expressed in millimeters of mercury (mmHg) because of the now obsolete manometers that used a column of mercury. Under normal circumstances, blood pressure is approximately 140 over 80mmHg, with an average blood pressure of 100mmHg.

During resuscitation, blood supply to the brain is less than 5 percent of its normal value, and during external heart massage the systolic pressure (the first number) usually reaches approximately 50mmHg, with an average of 20mmHg because of the low diastolic pressure (the second number). The maximum average blood pressure during proper resuscitation is 30 to 40mmHg, which is still far too low for the blood to deliver enough oxygen and glucose to the brain. Giving certain medications during resuscitation can increase blood pressure a little, but it will remain well below normal.[19] Furthermore, in the absence of a normal blood supply, the brain cells are likely to swell (edema), which results in increased pressure in the brain and actually requires higher than normal blood pressure to supply the brain with well-oxygenated blood and to remove carbon dioxide. Within seconds, a cardiac arrest results in serious oxygen deficiency and a buildup of carbon dioxide in the brain. This situation cannot be remedied during the resuscitation procedure itself, but can be addressed only by reestablishing cardiac rhythm through defibrillation (an electric shock).

Proper resuscitation, with adequate heart massage and mouth-to-mouth respiration or respiration via a mask, produces a low flow of blood to the brain, which increases the chances of recovering brain function after the cardiac arrest has been treated. There have been some cases in which the electrical activity of the brain was measured (EEG) during a cardiac arrest, for example during surgery. Following the cardiac arrest (no blood flow), this EEG flatlined after an average of fifteen seconds and remained flat during resuscitation.[20] The EEG does not recover until both heartbeat and blood pressure have been restored, and the more prolonged the cardiac arrest and resuscitation, the longer the EEG will remain flat (hours or days). In other words, after a complicated but successful resuscitation, the patient will remain in a coma longer. If no resuscitation is begun, the brain will usually be irreparably damaged within five to ten minutes and the patients will nearly always die.

Temporary and Permanent Brain Damage After a Cardiac Arrest

The ultimate severity of the brain damage depends on how long the brain was completely deprived of blood during the cardiac arrest and on how long it received minimal blood supply during resuscitation with external heart massage and artificial respiration. The severity of the brain damage also depends on temperature. The lower the temperature, the later permanent brain damage occurs because a lower temperature reduces cell demand for oxygen and increases the chances of survival.[21] Some people remain in a coma after a delayed resuscitation. A potential treatment for coma patients is hypothermia, which involves lowering the temperature of the head. A coma, both after a traffic accident (trauma) and after a delayed resuscitation, involves a cerebral edema, a swelling of the neurons, resulting in increased pressure in the brain. This means that despite normal blood pressure, blood supply to the brain decreases and patients remain in a coma for longer. The brain cells go into pilot-light state, otherwise known as hibernation of the brain.[22] When the brain again receives a normal supply of well-oxygenated blood, brain function is sometimes restored. Therapeutic hypothermia reduces cerebral edema, thereby slightly improving blood supply and the survival chances of cells in the pilot-light state. The chances of waking from coma increase a little while the risk of brain death decreases a bit.

When animals go into hibernation, their body temperature plummets and their metabolic rate slows to a near halt. These animals can survive for months without food by putting their body in a kind of pilot-light state with barely perceptible breathing and pulse. The principle of hibernation is known not only in animals and the human brain; cardiologists have identified it also in the heart. A myocardial infarction leaves the heart scarred because heart muscle cells die and are replaced by scar tissue. However, examinations with ultrasound (echo) show that the loss of function of the cardiac muscle extends beyond the actual infarction. The peripheral areas around the infarction go into hibernation (pilot-light state) because the surrounding small blood vessels, the capillaries, maintain a low blood flow. Ultrasound or nuclear imaging can establish how likely it is that this tissue will functionally recover because it is still viable. If patients receive treatment such as bypass surgery or angioplasty (inserting a small balloon into a coronary artery), the hibernating part of the cardiac muscle will recover fully, even when this period of hibernation lasted several years.

There is evidence that tissue in pilot-light state can survive prolonged cell dysfunction. The same is true for the brains of coma patients with a flat EEG.

Pamela Reynolds's NDE

Occasionally, a patient waking from a coma reports an exceptionally clear consciousness during his or her coma, including lucid thoughts and memories, emotions, a sense of identity, and verifiable perceptions from a position outside and above their unconscious body, despite the total absence of demonstrable brain activity. There are only a few known cases in which this loss of function has been carefully documented.

I therefore conclude this chapter with a comprehensive account of Pamela Reynolds's NDE, as described by cardiologist Michael Sabom.[23] Pamela also appeared at length in the BBC program *The Day I Died*. Because she had her NDE during brain surgery, when the activity of the cerebral cortex and brain stem were constantly monitored, hers is a good example of an NDE during carefully documented loss of brain function.

Pamela Reynolds was a thirty-five-year-old busy working mother who had carved out a name for herself as a singer-songwriter. In 1991 she became seriously ill. She experienced extreme dizziness, loss of speech, and difficulty in moving her body. Her physician recommended a CAT scan, which revealed a giant aneurysm in one of her cerebral arteries close to the brain stem. An aneurysm is a kind of balloonlike bulge at a weakened spot in a blood vessel, not unlike a bubble on the inside of a bicycle tire. If this aneurysm burst, and the risk of this happening was considerable, a cerebral hemorrhage would be immediately fatal. She was referred to a neurologist who told her that her chances of survival were minimal. But there was one last hope for Pamela. She contacted the Barrow Neurological Institute in Phoenix, Arizona, more than two thousand miles from her hometown.

Neurosurgeon Dr. Robert Spetzler at the Barrow Neurological Institute decided to operate on Pamela, even though her chances of survival were slight. Everything that happened during her operation was carefully recorded. During the operation her body temperature was lowered to approximately 50 degrees Fahrenheit. She was on a heart-lung machine because of the loss of all cardiac electrical activity (cardiac arrest), which always occurs during severe hypothermia. All the blood had been drained from her head. The electrical activity of her cerebral cortex (EEG) and of her brain stem ("evoked potentials" through 100-decibel clicks emitted by small molded speakers inserted into her ears) was under constant observation; in both cases, there was no activity whatsoever. During the interview in the BBC documentary, Spetzler explained:

> *What we're looking at is the aneurysm that she had, which is at the very base of the brain. This is the balloon that can burst and cause this incredible catastrophe in the patient's brain. This is why it was so difficult in this particular case. . . . What we want to do is we want to bring that brain to a halt. We don't just want the brain to be asleep. We want the metabolic activity of the brain to stop. Every measurable output that the body puts out really disappears completely so that you have no measurable neuronal activity whatsoever.*
>
> *Prior to the operation starting, a lot of activity goes on. The patient is put to sleep, the eyes are taped shut, and there are little clicking devices put in each ear in order to monitor the brain. The patient is then completely covered; the only thing that's really exposed is the area of the head where we work.*

And Sabom emphasizes:

During standstill, Pam's brain was found dead by all three clinical tests—her electroencephalogram was silent, her brain-stem response was absent, and no blood flowed through her brain. . . . Her eyes were lubricated to prevent drying and then taped shut. Additionally, she was under deep general anesthesia.

Pamela's account below is a composite of the written account of her experience in Sabom's book and her interview in the BBC documentary:

I don't remember an operating room. I don't remember seeing Doctor Spetzler at all. I was with a fellow; one of his fellows was with me at that time. After that . . . nothing. Absolutely nothing. Until the sound . . . and the sound was . . . unpleasant. It was guttural. It was reminiscent of being in a dentist's office. And I remember the top of my head tingling, and I just sort of popped out of the top of my head. The further out of my body I got, the more clear the tone became. I remember seeing several things in the operating room when I was looking down. I was the most aware that I've ever been in my entire life. And I was then looking down at my body, and I knew that it was my body. But I didn't care. I thought the way they had my head shaved was very peculiar. I expected them to take all of the hair, but they didn't.

I was metaphorically sitting on Dr. Spetzler's shoulder. It wasn't like normal vision. It was brighter and more focused and clearer than normal vision. There was so much in the operating room that I didn't recognize, and so many people. I remember the instrument in his hand; it looked like the handle of my electric toothbrush. I had assumed that they were going to open the skull with a saw. I had heard the term *saw*, but what I saw looked a lot more like a drill than a saw. It even had little bits that were kept in this case that looked like the case that my father stored his socket wrenches in when I was a child. I saw the grip of the saw, but I didn't see them use it on my head, but I think I heard it being used on something. It was humming at a relatively high pitch. I remember the heart-lung machine. I didn't like the respirator. . . . I remember a lot of tools and instruments that I did not readily recognize. And I distinctly remember a female voice saying: "We have a problem. Her arteries are too small." And then a male voice: "Try the other side." It seemed to come from further down on the table. I do remember wondering what are they doing there [laughs] because this is brain surgery! What had happened was that they accessed the femoral arteries in order to drain the blood, and I didn't understand that. . . .

I felt a "presence." I sort of turned around to look at it. And that's when I saw the very tiny pinpoint of light. And the light started to pull me, but not against my will. I was going of my own accord because I wanted to go. And there was a physical sensation to the point where . . . and I know how that must sound . . . nonetheless it's true. There was a physical sensation, rather like going over a hill real fast. It was like *The Wizard of Oz*—being taken up in a tornado vortex, only you're not spinning around. The feeling was like going up in an elevator real fast. It was like a tunnel, but it wasn't a tunnel. And I went toward the light. The closer I got to the light, I began to discern different figures, different people, and I distinctly heard my grandmother calling me. She has a very distinct voice. But I didn't hear her call me with my ears. . . . It was a clearer hearing than with my ears. And I immediately went to her. The light was incredibly bright, like sitting in the middle of a lightbulb. I noticed that as I began to discern different figures in the light— and they were all covered with light, they *were* light, and had light permeating all around them—they began to form shapes I could recognize and understand. And I saw many, many people I knew and many, many I didn't know, but I knew that I was somehow and in some way connected to them. And it felt . . . great! Everyone I saw, looking back on it, fit perfectly into my understanding of what that person looked like at their best during their lives.

I recognized a lot of people. And one of them was my grandmother. And I saw my uncle Gene, who passed away when he was only thirty-nine years old. He taught me a lot; he taught me to play my first guitar. So was my great-great-aunt Maggie. On Papa's side of the family, my grandfather was there. . . . They were specifically taking care of me, looking after me.

They wouldn't permit me to go further. . . . It was communicated to me—that's the best way I know how to say it because they didn't speak like I'm speaking—that if I went all the way into the light something would happen to me physically. They would be unable to put (this) me back into the body (me), like I had gone too far and they couldn't reconnect. So they wouldn't let me go anywhere or do anything.

I wanted to go into the light, but I also wanted to come back. I had children to be reared. It was like watching a movie on fastforward on your VCR: You get the general idea, but the individual freeze-frames aren't slow enough to get detail. . . . Sparkles is the image that I get. I asked if God was the light, and the answer was: "No, God is not the light, the light is what happens when God breathes." And I distinctly remember thinking: I'm standing in the breath of God. . . .

At some point in time I was reminded that it was time to go back. Of course I had made my decision to go back before I ever lay down on that table. But, you know, the more I was there, the better I liked it [laughs]. My grandmother didn't take me back through the tunnel or even send me back or ask me to go. She just looked up at me. I expected to go with her. My uncle was the one who brought me back down to the body. But then I got to where the body was, and I looked at the thing, and I for sure didn't want to get in it because it looked pretty much like what it was: void of life. I believe it was covered. It scared me, and I didn't want to look at it. And I knew it would hurt, so I didn't want to get in. But he kept reasoning with me. He says: "Like diving into a swimming pool, just jump in." No. "What about the children?" You know what, the children will be fine [laughs]. And he goes: "Honey, you got to go." No. He pushed me; he gave me a little help there. It's taken a long time, but I think I'm ready to forgive him for that [laughs].

I saw the body jump. . . . And then he pushed me, and I felt it chill me inside. I returned to my body. It was like diving into a pool of ice water. . . . It hurt!

When I came back, and I was still under general anesthesia in the operating theater, they were playing "Hotel California," and the line was "You can check out anytime you like, but you can never leave." I mentioned [later] to Dr. Brown that that was incredibly insensitive, and he told me that I needed to sleep more [laughter]. When I regained consciousness, I was still on the respirator.

Pam concludes her account by saying,

I think death is an illusion. I think death is a really nasty bad lie.

Sabom commented on Pam's story:

I found that what she saw from her out-of-body experience seemingly corresponded very accurately to what had actually occurred. She looked at the bone saw that was being used to cut open her skull. It indeed does resemble an electric toothbrush. . . . And there was some conversation at the time between the doctors and Pam accurately recalled hearing that conversation. . . .

These are the words of neurosurgeon Spetzler:

I don't think that the observations she made were based on what she experienced as she went into the operating theater. They were just not available to her. For example, the drill and so on, those things are all covered up. They aren't visible; they were inside their packages. You really don't begin to open until the patient is completely asleep so that you maintain a sterile environment. . . . At that stage in the operation nobody can observe, hear in that state. And . . . I find it inconceivable that the normal senses, such as hearing, let alone the fact that she had clicking devices in each ear, that there was any way for her to hear those through normal auditory pathways. . . .

I don't have an explanation for it. I don't know how it's possible for it to happen, considering the physiological state she was in. At the same time, I have seen so many things that I can't explain that I don't want to be so arrogant as to be able to say that there's no way it can happen.

I would like to offer some additional comments on Pamela Reynolds's NDE and the medical circumstances under which she had her experience.

The kind of brain surgery that she underwent lasts at least four to six hours, just as long as heart surgery. While she was under general anesthesia and surgery on her skull had begun, but her body had not yet significantly cooled down and the blood had not yet been removed from her head, she had an out-of-body experience. She was able to see and hear a great many details despite the fact that she was under general anesthesia, her eyes had been taped shut, and small speakers emitting loud clicks had been inserted into her ears. She could see the equipment and the people in the operating theater; she could see the instrument with which her skull was opened up. During the operation she could hear the conversation between Spetzler and the female cardiovascular surgeon operating in her groin to link her up to the heart-lung machine. When the cardiovascular surgeon made an incision in her right groin, she found that Pamela's veins and arteries were too small, so she had to switch to the left groin. The doctors had a brief exchange on this matter. Pamela heard these remarks and repeated them word for word. After her out-of-body experience, she was pulled into a tunnel. The rest of her NDE, which included an extremely lucid consciousness, the recognition of and communication with deceased relatives, and an encounter with the light, took place during a time when induced hypothermia and anoxia (no oxygen due to lack of blood) had rendered her brain completely nonfunctional. Some time into the operation, the hypothermia induced a cardiac arrest, which is why Pamela had to be connected to the heart-lung machine. Once her body had cooled down to 10 degrees C (50 degrees F) and had been linked to the heart-lung machine, the top of the operating table was raised briefly to remove all the blood from the brain. All these measures were necessary for the operation on the aneurysm to have any chance of success.[24]

Few medical centers in the world would be prepared to carry out such a risky procedure. Extreme hypothermia enables the brain cells to survive for a maximum of sixty minutes during an operation of this kind. The low temperature slows the cellular metabolic rate to such an extent that the brain cells can remain in the pilot-light state for longer without dying. Toward the end of her NDE, Pamela had another out-of-body experience while she remained under general anesthesia, her eyes were still taped shut, and she was still in cardiac arrest. She saw her body jump, which was the result of the electric shock (defibrillation) that was administered to get her heart going again. This only happens once the operation has ended and the body has been warmed up. She felt a chill upon returning to her body because it had not yet reached its normal temperature of 98.6 degrees Fahrenheit.

The prospective NDE studies showed that an enhanced and extremely lucid consciousness can be experienced during a cardiac arrest. We know from other studies that during a cardiac arrest the cerebral cortex and brain stem show no measurable activity while the clinical picture shows a complete loss of all brain function. To our utter surprise, we were forced to conclude that oxygen deficiency alone does not explain the experience of an enhanced consciousness. If an NDE arose in response to oxygen deficiency in the brain, all patients in the Dutch study should have reported one. Nor did the severity of the medical situation, such as a lengthy coma after a complicated resuscitation with prolonged oxygen deprivation, explain whether or not patients reported an NDE. Near-death experiences also take place under circumstances that do not involve oxygen deficiency.

These findings made me all the more curious about brain function under normal, everyday circumstances. How can we reconcile the experience of an NDE during the loss of all brain function with our commonly accepted ideas about brain function? What do we know about the way our brain functions, and how do we know this? What theories do we have about the relationship between the brain and consciousness? I will consider these questions in more detail in the next chapter.

Chapter Nine

What Do We Know About Brain Function?

Consciousness, the subjective experience of an Inner self, poses one of the greatest challenges to neuroscience. Even a detailed knowledge of the brain's workings and neural correlates of consciousness may fail to explain how or why human beings have self-aware minds.
—DAVID J. CHALMERS

I am well aware that this chapter may not be easy for everyone. But my exploration of the anatomy, function, and imaging techniques of the brain is aimed at improving our understanding of the complexity of the brain while at the same time recognizing how little we still know about brain function and the origins of consciousness. Most neuroscientists take a materialist approach, which is based on the premise that the content of thoughts, feelings, and memories can be accounted for on the basis of measurable brain activity. However, the hypothesis that consciousness and memory are produced and stored exclusively in the brain remains unproven. There is no direct evidence to prove if and how neurons in the brain produce the subjective essence of our consciousness. This chapter will draw on scientific studies to prove that the materialist approach falls short in many respects and can no longer be maintained in its current form. It is now becoming increasingly clear that brain activity in itself cannot explain consciousness.

The Search for Consciousness

The previous chapters revealed that people can experience a clear consciousness with memories, lucid thoughts, and emotions during a cardiac arrest. But a period of clinical death, precipitated by the loss of blood flow to the brain, is characterized by the absence of all measurable and clinical brain activity. How can this be? What does science tell us about the relationship between the brain and consciousness, and where and how can consciousness be localized in the brain? How can matter produce consciousness? After all, the brain is made up of pure matter, of atoms and molecules, which in turn are the building blocks of cells with chemical and electrical processes. Composed of "unconscious building blocks," the brain is certainly capable of facilitating consciousness. But does the brain actually "produce" our consciousness? And where in the brain can this consciousness be produced and stored?

Another question is how a nonmaterial activity such as mindfulness or thinking corresponds with a visible reaction in the form of measurable electrical, magnetic, and chemical activity in a certain part of the brain. These activities can be measured with the help of (1) an electroencephalogram (EEG), which registers the electrical activity in the cerebral cortex, or with (2) a magnetoencephalogram (MEG), which registers the brain's magnetic activity. Differences in activity are measured (indirectly) with (3) functional magnetic resonance imaging (fMRI), which can map differences in the brain's blood flow because blood contrasts with surrounding tissue. The scan produces a blood-oxygen-level dependent (BOLD) contrast, which indirectly reflects the metabolic activity of neural networks. It does not register neural activity directly. Brain activity can also be registered with the help of (4) a positron emission tomography (PET scan), in which an injected radioactive substance can provide more direct information about differences in the metabolic activity of brain cells. It is possible to measure a 30 percent increase in blood flow to the brain during thinking or mindfulness because the neurons use more energy for such processes.[1]

All of these forms of research measure changes in blood flow and activity at certain locations in the brain. These activities may vary for individual subjects while changing thoughts and emotions also activate new locations. While this suggests that neural networks play some role in the manifestation of thoughts, feelings, and memories, it does not necessarily imply that these cells actually produce and store our thoughts and emotions. We have no direct evidence to prove if and how neurons in the brain produce the subjective essence of our consciousness. What we do know is that the following three structures and the close connections between them enable us to experience consciousness: (1) the ascending reticular activating system (ARAS) in the brain stem; (2) the cerebral cortex, especially the frontal lobe, temporal lobes, and parietal lobes; and (3) the connections between the cortex and the stem, by way of the thalamus and the hippocampus (see figure). These centers show distinct activity during consciousness, and impairment of these centers leads to unconsciousness or coma. The logical conclusion is that collaboration between these brain centers plays a role in enabling the experience of everyday (waking) consciousness.[2]

Anatomy of the Brain

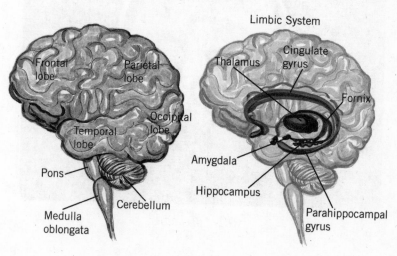

The anatomical structures in the brain.

The Reliability of Contemporary Brain Research

How accurate is contemporary brain research? Which brain activities can be measured, what causes these activities, and what do these measured activities tell us about what actually happens in the brain? And of course we must ask ourselves what it means if we can no longer measure any of these activities in the brain.

An fMRI (see figure) shows blood flow activity in the brain at a resolution of approximately a grain of rice (a "voxel"). Millions of neurons must be fired simultaneously to illuminate such a small region. An fMRI cannot be used to determine the sequence of, and hence the connections between, the measured activities in the various regions. The fastest possible speed is currently only one scan per two seconds, which is far too slow to trace and map cerebral processes that take place in mere milliseconds. This could be compared with reading a book by reading only one of each thousand words.

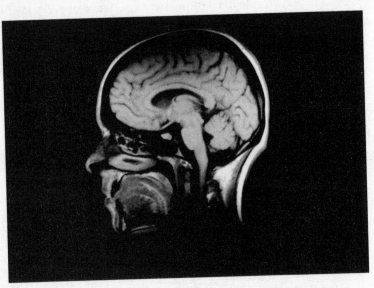

MRI of the brain.

At present, scientific research methods appear to be incapable of accurately studying the neural processes associated with our experience of consciousness. If changing forms of cooperation between large groups of neurons are indeed the neural correlates of consciousness, our current research techniques are not up to the task of mapping this process. Even the most modern and detailed fMRI scan shows us no more than the physical basis of an observation or of mindfulness and fails to offer any explanation for what happens in our mind. Besides, an fMRI scan says nothing about the content of our thoughts and feelings. It is alarming therefore that fMRI research has prompted some scientists to trace a causal link between certain brain functions and specific mental processes.[3]

This concern is shared by the Danish neuroscientist and anthropologist Andreas Roepstorff in a recent interview.[4] The interviewer noted that "Roepstorff has a difficult message for his peers": "Complete, objective knowledge about the nature of the human mind is impossible, despite all brain scans." Roepstorff describes his experiences as a volunteer in an experiment. As part of this experiment, subjects were placed in an fMRI scanner and had the soles of their feet tickled. Sometimes they could only feel the tickling, and sometimes they could also see it happening in a mirror. The experiment sought to identify differences in the way the brain processes information. But Roepstorff was annoyed by the experiment leader, who for no apparent reason left him in the scanner for a very long time. He decided to play a trick on the leader. If he was tickled, he would think about football, but if he could also see himself being tickled, he would think about his cat's funeral. He comments: "Because I was thinking about different things the brain scans should, in theory, have shown activity in different parts of the brain."

Given the fact that Roepstorff's thoughts fell outside the scope of the experiment, the test leader should not have been able to understand the findings. But the leader failed to notice anything strange about the scans. They were no different from the scans of any of the other subjects.

Because I had decided not to do as the test leader had asked, my state of consciousness was, by definition, different from that of a subject who had followed instructions. . . . But the test leader has no way of interpreting such a difference objectively because I did not behave

noticeably different from an obedient subject. Suppose that the
measurement had been precise enough and that my brain scans had
struck him as unusual; the only way he could have made sense of
the difference was by asking me what went on in my head. In which
case I could either lie or tell the truth or, if I had forgotten what I
had been thinking about during the scan, be useless. . . . Thoughts
are subjective. Somebody's thoughts can often be deduced from his
or her behavior. . . . But only the subject himself has direct access
to his thoughts. This discrepancy between the first-person perspective (subjective) and the third-person perspective (objective) presents the scientist with seemingly insurmountable methodological
problems. How can the neuroscientist obtain objective knowledge
about consciousness when direct access to the brain is purely subjective, via introspection? Consciousness is fundamentally unverifiable, and thus fails to meet scientific criteria. . . . This evaporates
the hope for completely objective knowledge about consciousness.
Sooner or later, you will have to talk to your subject, so there will
always be a subjective link.

In fact, Roepstorff is curious to know whether in five to ten years'
time we may have to concede that we were stupid to think that we
could explain us to ourselves by way of the brain.

Both Roepstorff and neuropsychologist Anthony I. Jack are convinced that it is impossible to produce objective evidence with which to
test the reliability or veracity of subjective reports.[5] The same applies, in
their view, to psychological research techniques. The fact that specific
parts of the brain are activated and thus involved in specific tasks does
not seem to explain cognitive processes. When researching the structure
of information processes in the brain, psychologists are not allowed to
use reports on thoughts and feelings in the same way as objective reports on behavior. But Roepstorff and Jack also reject the assertion that
the accuracy of reported thoughts and feelings must always be called
into question. They believe that there should be room for faith in reported ideas and emotions, provided of course that the subject has been
proven to be reasonably reliable. Needless to say, the same applies to the
subjective reports of near-death experiences, the subject of this book.

In their scientific publications Jack and Roepstorff write that the validity of the conclusions drawn from registrations of brain activities is both the most important and theoretically most complex and thorny subject in the cognitive sciences. A measurement is deemed reliable only if there is evidence that it accurately reflects the phenomenon it set out to examine. Reliability is all the more complex because scientific measurements are often used to prove phenomena straddling entirely different levels. By *different levels* they mean which aspects or conclusions of measurements have definitely been proven, which may have been proven, or which are unlikely.[6]

They clarify this with a few examples:

1. Evidence has shown that the results of fMRI scans can be used to track blood flow in the brain. But these scans do not yet allow positive conclusions about neural activities because scientists are still in the process of establishing these activities with some degree of accuracy. Besides, an fMRI scan probably says (almost?) nothing about the content and location of cognitive functions because these have not yet been established with any degree of certainty.

2. Measurements of observed behavior, such as the measurements of reaction times, are valuable as direct proof of stable behavioral patterns but less valuable for assessing the development of information processes, and least of all as proof of the existence and mechanism of certain cognitive functions.

3. Reports of subjective thoughts and feelings serve primarily as proof of people's ideas about their own experience but are less direct proof of the existence of experimentally induced phenomena, while reports of subjective experiences cannot provide any direct proof of the origins of specific cognitive functions.

Unproven Hypotheses

So far we can locate no single region in which the neural activity corresponds exactly to the vivid picture of the world we see in front of our eyes.

—Francis H. C. Crick

The hypothesis that consciousness and memory are produced and stored exclusively in the brain remains unproven. For decades, scientists have tried unsuccessfully to localize memories and consciousness in the brain. It is doubtful whether they will ever succeed. At present science cannot explain how certain neural networks produce the subjective essence of thoughts and feelings because so far no neurophysiological study has identified any exact correspondence between specific neural activities and the specific content of memories, experiences, feelings, or thoughts. The assumption was that an activity in specific neural networks would always result in the same thoughts and feelings. Some studies spoke of a "matching content doctrine" because it was thought that the sight of certain images would always prompt the same visual perception with associated thoughts and emotions on account of this perception triggering activity in specific neural visual networks. These days scientists simply speak of neural correlates of consciousness, which means that there is a correlation (a relationship or connection) between registered activities in the brain and experiences in consciousness, and various imaging techniques (EEG, MEG, fMRI, or PET scan) have shown that a specific conscious experience can activate many, sometimes quite remote, brain centers.[7]

But a connection says nothing about cause or effect. A conscious experience can be the result of brain activity, but a brain activity can also be the result of consciousness. Likewise, a connection says nothing about the content of a subjective experience. An exact match between measurable brain activity and the subjective content of conscious experiences seems highly unlikely because neural activity is no more than neural activity: a way of coding information. Evidence of neural activity only means the presence of active structures. You might think of it as a radio: somebody can activate a radio by switching it on and then search for a certain wavelength to receive a particular station, but doing this does not affect the content of the broadcast. In other words, tuning in to a radio station has no influence on program content. Equally, switching on your computer, connecting to the Internet, and navigating to a Web site does not determine the content of this Web site.

The activation of certain areas of the brain cannot explain the content of thoughts and emotions. A correlation between activities in certain areas of the brain and certain conscious experiences fails to explain the origins of either consciousness or the subjective content of consciousness. The explanatory gap between the brain and consciousness has never been bridged because a certain neuronal state is not the same as a certain state of consciousness. It looks as if scientific research methods are not accurate enough for studying the neural processes underlying our conscious experiences or for demonstrating how neurons or neural networks might produce the essence of our private thoughts and feelings because, as I have explained before, what we can measure is only a correlation between registered activities in the brain and experiences in consciousness. It seems fair to conclude that current knowledge does not permit us to reduce consciousness only to activities and processes in the brain.

Interestingly, this view is fundamentally in agreement with the ideas of the philosopher and neuroscientist Alva Noë, who, based on entirely different neuroscientific research, writes in his recent book:

All scientific theories rest on assumptions. It is important that these assumptions be true. I will try to convince the reader that this startling assumption of consciousness research that consciousness is a neuroscientific phenomenon and that it happens in the brain is badly mistaken. . . . Contemporary research on consciousness in neuroscience rests on unquestioned but highly questionable foundations. Consciousness does not happen in the brain. . . . What determines and controls the character of conscious experience is not the associated neural activity. It is misguided to search for neural correlates of consciousness: There are no such neural structures. That is why we have been unable to come up with a good explanation of its neural basis. . . . The idea that we are our brains is not something scientists have learned; it is rather a preconception. That consciousness arises in the brain goes unquestioned. It is an unargued-for starting assumption. . . . It is just prejudice. We are not entitled to conclude that consciousness depends only on actions of the brain itself. And in fact we have every reason to reject it now. . . . Experience and cognition are not bodily by-products. It is a hard conclusion, but one that is hard to avoid. . . . Moreover, the

mere absence of the normal behavioural markers of consciousness
does not entail the absence of consciousness.[8]

Noë proposes that consciousness is not a by-product of the brain but that the brain's job is that of facilitating a dynamic pattern of interaction among brain, body, and world.

Neuroscience has so far been unable to explain how neuronal behavior might account for the cause and content of thoughts and emotions, but still most scientists continue to support the view that cerebral processes underpin all aspects of consciousness. An article by Jeffrey Saver and John Rabin about the neural substrate of religious experience illustrates just how extreme this view is: "All human experience is brain-based, including scientific reasoning, mathematical deduction, moral judgment, and artistic creation, as well as religious states of mind. . . . There are no exceptions to this rule."[9]

Neuropsychiatrist Jeffrey Schwartz writes, "Mainstream philosophical and scientific discussions may remain strongly biased toward a materialistic perspective, because one restricts one's questions to the domain where materialism is unchallenged."[10]

For decades, scientists have tried unsuccessfully to localize memories and consciousness in the brain, and it seems doubtful that they will ever succeed. So despite the fact that a majority of contemporary scientists specializing in consciousness research still espouse a materialist and reductionist explanation for consciousness, the hypothesis that consciousness and memory are produced and stored exclusively in the brain remains unproven.

Neurons and Electromagnetic Fields

The brain consists of a hundred billion neurons, twenty billion of which are located in the cerebral cortex.

Several thousand neurons die every day, but in the course of days and weeks the fats and proteins that constitute the neurons' cell membrane undergo constant regeneration.[11] The composition and cohesion of all cerebral structures, from molecules to neurons, are in constant flux, which raises a question about long-term memory.

Neurons process and transmit information through electrical charges across their cell membranes, and each neuron has at least a thousand and sometimes up to ten thousand synapses, which can both excite and inhibit other neurons. Synapses are the junctions between neurons (see figure).[12]

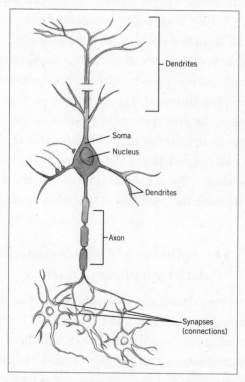

A neuron with dendrites and synapses.

Neurons work together in a highly complex network. Complexity involves a high level of integration (mutual cooperation) and differentiation (mutual differences). It means that there is a system of different neural networks (subsystems) that are both dynamically connected and differentiated.[13] The result is "organized chaos." Such processes are known as a self-organizing system, producing patterns and structures that interact with their environment but are not directly caused by external factors. A good example of self-organization is a vortex in flowing water, in which the shape of the vortex is determined by the rate of flow and the quantity of water, but the vortex itself is spontaneous and self-regulating.

We also find such self-organization in the electrical phenomena in the brain. Neurons communicate through changes in voltage, which release neurotransmitters in the synapses, or junctions between cells. The sum total of all the voltage changes produces constantly changing electrical fields.[14] This also results in constantly fluctuating magnetic fields along the simultaneously activated dendrites, as the branched projections of the neurons are called. During each activity in the brain, all the electrical and magnetic patterns of millions or billions of neurons change every millisecond. Yet neither the number of neurons nor the exact shape of the dendrites nor the individual electrical pattern of individual neurons appears to play a decisive role in the information exchange. Instead the exchange is determined by the constantly changing patterns—caused by self-organization—of the electromagnetic fields that form along the dendrites in specialized neural networks.

The Influence of Electromagnetic Activity on Brain Function

The constantly fluctuating electromagnetic field in and around the brain is registered by the EEG, and from the EEG even the electrical activity of the heart can be deduced (ECG). The question now is what role the electromagnetic activity of neural networks might play in the working of the brain and in the experience of consciousness. Interfering with the brain's electromagnetic field appears to have an effect on brain function because several studies have shown a clear change in the function of neural networks when external magnetic or electrical fields were aimed at the brain. Stimulating or inhibiting neural networks through electrical or magnetic stimulation makes it possible to study the function of these networks, while it can also trigger certain experiences in the mind and can offer a therapeutic potential.

Magnetic Stimulation

When magnetic fields are aimed at the brain, as in transcranial magnetic stimulation or TMS (see figure), this can, depending on the duration and intensity of the administered magnetic energy, either inhibit or excite certain parts of the brain. Targeted magnetic fields are thus capable of temporarily exciting or inhibiting local brain function by influencing the neurons' constantly changing electromagnetic fields, sometimes beyond the time of stimulation, but apparently without any lasting effect.[15] Transcranial magnetic stimulation thus makes it possible to map the function of certain areas of the cerebral cortex. The function of small areas in the cortex can be studied to within milliseconds, enabling scientists to research the contribution of neural networks in the cerebral cortex to specific cognitive functions.

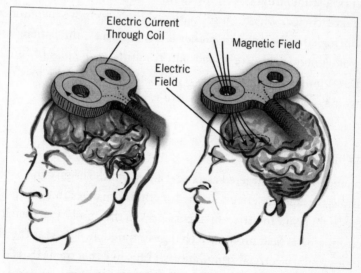

Transcranial Magnetic Stimulation (TMS).

But at a higher intensity TMS can also cause temporary impairment of brain function. Interrupting the electromagnetic processes in the cerebral cortex can momentarily disrupt vision or the experience of physical movement. Stimulating the occipital lobes, the visual processing center, can cause temporary blindness. There appears to be a direct link between the presence of an electromagnetic field and the function of neural networks. The loss of this field causes the loss of function.

Electrical Stimulation

Electrical stimulation of local neuronal networks also disrupts normal brain function, as described in 1958 by neurosurgeon Wilder Penfield and in 2004 by neurologist Olaf Blanke. Local electrical stimulation of epilepsy patients sometimes triggers images from the past (but never a panoramic life review), flashes of light, sounds, and (very rarely) a sense of detachment from the body. These artificially induced experiences are never identical to a typical NDE or to an out-of-body experience with verifiable components, nor are they life-changing. The use of low levels of electrical energy would occasionally produce either no effect or a stimulating effect, for example in the case of stimulation of the motor cortex, which causes patients' limbs to move involuntarily. But during stimulation with higher levels of energy, the patient's own electromagnetic fields are wiped out, resulting in the loss of function of the stimulated area in the cerebral cortex. Again, the loss of the electromagnetic field leads to a loss of function.[16]

Therapeutic Effects

Applying local and targeted electrical energy to the brain can also have a lasting therapeutic effect, as the functional changes in certain areas of the brain result in different experiences in the mind. A change in the electromagnetic field prompts a change in function. Whereas the effect of transcranial electrical stimulation (TES), like that of TMS, is short-lived, transcranial direct current stimulation (tDCS) causes permanent functional change in some parts of the brain because of its effect on the cerebral cortex.[17] This can be used to help seriously depressed patients who do not respond well to antidepressants. The brain state of major depression in such therapy-resistant patients has been demonstrated with the help of fMRI and PET scans, which show that some parts of the brain (such as the subgenual cingulate cortex) are overactive and other parts (such as the prefrontal cortex) are seriously underactive.

Antidepressants can improve these impaired activity patterns in serious depression, but so can various forms of electrical therapy, such as electroconvulsive therapy (ECT), in which a powerful electrical current induces epileptic seizures (convulsions); stimulation of the vagus nerve; and more recently the implanting of deep electrodes in over- or underactive areas of the brain, known as deep brain stimulation (DBS).[18] Although the precise mechanism is unknown, it has been found to produce clinical benefits that were confirmed by fMRI. An article in *Nature* recently described how a man who had been in a form of coma for more than six years after sustaining a traumatic brain injury regained consciousness after DBS in the thalamus.[19]

Targeted magnetic energy, as administered during TMS and magnetic convulsive therapy, sometimes achieves an equally positive effect. But what is even more interesting is that placebo treatment has been found to produce the same neurological improvement in the brain.[20] The belief that one is receiving proper treatment thus appears to have the same effect on brain function as medication or electrical and magnetic stimulation therapy. More on how the mind can influence brain function can be found in the section on neuroplasticity below.

Consciousness Research Using TMS

A recent study published in *Science* drew on a combination of TMS and high-density electroencephalography (EEG) to see if changes in the cerebral cortex might play a role in the loss of consciousness during deep, dreamless sleep (non-REM sleep) while the brain remains active.[21] People have no memories of this phase of deep sleep, whereas they do remember dreams from the REM phase of sleep. Despite measurable brain activity, people do not usually experience consciousness during non-REM sleep.

The study found that during such deep, dreamless sleep the initial response to TMS was heightened but that the signal was rapidly extinguished a few millimeters from the crown of the skull. The electromagnetic signal did not propagate beyond the stimulation site. In contrast, when the same study was conducted in the daytime during wakefulness, the initial response (15 milliseconds) was followed by a sequence of waves that moved to other cortical areas and to some deeper structures several centimeters away. The study concluded that despite electromagnetic activity in the brain during deep sleep, communication between various cortical areas breaks down. This breakdown in communication between neural networks causes consciousness to fade. But when the connections between the various parts of the cerebral cortex and between the cortex and the thalamus function properly, information exchange is possible thanks to the system's integrating and differentiating properties. Such information exchange seems to be a condition for the experience of consciousness.[22]

The latter has also been demonstrated in research using PET scans to explain unconsciousness during general anesthesia, during which brain activity is registered but no (waking) consciousness is experienced. This research, recently published in *Science* and other journals, also shows that a functioning communication system among various neural networks with integration of information is a prerequisite for the experience of (waking) consciousness because during general anesthesia the pathways between thalamus and cortex, in particular, were found to be impaired.[23] And this essential condition is lacking during a cardiac arrest, during general anesthesia, and during deep sleep.

That proper communication within the brain is essential for the experience of consciousness has also been demonstrated by a study of people waking up from deep sleep. The process of deep sleep, which, as shown by TMS, includes the loss of communication between certain neural networks, is reversed upon waking. The study in question looked at the order in which brain centers are reactivated after sleep, during the first five and the first twenty minutes after waking. During the first few minutes, activity in the brain stem and thalamus increased, followed a little later by activity in the prefrontal cortex. The authors conclude that the process of regaining awareness of oneself and one's surroundings after sleep rests on a reorganization process in the brain that involves the recovery of working connections between the aforementioned centers. These centers need to function like a network to enable the experience of consciousness.[24]

During a cardiac arrest the cerebral cortex, thalamus, hippocampus, and brain stem as well as all connections between them stop functioning, as we have seen, which prevents information from being integrated and differentiated—a prerequisite for communication and thus for the experience of consciousness. The experience of consciousness should be impossible during a cardiac arrest. All measurable electrical activity in the brain has been extinguished and all bodily and brain-stem reflexes are gone. And yet, during this period of total dysfunction, some people experience a heightened and enhanced consciousness, known as an NDE.

The Brain, Information Storage Capacity, and Memory

According to current knowledge, consciousness cannot be reduced to activities and processes in the brain. It is highly unlikely that thoughts and emotions are produced by brain cells. Above, we looked at the influence of electromagnetic fields on consciousness as well as the fact that information exchange between brain stem and cerebral cortex is a prerequisite for the experience of consciousness. The next logical question is how all the memories from a person's life can be stored and then recalled again together with their associated emotions. How do we explain short-term and long-term memory? How and where in the brain is this virtually unlimited amount of information stored? And how can this information be accessible at all times?

A single cubic centimeter of the cerebral cortex contains no less than a hundred million neurons, and because each neuron has at least a thousand synapses connecting it with surrounding neurons, each cubic centimeter has approximately $100,000,000,000$ (10^{11}) synapses of dendrites that originate largely in other parts of the cerebral cortex. This means that the brain contains a total of about 10^{14} synapses. If one synapse contained one bit of information, brain function would require more than $100,000,000,000,000$ (10^{14}) bits of information processing, which is far more information than the human DNA, our genetic code, can handle according to current knowledge. For this reason consciousness cannot be stored in our DNA, rendering a cell in our body and brain a highly unlikely producer for consciousness.[25]

Simon Berkovich, a computer expert, has calculated that despite the brain's huge numbers of synapses, its capacity for storing a lifetime's memories, along with associated thoughts and emotions, is completely insufficient. At any waking moment during the day, there are approximately 10^{24} actions per second in the brain. Add to this the required capacity for long-term memory storage, and the total data storage capacity would have to be 3.10^{17} bits/cm^3, which, based on our current understanding of neuronal processes in the brain, is inconceivable. Neurobiologist Herms Romijn, formerly of the Netherlands Institute for Neuroscience, also demonstrated that the storage of all memories in the brain is anatomically and functionally impossible.[26]

On the basis of these findings, we are forced to conclude that the brain has insufficient capacity for storing all memories with associated thoughts and feelings or retrieving capacity for stored information. Neurosurgeon Karl Pribram was equally certain that memories cannot be stored in brain cells, but only in the coherent patterns of the electromagnetic fields of neural networks. In his view the brain functions like a hologram. This hologram is capable of storing the vast quantity of information of the human memory. According to Pribram's holographic hypothesis, memories are stored not in the brain itself but in the electromagnetic fields of the brain. Pribram's hypothesis was inspired by the extraordinary experiments of psychologist Karl Lashley, who proved as early as 1920 that memories are stored not in any single part of the brain but throughout the brain as a whole. His experiments on rats showed that it did not matter which parts or indeed how much of the rats' brains were removed. The animals were still capable of carrying out the complex tasks that they had learned to do before the brain operations.[27]

Earlier in this chapter I mentioned that the composition and cohesion of all brain structures, from molecules to neurons, is in constant flux, which raised a question about long-term memory. The debate about information storage and memory is further complicated by an article in Science with the provocative title "Is Your Brain Really Necessary?" This article was written in response to English neurologist John Lorber's description of a healthy young man with a university degree in mathematics and an IQ of 126. A brain scan revealed a severe case of hydrocephalus: 95 percent of his skull was filled with cerebrospinal fluid, and his cerebral cortex measured only about 2 millimeters thick, leaving barely any brain tissue. The weight of his remaining brain was estimated at 100 grams (compared to a normal weight of 1,500 grams), and yet his brain function was unimpaired. It seems scarcely possible to reconcile this exceptional case with our current belief that memories and consciousness are produced and stored in the brain.[28]

The question is not just how short-term and long-term memory can function properly given the constantly changing synaptic connections in neural networks, but also how memory loss arises. As we get older our brains can atrophy as a result of Alzheimer's disease or arteriosclerosis. Brain volume decreases when brain cells die and are no longer replaced, giving rise to damaged and less effective neural networks and slowly worsening dementia. Whereas long-term memory can remain intact for some time, short-term memory deteriorates, cognitive functions gradually decline, relatives are no longer recognized, and speech becomes more difficult or altogether impossible. These functions can also be lost after brain damage brought on by a cerebral hemorrhage, serious head trauma with permanent brain damage, long-term alcohol abuse, or encephalitis. The obvious and correct conclusion must be that the brain has a major impact on the way people show their everyday or waking consciousness to the outside world. The instrument, the brain, has been damaged, whereas "real" consciousness remains intact. Consciousness and the brain are interdependent, which is not to say that mental and emotional processes are identical with or reducible to cerebral processes. How else can we explain the fact that people with a severe form of dementia, or patients with chronic schizophrenia, sometimes can experience brief lucid moments ("terminal lucidity") shortly before they die?[29]

Neuroplasticity

Throughout life a process of constant adapting is taking place in the cerebral cortex because our mental, intellectual, and physical activities affect both the number and the location of the connections between neurons. This process of ongoing adaptation is called neuroplasticity. Under the influence of mindfulness, emotions, active thought processes as well as movement, the neural networks and electromagnetic activity of the brain undergo constant change. The term *mental gymnastics* or *mental training* speaks volumes. If we remain mentally (and physically) active until late in life, our brains will continue to function better thanks to a more extensive network of synapses. "The power of the mind" can change brain function.[30]

In his book *The Brain That Changes Itself,* the psychiatrist and psychoanalyst Norman Doidge provides an excellent survey of the many scientific studies offering convincing evidence of neuroplasticity. He also writes at length about the many patients who benefited from the therapeutic use of plasticity of the brain because "our thoughts can change the material structure of our brains at a microscopic level, because the brain is constantly adapting itself. So even talking therapy or imagination can change our brains."[31]

At a young age, up to about four, the brain is remarkably plastic. There is evidence that during this period, some hundred thousand synapses are lost and regenerated every second.[32] An extreme example of neuroplasticity is the case of a three-year-old girl whose left brain needed to be surgically removed because of serious chronic encephalitis with symptoms of epilepsy (see figure). Doctors at Johns Hopkins Hospital in Baltimore have performed this kind of major operation on at least a hundred young children, many of them suffering from intractable epilepsy precipitated by serious neurodevelopmental disorders. If adults were to undergo this kind of intervention, the consequences would be disastrous: the patients would be unable to speak or understand language, would be paralyzed on the right side, and would lose sight in one eye. But a year after her operation this girl showed almost no more symptoms. The one-sided paralysis was as good as gone, and she could think clearly. She is now developing normally, fluent in two languages, running and jumping about, and doing well in school.[33]

The only possible explanation for this remarkable adaptability is that the new connections forged by plasticity allowed all brain function to be assumed by the remaining right half of the brain. The girl can do as much with only half a brain as other people do with both

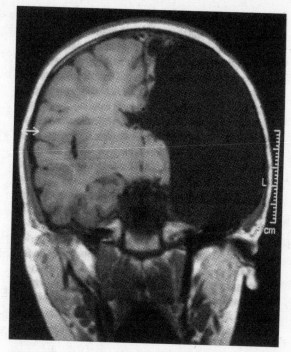

*fMRI scan of a three-year-old girl, following removal
of the left part of her brain (right on picture).*

halves. With practice and the will to get better, she was able to completely reprogram her brain because with only half a brain she had regained the same capacities as people with a normal functioning brain.

The Placebo Effect and Psychotherapy

Several scientific studies have shown that the mind can influence or
determine brain function to a considerable degree. In a study of cognitive behavioral therapy and placebo treatment for depression, fMRI
studies and PET scans found a permanent change in activity distribution in certain regions of the brain.[34] The brain scans of depressed
patients receiving placebo treatment showed neurological improvements in certain parts of the brain that were identical to those seen in
depressed patients receiving cognitive therapy or antidepressants. The
mere thought of receiving proper treatment triggered a clear objective
change in brain function among the depressed patients in the placebo

group. The placebo effect has been studied not only in patients suffering from depression but also in patients with Parkinson's disease, during the administration of pain stimuli, and during the measurement of changes in immune response.[35] In all of these studies, the changed expectations triggered by the placebo effect produced demonstrably different response patterns in both body and brain. Placebo treatment and positive pain manipulation had a favorable impact on some brain centers thanks to the release of endorphinlike substances, and the fMRI showed increased activity in the prefrontal cortex thanks to the raised expectations and changed attention processes. In Parkinson's patients who received placebo treatment, certain brain centers released more dopamine, which significantly reduced muscular stiffness.

In a recent article neuroscientist Mario Beauregard provides a comprehensive overview of everything we know from fMRI research and PET scans conducted during emotional self-regulation (the repression of emotions), psychotherapy, and placebo treatment. He concludes that trust and positive expectations can influence neurophysiological and neurochemical activity in areas of the brain that play a role in perception, movement, pain, and various emotional processes. This means that mental processes (thoughts, feelings, convictions, and volition) can have a significant influence on various levels of brain function. He calls this the psychoneural translation hypothesis. He concludes that when somebody's expectations are manipulated intentionally (through stimulation or self-regulation) or unintentionally (through placebo), this not only results in a positive impact on their (subjective) sense of well-being and in an (objective) reduction of symptoms but also brings about an actual biological change in the brain. His recent book, *The Spiritual Brain*, contains an extensive review of the many studies that have shown that the mind can indeed change brain function. The inevitable conclusion of all these well-designed studies seems to be that the mind is capable of changing the anatomy and function of the brain.[36]

Cognitive Therapy

Cognitive behavioral therapy can have the same effect as a placebo. Psychiatrist Jeffrey Schwartz carried out extensive neurological research in patients with obsessive-compulsive disorder, and with the help of PET scans he found abnormalities in some brain circuits. Intensive cognitive behavioral therapy, which taught these patients to harness the positive power of the mind to change abnormal compulsive thoughts, resulted in subjective and objective improvement of clinical symptoms, while a repeat brain scan showed clear neurological improvements. A new practical application is mindfulness-based cognitive therapy (MBCT) for patients with depression, stress, fear, pain, and physical ailments such as psoriasis, whereby a combination of cognitive therapy and meditation with mindfulness produces noticeable improvements and fMRI registers clear changes, especially in the prefrontal cortex. These cognitive therapeutic changes are a result of neuroplasticity. MBCT also boosted these patients' immune function after an influenza vaccination.[37]

Meditation

Meditation can produce similar changes in the brain. This is known as "spiritual neuroscience." A study showed that the quantitative EEG (or qEEG) of meditating volunteers displayed more gamma waves than normal while the EEG of meditating Buddhist monks, who have spent tens of thousands of hours engaged in meditation, displayed a much higher gamma activity (25–42 Hz), especially in the forehead and sides of the head, which did not disappear after the monks had stopped meditating.[38] Other researchers have found a shift toward theta waves (4–7 Hz) in the EEG during deep meditation. The results of these studies indicate both an acute change during meditation and a permanent change in brain activity as a result of neuroplasticity cultivated by many years of meditation. Some fMRI research comparing meditating monks and a control group also found clear differences in many areas of the brain, but especially in the frontal, temporal, and parietal regions (forehead, temporal bone, and parietal lobes). The centers that appear to be correlated with empathy and compassion showed a particular increase in activity.[39] These studies too show evidence of an acute change for the duration of meditation coupled with permanent functional change in certain areas of the brain. Long-term meditation appears to be the only explanation. We must therefore conclude that thanks to neuroplasticity, positive thinking, and an inward focus during meditation, even adults can bring about a permanent change in brain function.

Volition

Neurophysiologist and Nobel Prize laureate Roger Sperry, who has done a great deal of research among "split-brain" patients, also reached the surprising conclusion that the mind directly determines neural activities.[40] This conclusion is supported by neuropsychologist Benjamin Libet, known for his theory of "readiness potential" (RP), an unconscious "ready" signal in the brain preceding a conscious decision to act ("free will") or not to act ("free won't") by 350 milliseconds. He concluded that the conscious mental field can consolidate subjective experiences but also has the potential to directly influence neural activities. There has also been research into voluntary self-regulation of emotions among men viewing sexually arousing film excerpts and among women and children watching very sad film excerpts. With the help of fMRI techniques, the researchers demonstrated this correlation between special aspects of consciousness and brain function because certain centers were activated for emotions, and the conscious and voluntary regulation of emotions specifically involved increased activity in the frontal lobes (such as the prefrontal cortex). Conscious and voluntary self-regulation of emotions has a very real effect on the activity of the various brain centers involved.[41]

In summary, the human mind is capable of changing the anatomical structure and associated function of the brain. The mind can change the brain. There is unmistakable interaction between the mind and the brain and not just in the sense of cause and effect. As such, it would be incorrect to claim that consciousness can only be a product of brain function. How could a product be able to change its own producer?

Our Brain Is Not a Computer

The brain is the messenger to consciousness.
—JOHN C. ECCLES

Some scientists like to compare the brain to a complex computer. A few of them, such as philosopher Daniel Dennett and psychologist Susan Blackmore, even posit that consciousness is no more than an illusion caused by activities in this "computer."[42] This theory throws a whole new light on concepts such as volition and responsibility for one's actions. Computerlike processes in the neurons of our brain are supposed to be responsible for our thoughts and actions whereas our consciousness, being an illusion, plays no role. But the question that these scientists really ought to be answering is this: how could an illusion cause demonstrable changes in brain structure and function, as consciousness has clearly been proven to do?

Other scientists are of the opinion that in the (distant?) future computers will be able to produce consciousness. But the brain–computer comparison does not hold because for theoretical reasons it seems highly unlikely that a computer could ever adapt and change its own hardware and software to new demands and circumstances, something that the brain is capable of doing thanks to neuroplasticity. Mathematician and physicist Roger Penrose claims that a computer's algorithms are incapable of simulating mathematical reasoning and can therefore never produce consciousness. A machine, no matter how cleverly constructed by human intellect, is not in a position to answer philosophical questions about the meaning of life. On the basis of quantum-mechanical theories, Penrose put forward a hypothesis about the mind–brain relationship that posits that our consciousness cannot be localized in the brain because for purely theoretical reasons the brain is unable to produce human consciousness. According to him, the brain can facilitate but not cause the experience of subjective reality. The well-known neuroscientists and Nobel Prize winners Charles S. Sherrington and John C. Eccles and the neurosurgeon Wilder Penfield were also of the opinion that the brain is more like a complicated organism that registers and transmits consciousness than one that produces it. In his recent book neuroscientist Beauregard demonstrates that a materialist approach to the mind–brain relationship is no longer tenable in neuroscience. On the strength of his own and other people's research, he shows that religious, mystical, spiritual, and near-death experiences cannot be the product of the brain. He too is convinced that the brain merely facilitates the experience of consciousness. As mentioned earlier in this chapter, Noë reaches an identical conclusion in his recent book. He brands the assumption in neuroscience that consciousness arises in the brain and is a by-product of neural activity a prejudice and unfounded hypothesis. In his view the brain, in conjunction with the body and the world, has a facilitating function: it enables the experience of consciousness.[43]

> *What is mind? No matter. What is matter? Never mind.*
> —THOMAS HEWITT KEY

This chapter outlined what we do and above all what we do not (or not yet) know about brain function in relation to thoughts and emotions. In many respects this relationship remains a huge mystery. The brain is an extremely complex and mysterious organ, weighing about three pounds and consuming nearly 20 percent of our body's energy. Our brain consists of one hundred billion neurons, which are all interconnected via thousands of synapses per neuron, thus forming an extraordinarily complex network. Awareness of oneself and one's surroundings and having thoughts and feelings do not depend on a single active site in the brain but instead require a functioning multicentered network. The same applies to mindfulness during meditation, which activates other centers in the brain than are activated in everyday waking consciousness.

It appears increasingly unlikely that consciousness is simply a product of the brain, not just because measured activities in the brain tell us nothing about the content of thoughts and feelings, but also because the mind is capable of changing the anatomy and function of the brain (neuroplasticity, placebo effect) and because there is evidence that consciousness can be experienced independently of brain function (an NDE).

Chapter Ten

A Comprehensive NDE: Monique Hennequin

The speaker has no value whatsoever, nor what he says. What has value is how you understand yourself in listening to what he says. He is like a mirror, in which you see yourself reflected. Your consciousness, your daily activity, your unconscious demands, pursuits and fears are exposed. When you so listen, then you begin to discover for yourself not the ideas, the conclusions, the assertions of the speaker, but rather you see for yourself what is true and what is false.

—KRISHNAMURTI

As an interlude between the many theoretical reflections on the cause, content, and aftermath of an NDE, I present here Monique Hennequin's account of her two near-death experiences. I have met her many times and was so impressed by her account that I have asked her to write down her comprehensive NDE to include in this book. She was thirty-one when she had her second cesarean section. She gave birth to a healthy baby boy, and everything seemed to be going well until several hours later she suffered abdominal cramps, which worsened in spite of pain medication. Her clinical condition deteriorated fast, another abdominal operation followed, her temperature soared, and she went into shock with extremely low blood pressure and developed multiorgan failure. The latter involves kidney dysfunction, plummeting potassium levels in the blood, renal failure, and potentially fatal cardiac arrhythmia, and it usually requires artificial respiration. The cause of Monique's life-threatening condition was later found to be an undiagnosed constriction and perforation of the intestines, a complication arising from the cesarean section, which had caused the contents of the intestines to leak into her abdominal cavity and wreak havoc there.

After a few days in an extremely critical condition, during which the doctors informed her family that she was unlikely to pull through, she suffered a cardiac arrest. At that moment she had an NDE, which she describes fully below. She was successfully resuscitated but remained in a critical condition and somehow became aware of her "hopeless" situation. She was desperate to return to the loving environment that she had just visited. In her desperation she managed to bite her breathing tube in half, thus precipitating an apnea. She was resuscitated a second time, and after the old, severed one had been removed with some difficulty, a new breathing tube was inserted into her trachea. At that point Monique had another NDE.

She begins her story at the moment of her cardiac arrest, which initiated the first NDE.

After a final, desperate attempt at moving my body, I gave up. My heart was beating like mad, and even my rapid breathing gave me no air; I felt as though I was suffocating.

This was the time and place; how on earth could my life end here in the intensive care unit? What was to happen to my children, my job, and the assignments that wouldn't be finished, my house? There was so much left unsaid and undone. My God, they didn't even know where my funeral insurance policy was! And the fridge! I had this vision of people coming in after the funeral and the whole rigmarole were over, to clear up, clean the house, open the fridge . . . the stench! (The night before I went into the hospital I had cooked cauliflower, and there were some leftovers in the fridge.)

This was my last conscious thought before I "went." The astonishment at being able to exist and perceive outside my body was something I had already experienced during the operation a few days earlier.

The last thing I heard was a long beep from the monitor and an alarm that apparently sounds when you flatline. Suddenly I was in the room next door and saw two nurses rush over to my room and my body. It seemed odd to me that this adjacent room had a monitor and alarm even though it was unoccupied. I felt a smile on the lips of the body I no longer ensouled or inhabited. At the same time I felt the nurses' panic, and I felt deeply sorry for them. I couldn't help it; I appeared to be taken away. And I took my smile with me. Picking up speed, I saw every single room in the hospital, including patients and staff, as well as the past, present, and future of everything that whizzed past me. I knew it! I had often thought to myself: What if life is like a dream, and just as you can wake from a dream you can wake from life? At that point I didn't know yet what lay ahead of me and that I was to see that life is a cycle, just like sleep is a cycle in life. What's more: that this in turn is a cycle in a dimension that would be well beyond my comprehension.

Everything went black, and somehow I was relieved because I felt no pain or fear whatsoever. I felt safer than ever, and I had a sense of nostalgia. A childlike innocence and naïveté came over me, as if I was going to the movies with Daddy, and at the same time I felt more mature than everything I was leaving behind (including my own life). I didn't feel alone, despite not seeing anything or anyone. Gradually a sense of sight developed around me, like a sphere that I myself was a part of. I seemed to have ended up in the omniverse, as another image formed below me; in fact, it formed around and through me. I continued to feel protected, by somebody rather than something. It became lighter, and I saw myself enveloped as it were by a situation from my past. I immediately recognized the time and place: underwater at the age of fourteen! I was shocked because I suddenly realized: I've been here before! Exactly there in 1974, when I nearly drowned, I had also seen myself! At the time I thought that I saw myself reflected in the water's surface (from below). What I didn't realize at the time—but did during this experience—was that at the age of fourteen I had seen myself as a six-year-old girl.

It was clear to me now that back then I had also briefly woken up from life. At the time I had regained consciousness on the edge of the pool, lying in a pool of vomit, with a lifeguard on my back. It was very embarrassing, and I repressed the beautiful colors, the peace and quiet, and the excerpts from my youth. Afterward I would occasionally tell people that I had seen myself underwater; I couldn't understand it myself let alone explain it to others. As usual I thought: the less I say, the easier it will be to fit in with the rest, so I never mentioned it again.

But now, in this situation, I understood that there was no reason whatsoever for my shame, and at the same time I sensed the fear, worry, love, and relief of the lifeguard and some of the bystanders. I also felt the shame and reticence I developed afterward. At that moment I'd felt a connection with "the truth," with my origins, and had trouble recovering this feeling in society. From then on I knew that there was something that I didn't know but was desperate to know. Now I was literally dying to know. The strong sense of separation from my source—which I'd had since the incident— was something I would come to understand even better during this experience.

The situation made sense to me now, and I proceeded to other situations that had raised question marks in my life. The how and why of my actions became clear to me because I saw, sensed, and knew how people had felt during (and frequently also after) contact with me. I viewed several episodes from my life. I recognized and felt everything as though I had gone back in time and completely in the actual moment.

What had I done with my life? My God, I was my own judge and executioner at the same time. When I realized that I had done something wrong, I wanted to go back to make amends. A bit like rushing out the door and realizing you've forgotten your bag. No problem, you fetch it, but then back in the street, damn, you've forgotten your purse. You think you're all set, get to the store, and sure enough, you've forgotten your credit card!

I hadn't really done anything wrong, but I felt the pain, the misgivings, the anger, the powerlessness, and the sadness of all the people who felt upset by my words and actions. Actually, it wasn't really something I did to others but [rather I did it] to myself. These people were also reflections of my own pain, misgivings, anger, and powerlessness. I hadn't taken full responsibility for my thoughts, words, and actions and had thus deprived myself of the chance to grow and become more aware. I also saw that sometimes I spontaneously overreacted to some people (usually to my own great annoyance) and that they actually needed it as well. And there I was, feeling unnecessarily guilty over my supposed overreaction, not realizing that people's thoughts and emotions had triggered my unintentional responses and that it was a mirror for them as well. Everything could have been so different by open and sincere communication.

I lingered at those incidents where I had trouble recognizing my responsibility until I was ready to accept it. To everybody I had ever hurt, intentionally or unintentionally, I wanted to explain why and express my sincerest apologies. Nobody condemned me, and at all times I felt this warm support. How could this support love me? Could it not see how naïve I had been in life? And that I had been motivated by ambition, selfishness, fear—and, yes—even by joy or euphoria?

Fortunately, I also saw and felt all the wonderful, happy, rewarding, and joyful moments that my thoughts, words, and actions had given others (and thereby myself). Everything was shown simultaneously—my entire life! Some things even made me laugh. I didn't spare a thought for my surroundings and was completely engrossed in my life.

I was the one taking stock, and I didn't come off too badly. I had personally come to terms with everything now; my guilt (or rather my lack of awareness) had become clear through the pain and the acceptance of responsibility. However, I did feel that I owed some people an explanation. If only I could reach them. Why had I said so little in life? Fear. But that fear was completely gone now. Never again would I be afraid of my thoughts and feelings. Never again would I be afraid of being a failure in other people's eyes or of being undermined. I was and always will be my own judge. From now on, I would always be responsible for my own thoughts, words and actions. Everything faded, and suddenly, with terrible pain, I was back inside my body. What had happened? For a moment I had some control over my fingers, and I tried to send an SOS in Morse code. I wanted to stay in my body, if only for a chance to say that I was sorry about certain things. Never again did I want to do and say anything other than what I really thought and felt. If I were given another chance, things would be different. Honesty! It would begin with being open and true to myself.

It was [to] no avail; even if anyone had picked up my signals, they couldn't have stopped or understood what I was going through. Apparently I hadn't understood enough myself. It felt as if I had been taken by the scruff of the neck and been confronted with the facts. The concept of the individual, the need for cooperation, growth, awakening, purpose, the cause and effect of thoughts, words, and actions in human interaction were all clear to me now. But did I really want to continue down this difficult path? Did I really think the world was a worthwhile place to be in? Why couldn't I keep myself alive? Questions, lots of questions, were going through my mind. And once again I felt myself leaving my body. As quickly as I had returned, I was gone again.

I appeared to be ascending through a spectrum of light, and I recognized colors not by sight but by sense. The colors became lighter, warmer, and brighter and then intermingled, as if to form one big umbilical cord. A soft tornado of all colors—more than I had ever known—came together and blended without any losing their individual identity. The tip of the tornado appeared to be pointed at the earth's atmosphere. I went along and ended up in its "eye." Did that mean that I was heavier than the light? Lighter? Every intensity and color of light appeared to have its own mass and movement. Was I part of this? Did it pass through me, or did I pass through the light? Everything felt diffuse, and I had the impression that I was soaring to the center of this column. I sensed and knew instinctively that this force had the shape of an hourglass and that it would expand at some point, become even bigger than the place I had come from. For a moment it was tight, and I thought about the inside of an umbilical cord, about the placenta connecting mother and child; I felt dizzy, and I seemed to be experiencing birth at different levels and in different dimensions. Ascending, descending, entering or exiting places—I could no longer tell the difference, nor did it matter.

Everything around me went black, a warm black, and I realized that pure, warm, soft black is also light, a kind of energy, palpable even without a body. I wondered if it was palpable because it wasn't me who had control over myself or a body, but something else that was a great deal bigger, stronger, and more wise than I could ever be.

I rose up, like being in an airplane without side panels (so I could feel the atmosphere around me). Strangely enough, I could neither see nor hear, and yet I perceived everything. I was a part of it, and at the same time I felt enveloped by a protective shell. I rose higher and higher, further away from life and closer to what feels like real existence.

I sensed a hierarchy of the regions or atmospheres I literally and figuratively went through. Every "layer" had its own atmosphere with distinct boundaries and restrictions. "Aha," it flashed through me, "atmosphere," and I understood that every "level" I moved through contained parts of both myself and others. Did this mean that I could have arrived at different levels of awareness during my life review and that I might have ended up at other layers? "Atmos," "individual," "I am," "am I still?" flashed through me. It didn't matter, but I was glad that I didn't have to "get off" just yet. I was hoping to arrive at the essence, the purest form of consciousness. No more inadvertent mistakes or misadventures. Gradually, the notion of "enlightenment" began to take shape in me. I had never thought about spirituality, and to my embarrassment I had been known to make fun of it. But from a young age I'd always had many sincere questions, such as, If there is a God, why are there so many diseases and disasters? Why do we make war? Is nature God or God nature? Are we nature? One of my most fervent wishes as a child was that my father's high blood pressure could be cured or that my mother had never contracted polio. Of course, when my worldview broadened a little, I wished for peace on earth and prosperity for all. But my most urgent question and source of anguish was always, Why do so many innocent children suffer? It never struck me that I didn't place adults among the innocent. Now I would understand my unconscious views and get answers to my heartfelt questions. In fact, I would receive more answers than I had questions, and the answers would trigger yet more questions.

As I approached my "level," I became lighter and happier and felt almost elated about what lay ahead. I seemed to be making a soft landing, and the envelope around me dissolved. The black light cleared like a fog and turned a color so delicate I still don't know how to describe it. Palpable and all-pervasive; even the finest filter couldn't capture this luminosity. I was amazed that I could see and feel this.

I don't know for how long or short a time I had "stood" there (and how I could stand there without a body) when I became aware that the light was shot through with another luminescence, emanating from a source somewhere. I was floating along the shore, as it were, and continued to marvel at everything around me, no longer aware that I had no eyes and couldn't really see anything. Slowly but surely I realized that this luminescence consisted of a kind of infinite river of brilliance, like the brilliance of a setting sun reflected in rippling water with little pinpricks of light like small stars. The brilliance was made up of beautiful little globules of light, extremely bright and quite unlike anything on earth. They looked like nuclei surrounded by a body of light. Not literal bodies, but more like celestial bodies or atoms with clearly visible electrons floating past me, close to the ground. I checked to see where they were coming from and wondered if the dark aperture I saw might be a so-called black hole. The deep black looked more a cave from which the light gushed like a waterfall and thus formed the river in this tranquil field. The "river," the field, the current, and the black formed a peaceful whole. Brilliant clusters of DNA appeared to be flowing right by my feet (even though I didn't have any). The particles were linked in complementary pairs, which in turn made up an enormous organic spiral. The spirals formed the clusters in the field. I sensed that I could sail or float along with any pinprick of light (particle, being, consciousness, atom, soul, or whatever). All

I had to do was "enter" or join the chain. All communication was wordless (suddenly, I understood the meaning of "soul to soul"). All particles, atoms, lights, or celestial bodies understood one another; they knew everything, they were everything! What one half of the pair knew was reinforced by the other half and vice versa; the same applied to pairs and even to strings. The current appeared to be expanding all the time by emitting pulses. All particles were a fraction of an all-encompassing sympathetic force, love, and consciousness (like the support I had felt during my life review or the process of accountability or awareness, except several degrees stronger). Now I understand why some call this God, Allah, Parinirvana, the happy hunting grounds, the Akasha field, Asgard, or whatever. I don't know what to call it. Perhaps simply nature.

A sense of happiness and grace washed over me and glowed deep inside me. I felt privileged to be a part of this. Why me? What good had I done to deserve this? I was very eager to join, be together. Be together with these particles in this current; be one with this immense, wise, creative force. Be together with this all-encompassing consciousness. This is where I belonged! This was no longer the dimension of earth or other spheres, this was more! This was the beginning and the end, this was the source. I recognized many particles as belonging to people I had known on earth, including my younger brother, who had died before I was born. However, what I recognized wasn't the complete individual. These were elements, beautiful elements, conscious elements. I didn't recognize them by sight because every single particle was the same—the same color, light, vibration, consciousness, and level of maturity. The recognition was purely instinctual, like an animal has an instinct. And these particles seemed to represent only the purest or most profound parts of them. Perhaps this is why they resembled elements of a larger whole. All particles possessed knowledge; everything I might want to know, they knew. There couldn't be a single question to which they wouldn't know the answer. And I would know and understand everything as soon as I had become one with this whole. Everything! I hesitated at the prospect of becoming one because I knew that once I decided to do so I would never be in a body again. The force of attraction was incredibly strong, but at the same time I realized that if I connected there would be no way back. I was so eager to know and be like this current of infinite wisdom. The questions I'd had in life resurfaced: the reasons behind sickness, disasters, children's suffering, the way the earth and the body worked, evolution, the conflict between Jung and Freud, the clash between religions, gravity, birth, incarnation, relativity—in a word, *life*.

I wasn't even aware that I'd had so many questions and that I'd felt excluded from learning, knowledge, and awareness. Physics was the first subject I'd dropped in school, because it was way beyond me, and philosophy had seemed an area I would definitely never get my head around. But here and now I had the opportunity to instantly know everything. I saw that knowledge, conscience, intuition, awareness, thoughts, emotions, and the physicality of everything are all connected.

Time did not exist here, and a loving tranquillity suffused the field, the current, and myself; suddenly, for just one moment, I was a fraction just like they were and completely one with them. I don't know if I touched them or they touched me. This moment of eternal omniscience is still indescribable to me. The surge of love and the explosion of information were overwhelming.

The answers I received far outnumbered all the questions I'd ever had. In fact, these were no answers but an all-encompassing oneness and convergence of absolute knowledge. My smile felt bigger than ever, and I was complete as well as perfectly happy and in the place where I belonged.

This time I had not just woken up from my body, my ego, and physical life on earth, but from something much, much more profound. I had woken from deep inside my individual consciousness in the source.

It was all so easy, natural, and logical within this completeness, but at the same time I knew that I would never be able to express it in words from an "I," an ego in a physical body. Yet this is what I wanted; I wanted to share this, try to explain how eternal existence and consciousness looks and feels. Impossible, and yet, even if I could reach only one single person on earth and let him or her feel this tranquillity, my return would be worthwhile. I realized that I couldn't stand around here much longer, or I wouldn't be able to return to achieve my goal. I knew everything I needed to know. Even my own "end." The options of staying and returning were clear to me. Past and future were one, as if all opportunities were shown to me at once. I had to take my leave of this conscious oneness, this source. Nothing that had ever been born or come to maturity or that had died or been destroyed was unfamiliar to me. I made a well-considered choice, and this determined what happened next. A preview, showing me all the opportunities of the life ahead of me, as if it had already been lived, flashed before me. My smile was palpable again, and I was pleased with the life ahead. I was sorry that I couldn't stay and felt a little guilty, although I knew there was no need, that this was an unnecessary feeling. Naïvely and intensely happy, I began to wave at the particles (without hands): see you later! I knew that I would be back here at the appropriate time, to become a permanent part of this wholeness. The particles paid no attention; everything was fine. As if this current knew that I had to go back because I made this decision. They did know, and all my future decisions would be right. The flashes of moments from this next stretch of life would come to me regardless of the paths I would travel.

And I was ready now; I really wanted to be on this logical and coherent earth, to be aware and live accordingly. Nature is perfect.

With a violent and painful jolt, I returned to an immobile body. It was full of tubes and hooked up to machines that were keeping it alive. Even the oxygen that this body needed was provided by a machine. I heard the regular rhythm of the respiratory pump and thought of the awareness and knowledge I had just come back from: iron! The earth's nucleus and the absence of a nucleus in the particles of the transport fluid that delivers oxygen to all of the body's cells. What a wondrous thing, a body! So this was the first thing I had to learn to stay on this earth: breathing! I tried to stay alive by being rational, and literally and figuratively forgot my heart. Passion, dedication, surrender-ness and love, that is, no selfishness or rational beliefs. The loving smile and the tranquillity of where I came from were obviously gone too. Nor was I grateful; what a sluggish, harsh reality this was. I didn't know how many hours or days I had been lying there, feeling frustrated in the body that I couldn't get going again, when I decided that it was pointless. I didn't realize that it was my "I" making this rash decision instead of trusting the experience I'd had within the oneness of that beautiful consciousness. I became angry and felt betrayed. I thought that I had been abandoned by the source I came from and failed to see that I was abandoning this source (and myself) by seeing it as part of where I came from instead of as part of me.

I wanted to return to this heaven and didn't realize that a single loving thought would let me be part of the whole again. I thought that the only way back was via death, and I began to wonder how I could rid myself of the machinery that kept my body alive. I felt imprisoned in this life and wanted to free myself at all costs. Again I heard the respiratory pump and felt a tube in my mouth, right down my throat. Unlike the rest of my body, my mouth I could move; I tried to swallow and felt the obstruction. There was some—be it minimal—movement. With my jaws and teeth, I bit down as hard as I could until the intubation tube was severed. I tried to swallow it as quickly as possible. The alarm, the oxygen, the nurses—they were gone in a flash. "Good," it went through me; "that will teach them not to keep me here. I'm off!" But it wasn't good; it was neither a good thought nor a good deed, not loving toward nature, God, Allah, Jahweh, the Source, toward anyone on earth doing their best for me, and most of all not loving toward myself.

This time I saw nothing: no colors, no warmth. The soft vortex now looked like a hard, cold funnel. I was confused; where was I supposed to go, what was happening? Contact with all earthly things vanished abruptly, and this was by no means a smooth transition. A sense of panic came over me. I didn't understand what was happening and what had to be done except that this was damn serious and that I was now at a different "layer" in the atmosphere than last time.

In the distance, very far away, I saw a pinprick of light. The end of the darkness? Was I supposed to go there? Did I want to go there? I looked around me; everything was black—black, quiet, cold, and lonely. I was completely alone. Alone with my thoughts and feelings. The expression "godforsaken" made sense to me now, or whatever name you want to give the all-embracing. It wasn't as if the all-embracing wasn't there, but I had obstructed it, as it were. I had built a wall between myself and my heart, between myself and trust, between myself and gratitude. And above all between myself and love. The love of the heart that enables a clear consciousness, peace, health—in one word, everything. My wall cast a shadow over me so I couldn't see the light. Love and the higher consciousness are in the light; they are the light. And in order to see this, I had to break down this wall. Fear—I was full of fear. I think there's no bigger wall than fear except perhaps anger, which was still there too. How could I escape this misery? Why did nobody help me? I got more anxious and felt nothing but pain, anguish, grief, and loneliness.

The regret intensified, and I understood that I had made a huge mistake. Why had I not had any faith and patience? I felt deeply ashamed before all the light of which I was part of and from which I had isolated myself. I knew that "they" knew, and I also knew that I wasn't condemned for it. I felt small but no longer naïve or innocent. Humble or unassuming, rather. From then on I knew that humankind is as strong as the weakest link; in fact, not just humankind, but humanity itself. I recognized my weakness and felt that love, trust, and faith were the greatest strengths I needed. Now, but also in the intensive care unit, in the life before this happened, and in the life that might have been.

I was of no use to anything or anyone by isolating myself like this, and I encountered immediate pain. The pain of having chosen the life that might have been and having given up so easily, having abandoned oneness with those clusters. I felt the pain of those who had resuscitated me, of those who had devoted their efforts to me and thought that I was going to pull through. I became angry with myself, but that too was an unhelpful emotion. I wouldn't reach the "heaven" where I belonged via this road, not via "death," but would reach it instead via a life filled with faith, trust, love, and gratitude. "Help!" went through me, and I felt like I was breaking.

All of a sudden my father [who had previously died] drifted around the corner, like a huge shadow. He didn't look up at all. He moved solemnly. I sensed that he knew his way around this darkness. His feet never touched the ground as if he was walking or floating in slow motion, and yet he walked ahead of me at quite a steady pace. I wanted to see him, his face, his eyes. Whatever I tried, he kept moving and never once looked at me. I was so desperate to see, touch, question, almost beg him what to do. Should I join him, go after him, or head in another direction? Somehow I sensed that if we made contact, there would be no going back. Not to my "heaven," my level, and not to my body and the life I had chosen. There was a reason he didn't look at me, a reason he didn't grab hold of me. He had only come to clarify everything for me. It felt like his final duty as a parent before he could proceed to the light. I had to understand everything now. And again, the decision would be mine. If I wanted to, he would accompany me to that pinprick of light, but I knew that this light didn't have the same intensity as before. To get to the source, I would have to move through yet another "level." So again (or still) the choice was mine: I could choose life or choose this other light without pain, cold, or lovelessness. It remained to be seen if I would suffer again in a physical body, and deep down I knew the answer.

I could tell from my father's body language that he didn't want me to accompany him. I loved him, even here and now, behaving like this. Suddenly I felt what he felt, I knew what he knew. He clarified things instantly, including what I had to do. This wasn't my "final destination," this was his world, this was his (and temporarily my) domain, and it was his level and his light over there in the distance. I had to go back, of my own free will. Surely I knew that next time I would end up at my own level and at the source? But I was anxious and kept following him (without legs, of course). I couldn't for the life of me catch up with him. I tried to scream, cry, beg, but nothing got through to him. I felt the pain in him: the pain of not being able to hug me and help me decide. I tried to stop and scream like a small child in a supermarket, in the hope that he would stop and help me. He continued to the light, which drew closer and closer. I had to make my decision now; the end (or the beginning of his light) was approaching. I could almost touch him. Almost . . . Right before my eyes he entered the overwhelming light. Such power! Such love! One more step and I'd be with him. With him and with many others. He had arrived; he had finally found peace. Now I knew that part of him, part of his energy that had been with me all those years after his death, was gone. Because of him I could move on, and vice versa.

What next for me? If I wanted to return, I had to go "backward" immediately or I would be received into the light. As an example of how I felt at that moment: imagine you're on one of those moving walkways that take you to your gate at the airport. You reach the end without doing anything, but you do have to "get off" to avoid taking a tumble. If you don't want to get off, you have to walk backward at what feels like a very brisk pace. When you do this you will hover at the point of disembarkation so you can delay your decision. You also know that if you stand still for only a second, you will fall off the moving walkway. Constant attention and effort are required. And if you really want to return, you have to walk backward as fast as possible (mind you, without legs), faster than the moving walkway moves forward.

A lot of words for a feeling that lasted a split second. That blasted language of ours! If only I could conscious express my feelings through feelings, then I wouldn't need all these words. Without the limitations of words and pictures, but with love and a lucid consciousness, everything can be expressed and conveyed as it really is, and not as people think it is. Everything will make perfect sense.

When I was about to get off (by simply letting myself go) I wasn't ready to decide, but the indecision was too much to sustain. I had no time to weigh up the two options. I felt like I was suffocating; I had to decide. I had reached the hardest part of both of my NDEs, and the decision this time to return to my body on earth was actually the only time that felt like I was "dying." The decision to walk backward and return as quickly as possible seemed inexplicable to me. It was the more painful of the two options, and I knew I would be suffering a lot of pain. Ahead of me everything was good, full of love, warmth, honesty, knowledge, everything I had always wanted here on earth. Then why return to that hell when I knew for sure that what I saw in front of me I would never have on earth?

Actually, it wasn't about having but about being. I myself should be full of love, honesty, warmth, and awareness without wanting or having it from someone or something else. Only when I am these things shall I be able to give and receive them; therein lies the greatest wealth.

I realized that I would have to fight not just to reenter my aching and weak body but also to rebuild that body to meet my next challenge. I had no time to rationalize my decision. But I knew that it would have been pointless to stay. What is the point of enlightenment if I can't reach out to others? Sharing knowledge, love, honesty, and awareness—that's it! I had to reach out to myself (my ego, my "I") and others.

The intention behind the earlier choice I made at my level, "my heaven," came back to me very clearly: "Even if I can only reach one person to feel this, it will be worth all the pain." As soon as I became aware of this, I felt the connection, the warmth, and the support of where I belonged, and I no longer felt isolated and alone. I had reconnected to the source and knew that I needed my body.

Suddenly I was back in my body, rudderless, and this time with a mask over my mouth and nose. Somebody leaned over me, manually respirating me. I was in pain and longed to be back, but I also knew that I needn't feel alone as long as I kept the connection with the love and gratitude. Strangely enough, this painful decision was also motivated by conscious love. Love for creation, the nature of everything, and consciousness as well as love for myself, because the decision to give up ran counter to nature and creation; in other words, it ran counter to me. Everything remained dark for days, and during the moments when I was aware of my comatose state, I knew that my first experience was a natural one and that my so-called negative experience was an unnatural one that sprang from lovelessness. However, the latter NDE taught me most about love and conscious choices because I had to feel right down to my toes what free will, faith, and love can achieve and that I'm always only one thought removed from the source, irrespective of the horrible state I'm in.

At no point during my NDE did I feel that somebody other than myself forced me to do anything. I made all the decisions. That's what made the crucial choice of staying or going back so intensely difficult.

I will never be able to let anyone see or feel what I perceived or where I was during my NDE. I can only describe what I felt, what I pictured, imagined, and what it signified to me. In other words: what I experienced. Everybody is free to make of this what he or she wants. And everybody will have a different interpretation because everybody attaches different meanings, images, and values to words. I use the words *picture* and *image* because that is literally what I do. When I describe something I have witnessed but that can't be seen or experienced on earth, I paint a picture, an image with words; in fact, unless I refer to earthly properties, none of it makes sense. How can I explain something that doesn't exist here or that does exist but cannot be seen? When my soul (consciousness or energy) perceived instantaneously everything that can't be perceived by earthly senses, the "picture" was clear and complete. There was no need for me to "experience" or discover it via the faculties of a body in space and time and to find a rational explanation for it.

Without a body and without time, I wasn't disturbed by my ego or by the function or dysfunction of my brain and memory. And during my pure perception within the source, I had no opinion either. I didn't even have an "I." An opinion is tied to the ego, and the ego to the body. Without any of this, there was only objective consciousness.

During my negative experience I was neither complete nor objective. I still had an "I." I actually saw that my father was wearing his old raincoat whereas I knew perfectly well that we had got rid of it a long time ago; the very fact that I saw him as a human figure made me doubt my objectivity later. My "I" only disappeared toward the end when I felt the oneness from my positive experience again.

Back inside my body, I found it difficult to put everything into words although this had nothing to do with what I had perceived. I often wonder whether everything remains pure when such an experience is put into words because of being back inside a body and dealing with time, the brain, memory, and an ego.

When I left the hospital after my NDEs, the search began. What I had perceived must surely be known here on earth? I knew that numbers and units have names—obviously! A one and a zero is called ten, the perfect number according to Pythagoras, everything and nothing, both full and empty, but who knew what a googol was? And why couldn't this be deduced from the Latin word, like the meaning of one and zero? Who knew that this "accidentally" coined number of ten to the power of a hundred takes up 333 bits in binary format and that this number is sometimes used to refer to the trinity or perfection, fullness in emptiness? Or who knew that the "universe" has numerical values too and is not merely a name? Or that even thoughts, letters, and words have vibrations and values, which have an effect on physical matter? And even form matter? Where could I find information on, for example, gematria? Where was I to look, who could I turn to? Where did science and spirituality meet? What is matter and what is reality? I practically lived at the library and in bookstores; I recognized all kinds of things, but none of it felt like real "life."

The greatest reality for me was there! There where I awoke again and again with a speed far greater than the speed of light. A pure life at spirit level, an energy that surges on and through the earth, through everything! Would I ever be able to feel it here—if only for a split second, so I could feel whole again and no longer isolated from real life? During this brief spell here on earth, I occasionally feel hindered by my body, my physical senses, my thoughts, time in general, and "my" time. Everything that is visible here on earth feels like a feeble reflection of reality. The tiniest piece of matter that is as big as the universe and the universe that permeates the smallest known unit—these I find difficult to grasp.

I feel like I'm retracing my steps in my search for what is known here while moving forward in time. Knowing that the more I encounter on my path, the more questions I will have, and knowing that when I'm briefly without a manufactured thought I know everything, I sometimes find it futile to carry on searching. But I continue down my path, see and feel intuitively where, when, and especially with what intentions I take the next step. I still have no scientific explanation for what I experienced in 1991, but within the exact sciences, the M-theory and the description of antagonist harmonic pairs seem to come closest to my picture of the clusters in the field and their effect.

The feelings are even harder to explain. I can't give any examples of the love, the acceptance, the awareness, and the source except perhaps something as simple as newborn lambs or ducklings in spring, bright reflections in the water, or the smile of a child with an even brighter sparkle in its eyes.

Reentering a body and staying there was difficult; waking up from earthly life and everyday consciousness happened spontaneously and was as natural as being born. Staying there, in the eternal and endless consciousness, would have been a choice, just like the choice to take the body that I ensouled or inhabited before my death and make it entirely "my" body again and getting it literally and figuratively going again. Every day I'm happy that I chose the difficult route and that I've been through this heaven and hell. On a wall in my living room I calligraphied "What you think matters; in fact, it forms matter" as a daily reminder of the opportunities offered by life. Whether the experiences will feel positive or negative depends on the intention of my heart.

Chapter Eleven

Quantum Physics and Consciousness

Anyone who is not shocked by quantum theory has not understood it.

—NIELS BOHR

The preceding chapters dwelt at length on the various theories seeking to explain all aspects of a near-death experience. We reached the conclusion that the scientific approaches outlined so far fail to offer a satisfactory, irrefutable explanation for either the occurrence or the content of an NDE. We still do not know how it is possible for people to experience an enhanced consciousness during a cardiac arrest, that is, during a period when the brain displays no measurable activity and all brain function, such as bodily and brain-stem reflexes and breathing, has ceased. Looking at the interaction between consciousness and the brain, we concluded that consciousness cannot be seen as the product of brain function. In fact, sometimes the opposite seems to apply: the mind influences brain function, both in the short and long term as a result of the empirically proven principle of neuroplasticity. Our current scientific knowledge cannot account for all aspects of the subjective experiences reported by some cardiac arrest patients with complete loss of all brain function.

Some NDE Elements Reconsidered

Let us therefore reexamine the contents of a detailed near-death experience, as described before. Some subjective aspects of this profound experience invite comparison with concepts from quantum physics. Quantum theory emerged at the start of the twentieth century when scientists began measuring the behavior of subatomic particles like electrons and protons and were surprised to learn that these particles did not follow the rules of classical physics. This chapter will feature a comprehensive account of such quantum behavior and the theory behind it and then compare that to aspects of NDEs.

We learned earlier that during a life review every single detail of one's past life can be relived. Everything appears to be connected to everything else, an interconnection similar to what in quantum physics is called entanglement; everything is one. All past events appear to be stored and available as soon as one's mind turns to them. Time no longer plays a role; everything exists in an eternal present. This is true for time as well as for place. NDErs report that during their experiences they can be anywhere in the past as soon as they think about or want to be in a particular place, be it as a baby in its crib, at a sporting event in elementary school, as a student on a study abroad program, or during a vacation in Australia. They instantly return to that situation and relive everything that mattered at that moment in time, including the emotional impact on themselves and others. The mind seems to contain everything at once in a timeless and placeless dimension. In quantum theory this timeless and placeless interconnectedness is called nonlocality (see later in this chapter).

Similarly, during a preview or flash forward the concept of time as we know it in everyday life appears to be nonexistent. We also experience this timeless aspect in dreams, in which everything appears to happen outside of time. But the vivid reality experienced during an NDE is quite unlike everyday reality or a dream. As was already described, an NDE seems to generate images from one's own future and from that of the world. In this timeless dimension everything seems possible and accessible. And years later the perceived events are found to be true when they are recognized as part of the earlier NDE or experienced as a kind of déjà vu.

Reports seem to confirm that a nonlocal experience can also occur during an out-of-body episode when an NDEr's consciousness, independently of the body, can provide instant access to whichever place he or she is thinking of. When somebody is comatose in a car wreck and thinks of their partner, they instantly join their partner at home; they can even see what the partner is doing or thinking. In retrospect, this observation is found to be true. In other words, it seems to be possible to have a nonlocal connection with other people's consciousness as well as with the thoughts and feelings of deceased friends and family and to communicate with them by way of thought transfer. To their utter confusion, NDErs often retain this ability for nonlocal connection. Without really wanting to, they can still communicate beyond time and space. This is known as heightened intuitive sensitivity. I already mentioned it in an earlier chapter and will come back to it later.

Additionally, people with a tunnel experience during their NDE seem to be making a conscious transition from our physical world, also known as space-time, to a multidimensional space in which time and distance no longer play a role. Theoretical physicist Stephen Hawking calls the instantaneous transition from space-time to a multidimensional space a wormhole.[1] Wormhole models look remarkably similar to hourglass-shaped tunnels.

Scientific studies of near-death experience seem to show that various aspects of an NDE correspond with or are analogous to some of the basic principles from quantum physics. Quantum theory might perhaps explain the reported connection between one's own consciousness and that of other living persons or deceased relatives. The same applies to nonlocal phenomena such as the life review and preview, in which past, present, and future can be experienced simultaneously and which elude our conventional embodied conception of time and space.

Some Concepts from Classical and Quantum Mechanics

Because discussions of quantum mechanics can get very technical, I provide here a synopsis of the concepts that are necessary for understanding this and the next few chapters. This synopsis does not contain any references to scientific literature. Readers who would rather not delve into quantum physics just yet can skip this chapter (for now).

According to *classical* physics, objective reality comes about according to certain fixed principles. Everything in our world occurs within an *unchanging structure of space and time* on the basis of *unchangeable laws* that can be accounted for with unambiguous ideas about *reality, causality, continuity,* and *locality*. Classical physics is based on the premise that perceived reality in the physical world equals objective reality. This chapter will first look at a few "classical" physics concepts such as waves, fields, and information.

We start with electromagnetic fields, which are fields that are produced by electrically charged particles whose patterns travel in waves. An electromagnetic field appears to possess an infinite capacity for storing or encoding information. Think of the more than one billion Web sites to which computers all over the world have wireless access. All this wireless information is encoded as differences in wavelength. In particular, information is coded within interference, which is a phenomenon whereby overlapping coherent waves create a specific pattern. When two waves oscillate regularly in a certain relationship they are said to be coherent, and that coherent relationship enables interference. Information in a coherent field is like a hologram, which is a picture of, for instance, a three-dimensional object encoded in two dimensions. The holographic principle makes it possible to retrieve information about an object as a whole from any location in a coherent field. So far we are still in the realm of classical physics, where particles and waves behave in predictable ways.

Quantum physics turned the classical scientific conception of our material, manifest world upside down. New concepts from quantum physics include *superposition, complementarity, the uncertainty principle, the measuring problem,* and *entanglement* or *nonlocality*. All of these concepts relate to the same problem: certain observations cannot be predicted absolutely. Unless a quantum object is observed, it has neither a definitive location in time and space nor any of the fixed properties that classical physics ascribes to objects. Instead, there is a range of possible observations, each with a different possibility. The different possibilities are called *probability waves*. Light behaves like either a particle or a wave, depending on the experiment design, but never like both at the same time. This phenomenon has been termed *complementarity*. Particles and waves are complementary aspects of light. What had already been proven for light—that it has both a particle and a wave aspect—was found to apply to matter as well. All matter, 99.999 percent of which is emptiness, can ultimately be regarded as a wave function and thus possesses wave–particle complementarity.

Experiments with isolated photons show that a photon sometimes behaves like a wave, which means that it is entangled with itself. Entanglement is a quantum phenomenon whereby spatially separated particles possess properties that are connected beyond time and place. They are linked together so that one object can no longer be adequately described without full mention of its counterpart. This is known as the *superposition* of wave functions, whereby a wave should no longer be seen as a real wave but as a *probability wave,* as this quantum phenomenon is called. It means that we can calculate only the probability that a particle will be found in a given location, not where it will actually end up; the range of probable locations is the probability wave. In other words, we can never know a particle's exact location at the same time as its momentum, which is an indicator of its proper velocity. This is the *uncertainty principle* of Werner Heisenberg, which posits that observation is impossible without fundamentally altering the observed object. Some quantum physicists champion the radical interpretation that observation itself literally creates physical reality, thereby ascribing consciousness a more fundamental role than matter or energy. I personally support this not-yet-widespread view that consciousness could determine if and how we experience (subjective) reality. I will come back to this later.

One of the most important principles of quantum physics is that two isolated, remote particles can have an *instantaneous* effect on one another because these two remote objects can become entangled. This is known as *nonlocality* and has given rise to the quantum physics concept of *nonlocal space:* a multidimensional space, with nothing but possibilities, also known as probability waves, and without certainties, without matter, and without a role for time and distance. Everything in this space is uncertain, and physicists can carry out neither measurements nor observations. The nonlocal space represents a hidden reality that, at the quantum level, exerts a continuous influence on our physical world, which is the *complement* of nonlocal space.

Another possible name for nonlocal space could be the absolute or true vacuum; it has no structure and is a timeless and empty space, in which quarks (elementary particles and fundamental constituents of matter), electrons, gravity, and electricity have all become one and as such no longer exist. This space forms the foundation for an infinite number of possibilities, and at a temperature of absolute zero the true vacuum possesses an infinite amount of energy. On the strength of these and other findings, some scientists like the Nobel Prize winners and physicists Eugene Wigner and Brian Josephson or the mathematician John von Neumann argue that this absolute vacuum, this nonlocal space, could form the basis for consciousness (see later in this chapter). This chapter also considers whether or not quantum physics applies to living systems.

Not everybody will be able to accept the ideas, concepts, and interpretations of quantum physics. We do not know yet if and to what extent quantum physics can help us find answers to all of our unanswered questions. But in my view the foundations of quantum physics, such as wave-particle complementarity, entanglement, and a nonlocal space with probability waves, which have been accepted by most quantum physicists, might be crucial for understanding the mind–brain relationship and the nonlocal aspects of consciousness itself.

Our Classical Worldview

We begin our in-depth look at what quantum physics might tell us about NDEs by looking more closely at our classical worldview. During and after an NDE people experience phenomena that suggest an instantaneous, intuitive connection with the thoughts and feelings of others. NDErs experience an enhanced consciousness in a dimension where time and distance no longer play a role. These phenomena cannot be accounted for with concepts from classical physics.

According to classical physics, it is impossible to be in two or more places at the same time or to move instantaneously to another time or another place. *Instantaneously* means immediately, much faster than the speed of light; it refers to a connection that, independent of distance, is timeless and ubiquitous. According to classical physics, we live in objective reality, which means that everything in our world is supposed to happen within a structure of absolute, fixed space and time. Adherents of classical physics assume that the perceived reality in the physical world equals objective reality. Under this assumption, reality exists independently of observation. The unchangeable laws of classical physics assume that everything in our natural world happens along orderly and predictable lines. As Albert Einstein said, "God does not play dice."

In classical physics, causality is paramount. This means that time is unidirectional and that the order of cause and effect is always a foregone conclusion. Classical physics assumes that reality is continuous, which refers to the fact that our physical world is free of discontinuity and that everything happens gradually and orderly in time and space.

Classical physics assumes locality, which means that objects are influenced only by direct (local) contact. This law rules out remote influence. That said, several centuries ago classical physicists were already engaged in a fierce debate about local versus nonlocal causality. Isaac Newton proposed a nonlocal model of gravity, which could exert a remote influence in, for example, our solar system, and which was fiercely disputed by contemporaries.

Finally, it should also be mentioned that in classical physics Einstein described time as relative, after proving that time is not an absolute constant in the universe. When his lifelong friend Michele Besso died, Albert Einstein wrote to his family, "He has departed from this strange world a little ahead of me. That means nothing. People like us, who believe in physics, know that the distinction between past, present, and future is only a stubbornly persistent illusion."

So even in classical physics remote influences like gravitation and the relativity of time were accepted ideas.

What Is a Wave?

Before I describe the many challenging and often incomprehensible aspects of quantum theory, let us try to familiarize ourselves with a few more important concepts from classical physics. What is a wave? A wave is a standing or traveling disturbance moving through air (as in sound waves) or water or space. Light is also a wave phenomenon, that is, an electromagnetic wave with a magnetic and an electrical component. What is true for light, namely that it possesses both a particle and a wave aspect, also pertains at subatomic level to matter (see figure).

Later in this chapter I will describe in more detail that according to the laws of quantum physics, we cannot determine the exact location of a quantum particle; we can only establish the particle's probable location. The equation expressing this probability is known as the particle's wave function.

Matter as a complex field of standing waves.
Extreme enlargement (x 700,000) of platinum
(Photo:Pennsylvania State University). The white dots are
individual atoms. At atomic level, matter behaves
like a field with standing waves.[2]

The Definition of a Field

What is a field? A field is a complicated concept: although a field cannot be perceived, it does have a visible effect. The magnetic field is a case in point; it has a penetrating, space-filling capacity and can exert an invisible, remote influence on metal objects such as a compass. The electromagnetic field is a physical field produced by electrically charged objects. A field requires no medium to exert its remote influence; it occupies the vacuum of empty space. A field is itself a form of space.[3] In classical physics (local causation), a field denotes coherence in a system, ensuring a correlation or rhythmic cohesion between all parts or components of that system. If something happens in one part of the field, it automatically happens in the system as a whole. A disturbance in a field always travels at a maximum speed, the speed of light. A gravitational field can be considered an immaterial and invisible domain that can instantaneously influence our visible, physical world.

Electromagnetic Fields

An electromagnetic field is a physical phenomenon that causes only charged particles to move. The electromagnetic field extends indefinitely throughout space and describes the electromagnetic interaction. It is one of the four fundamental forces of nature. Electromagnetic fields are an integral part of the organization of all material systems, from atoms to galaxies. As well as the basis of the function of our heart, brain, and body, they are at the core of all the electrical equipment that contemporary society depends on. All the information we receive on a day-to-basis is encoded in waves or wave functions of the electromagnetic field. It is encoded in these waves as differences in frequency or wavelength. An electromagnetic field has a near-infinite capacity for storing information in frequencies, or phase speeds, without causing any disturbance or interference. Imagine the vast quantity of information that reaches us on the fiber-optic cable network that nearly all homes are hooked up to these days as well as the information transmitted between continents by way of cables on the ocean bed. Or imagine the global reach of the Internet with more than a billion Web sites or the information that is available worldwide via GPS satellites for use in such diverse applications as car navigation systems and cell phones.

Fields, Frequencies, and Information

Wavelength is inversely proportional to frequency. Hertz (Hz) is the unit of frequency, and 1 Hz equals 1 cycle per second. The electromagnetic spectrum is very broad: the ultraviolet (UV) light of a tanning bed or x-rays have a wavelength of less than 100 nanometers at a frequency of over 3×10^{15} Hz. Visible light has a wavelength ranging from 300 to 800 nanometers while a radar or satellite TV uses a wavelength of 1 mm to 3 cm, a microwave oven a wavelength of 10 cm, mobile telephone a wavelength of 30 cm (at a frequency of 1 GHz), TV a wavelength of 1 m (300 MHz), medium-wave radio a wavelength of 300 m (1 MHz), and submarine communication a wavelength of more than 3,000 km (less than 100 Hz). The greater the wavelength, the lower the frequency and the better the reach or reception. The alternating current of our domestic electricity is 50 Hz. The sound waves that we can hear (at a young age) have a frequency ranging from 20 to 24,000 Hz. All sensory perception is based on information obtained from waves: we can see colors thanks to the information from light waves, and we can hear different sounds and tones thanks to sound waves with different frequencies. We feel warmth on our skin thanks to thermal waves: the infrared light of the sun, which has a frequency of approximately 10^{13} Hz.

The information that astronomers draw on for their theories about the origins of the universe is largely based on images obtained by the Hubble space telescope. With the help of this telescope astronomers have recorded images of galaxies at a distance of 5 billion light-years, and they have seen exploding stars 42 million light-years away. Because information about these extremely remote events has been retained in light waves, we now have crystal clear images of them. Information encoded in light waves is retained, unchanged, for at least 5 billion light-years. The capacity for storing information in wave functions seems potentially infinite and eternal.

Our worldwide communication is based on the encoding and decoding of information stored in particular frequencies (wavelengths) of the electromagnetic field, which is not immediately visible to our senses. In order to receive and retransmit this information, we use radio, TV, mobile phone, and wireless Internet technology. Our entire worldview is constructed on the basis of all the information we receive into our consciousness by way of the senses. Our consciousness uses this information to form our conception of the world and of ourselves. In order to receive this vast quantity of information into our consciousness, we need a receiver to pick up, or decode, the factual information encoded in waves: cell phone, radio, TV, and wireless computer.

Quantum Theory and the Particle–Wave Complementarity

As said before, quantum physics emerged at the start of the twentieth century because certain natural phenomena could no longer be accounted for with classical physics. Scientists had known for some time that when metal is heated, the actual increased intensity of the light, especially in the ultraviolet spectrum, does not correspond with the predicted increase. In 1900 Nobel Prize–winning physicist Max Planck came up with the mathematical description of a discontinuous interaction between light and matter that he called quanta. This discontinuity is reminiscent of a ball that bounces down a staircase and lingers briefly on each step but can never be observed between two steps. This discontinuity was called a quantum leap. A few years later Albert Einstein developed the hypothesis that light also moves in packets (light quanta), and in 1905 he gave this energy packet the name *photon*. In 1926 an experiment confirmed his photon hypothesis.

For centuries the properties of light have been one of the biggest problems in physics. According to the seventeenth-century Dutch mathematician, astronomer, and physicist Christiaan Huygens, light behaved like a wave, whereas Newton believed it consisted of particles. In the famous double-slit experiment, first conducted in 1801 by the English physician and physicist Thomas Young, light is passed through either a double narrow slit or, after one of the slits has been closed off, through a single narrow slit. When the light passes through both slits, it behaves like a wave, with interference creating dark and light bands (see figure). Interference is the phenomenon we see when we throw two pebbles into a pond and the ripples intersect. Interference patterns create some bigger waves while making other waves disappear; these waves are the equivalent of the light and dark bands in the double-slit experiment. When Young published his double-slit experiment in 1802 and concluded that light behaves like a wave, he was showered with scorn and hostility because his results ran counter to Newton's particle theory of light. Young's publication, critics at the time argued,

> *contains nothing which deserves the name, either of experiment or discovery, and . . . is destitute of merit. . . . We wish to raise our voice against innovations, that can have no other effect than to check the progress of science, and renew all those phantoms of the imagination which . . . Newton put to flight from her temple.*[4]

But things proved to be even more complex. If a very faint light travels through both slits, with only a single photon passing the slits at a time, there is a possibility that the light will also behave like a par-

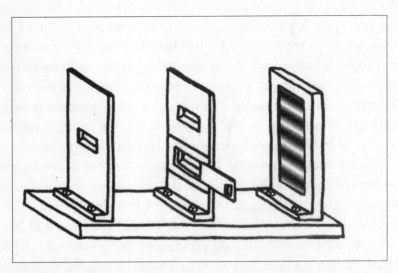

The double-slit experiment, with interference creating bright and dark bands.

ticle; in this case the light will be distributed evenly across the entire projection plane (a photographic plate), and the interference pattern of light and dark bands will disappear. However, this happens only when scientists want to know the exact position of the photon and record which slit the photon passed through. Only if an instrument positioned in front of or behind the slits measures if and where a photon has passed do we know the photon's exact route and if the light continues to behave like a particle. The same applies when a measurement is carried out behind the slits and the measuring instrument is not switched on until the photon has passed the slits but has not yet reached the photographic plate. Because of the measurement, the photon still behaves like a particle. If we do not take any measurements during the experiment, the light continues to behave like a wave.

Nobel Prize–winning physicist Niels Bohr called this phenomenon *complementarity*. The light behaves either like a particle or like a wave, depending on the setup, but not like both at the same time. Particles and waves are complementary aspects of light; they are incompatible and never visible at the same time. The problem for physicists was that, depending on the experiment design, light could behave like either a wave or a particle. With the double-slit experiment, an astounding thing was discovered: the behavior of light depends on the researcher's decision to install extra measuring instruments or to open one or two slits. The researcher's deliberate choices regarding the experiment design determine whether the light will behave like a wave or like a particle. This brought about a profound transformation of the core structure of the basic general physical theory: the connection between physical behavior and human knowledge was changed from one-way traffic to a mathematically specified two-way interaction that involves selections performed by conscious minds. As Bohr put it, "In the great drama of existence we ourselves are both actors and spectators."[5]

The revolutionary message of quantum physics is that while there is a certain order in the universe (waves or particles), there is more to it than the physical aspect alone. Matter can be measured, but the mind determines what we know. Our thoughts and feelings play a part in determining how the universe functions and how we perceive the universe. The way we think has a physical effect on what we perceive, and this has brought about a revolution in physics as well as in philosophy and in consciousness research. Nobel laureate and quantum physicist Max Born said, "I am now convinced that *theoretical physics is* actually *philosophy*."[6]

The slow and reluctant acceptance of the new insights from quantum physics can be ascribed in part to the materialist worldview with which most of us were raised. In this view the objective, material world functions only according to the unchangeable laws of classical physics outlined in the previous section.

Entanglement

In the double-slit experiment, if so little light is emitted that only a single photon passes through the two slits at a time, and no other measurements are carried out to determine the position of this photon, the photosensitive plate will eventually exhibit an interference pattern again, showing light behaving like a wave. Even when it consists of isolated photons, the light behaves like a wave, which can mean only that each photon passes through both slits at the same time. The photon becomes entangled with itself, as it were. This is known as a *superposition* of wave functions, whereby a wave should no longer be seen as a real wave but, in Born's term, as a *probability wave*. A probability wave is an equation that describes the probability with which a particle can be found in a certain position; it is also known as the wave function of a particle. When the intensity of the light dwindles from a massive bombardment to an isolated photon emission, the light is no longer described as an electromagnetic wave but as a probability wave. Light is normally defined as an electromagnetic field that behaves like a disturbance in an empty space or a vacuum. In large numbers, photons behave like an electromagnetic wave packet. But when a single photon passes and no electromagnetic wave can be measured, the immeasurable probability wave is used to statistically predict where the photon will hit the photographic plate. At that moment an isolated photon behaves like a probability wave. Unobserved, the photon has no location because it has an infinite number of possible locations.

Quantum physicist Erwin Schrödinger formulated the equation for these quantum-mechanical waves.[7]

An atom's electrons occupy a probability field around the nucleus, and each time they are observed they occupy a different position in this field. But matters are complicated by the fact that the position and the momentum of an electron cannot be measured at the same time. As a result, we never really know where the electron is. This is the *uncertainty principle* of Nobel Prize–winning quantum physicist Werner Heisenberg: when we try to measure the momentum of an electron, we become unable to locate its position at the same time. Observation is impossible without fundamentally altering the observed object. An observation reduces the countless possibilities (probability waves) to a single fact, the particle's position at that moment in time. Mathematician and physicist Roger Penrose calls this *objective reduction.*[8] Scientists have come to the conclusion that the observer determines where and how a particle will be perceived. Observing the process affects the results because everything is connected to everything else. This rules out any chance of objective observation. And this applies to both experiments and everyday life. All (observation of) reality is subjective because the observer's mind determines what will be observed. And if two or more observers are in agreement, we ought to speak of the intersubjectivity rather than the objectivity of perceived reality.

Some prominent quantum physicists, including Eugene Wigner, Brian Josephson, and John Wheeler, as well as mathematician John von Neumann support the radical interpretation that observation itself literally creates physical reality, a position that regards consciousness as more fundamental than matter or energy. Von Neumann writes, "The world is built not out of bits of matter, but out of bits of knowledge—subjective, conscious knowings."[9]

Quantum physicist Henry Stapp writes that including human consciousness in the basic structure of physical theories is one of the most crucial developments in quantum physics. He regards the idea in classical physics that our thoughts are completely irrelevant as a serious problem. Quantum physics allows us to view ourselves as people who seek and use knowledge and who, thanks to our investigative activities, are able to exert some influence over our environment and therefore cannot be reduced to automatons. This is why Nobel laureate Eugene Wigner claims that quantum physics is concerned with observations and not with the observable. Books such as *The Non-Local Universe: The New Physics and Matters of the Mind; The Self-Aware Universe: How Consciousness Creates the Material World;* and *The Spiritual Universe* also elaborate the key role of consciousness in relation to quantum physics and the consequences for our worldview. Many physicists and philosophers struggle to accept this interpretation of quantum physics.[10]

Nonlocality

How can two separate particles have a remote and instantaneous influence on one another? How do we explain the entanglement of two (or more) remote objects? This constitutes one of the key principles of quantum physics and one of the most profound and astonishing discoveries in the history of physics. It is based on Bell's theorem, which was proven by physicist Alain Aspect and his colleagues in 1982.[11] In their experiment they measured the change in spin of two particles.

But what exactly is spin? Spin is a fundamental property of nature, akin to electric charge or mass. All the smallest particles, such as protons, neutrons, and electrons, have a spin that is either positive or negative and that is always a multiple of one-half. Individual, unpaired particles have a spin of one-half.

If a local measurement is performed on the spin or rotation of a particle, classical physicists assume that the measurement has a local effect. But if the experiment is performed with two particles emanating from the same source but fired in two different directions, and measurements are then conducted in two separate places, what scientists found is that measuring the first particle also gives us the results of measuring the second particle. In other words, there is a correlation, an entanglement of the two particles, which allows us to predict the outcome; there is no local, or direct, influence between the two particles that would cause the outcome of the measurement of the second particle to match that of the first. This was a revolutionary finding because up until that point the consensus had been that only local, or direct, causes could determine the outcome of a measurement. Not so, according to quantum mechanics.

Initially, many struggled to accept such an instantaneous, remote effect; even Einstein had tremendous difficulty with nonlocal effects in quantum physics. However, experiments in 1982 produced definitive proof that entanglement between two particles creates a nonlocal relationship. The physicist Nicolas Gisin repeated these experiments with photons eleven kilometers apart via a fiber-optic cable at CERN, the European Organization for Nuclear Research near Geneva, Switzerland. The same nonlocal entanglement was later demonstrated across a distance of fifty kilometers. Nonlocality has even been proved in three entangled systems (the Greenberger-Horne-Zeilinger paradox).[12]

The New Worldview Based on Quantum Physics

According to quantum theory, everything is interconnected, there is no local cause for an event, and when an event takes place it instantly changes the entire universe. As early as 1923, Nobel laureate Louis de Broglie wrote that ultimately all matter in the universe can also be seen as a wave function.[13] This means that matter too has a wave-particle complementarity. What had already been proven for light—that it exhibits both particle and wavelike properties—was found to apply to matter as well.

In 1930 Einstein wrote, "We have now come to the conclusion that space is the primary thing and matter only secondary." And a few years later Schrödinger claimed, "What we observe as material bodies and forces are nothing but shapes and variations in the structure of space." Physicist Steven Weinberg recently expressed the current position in quantum physics quite succinctly: "Matter thus loses its central role in physics."[14]

But what is matter? Does matter actually exist? What can materialist-minded scientists still believe in?

As mentioned earlier in this chapter, the new and revolutionary concepts from quantum physics include superposition, complementarity, the uncertainty principle, the measuring problem, and entanglement. All of these concepts revolve around the same problem: when it is not observed, the quantum object has neither a definitive location in time and space nor the kind of fixed properties that classical physics ascribes to objects. This is known as the "quantum measurement problem." It is difficult to gauge the consequences for our worldview if we accept that something can exist without a location in space, a place in time, or any properties. If fundamental properties can be established only after an observation has taken place, the big question becomes: What kind of reality could exist without observation? "Does the moon exist when nobody is looking?"[15]

Before you continue reading, I would like you to stop and close your eyes for a moment. Now open them again and ask yourself: What did the world look like while you had your eyes closed and you were unable to see the world around you? How can you know what the world looked like at that moment? And what reality existed last night while you were asleep? Where was the world when you were sleeping? How can you be sure that the world exists while you are asleep? It may seem implausible, but some renowned quantum physicists maintain, on theoretical grounds, that the world does not exist when nobody is looking because without observation we cannot be certain that it really exists. These quantum physicists claim that an observation creates a personal subjective world from an infinite number of unlimited possibilities.

We can build on this thought experiment by manipulating a person's consciousness: if a person is hypnotized and told that everybody present is bald, he or she will actually see people without hair on their heads. Or if somebody under hypnosis is told that he will be touched with an extremely hot object but is really touched with, say, a pencil, his skin will still blister.[16] The mind is primed by hypnosis to perceive its surroundings in a particular way. The expectations imposed by the mind even trigger a visible reaction in the body. The mind, thus primed, determines how reality will be experienced.

In other words: expectations shape our reality. So what about people with prejudices or materialist views? Will these people have a different view of reality because of their expectations? I will come back to this intriguing question later in the book.

Unobserved objects are instantaneously connected or entangled in a timeless, nonlocal way. The concept of nonlocality is now a commonly accepted aspect of quantum physics, but about a hundred years ago Einstein still spoke of "spooky action at a distance." Actually, Newton's laws of gravitation were viewed in a similar light by his contemporaries. The following remark encapsulates the unimaginable consequences of quantum theory: "Quantum mechanics is magic."[17]

Quantum mechanics also exposes the concept of causality, the fixed relationship between cause and effect, as an illusion. Events happen only in the presence of an observer. In classical physics, by contrast, reality consists of separate elements that can be individually examined and measured. But since the advent of quantum physics we know that everything is interconnected, that everything operates like a holistic system and not in isolation, and that analysis of these separate elements will never uncover a so-called objective reality. In fact, the conclusion goes one step further: there is no such thing as objective reality, only intersubjective reality. As Schrödinger put it in his influential book *What Is Life,* "The world is a construct of our sensations, perceptions, and memories."[18]

Based on the empirical data produced by scientific research into NDE and on the purely theoretical assumptions of quantum physics, as formulated by aforementioned scientists such as von Neumann, Wigner, Josephson, Wheeler, and Stapp, I support the not yet commonly accepted interpretation that consciousness determines if and how we experience reality.

The Nonlocal Space of Probability Waves

Most contemporary quantum physicists believe that the nonlocal space of Schrödinger's probability waves is a purely mathematical concept and cannot be ascribed any reality. In other words, it is purely hypothetical. It cannot be measured because it is only a series of probability waves that have not collapsed, through observation, into measurable results. The velocity of the probability waves ranges from the speed of light to infinity (or instantaneous).

In 1901 the American physicist Josiah W. Gibbs was probably the first to call this nonlocal space of probability waves the *phase space*. In 1924 the German physicist Arnold Sommerfeld described the phase space as a six-dimensional space with only wave aspects, which, for readers familiar with modern string theory, is somewhat comparable to the many dimensions called for by certain recent versions of that theory. The waves of phase space have measure but no direction because they occupy a nonlocal dimension. This phase space is difficult to visualize, but such a multidimensional space can be constructed with the help of mathematical formulas.[19]

The nonlocal space of probability waves, the so-called phase space or nonlocal space, contains no matter; everything inside it is uncertain, and physicists cannot carry out any measurements or observations. However, nonlocal space can be influenced from outside. After a deliberate measurement or observation, the probability waves in nonlocal space collapse statistically into physically measurable particles.[20] If and how this collapse occurs remains a moot point. Quantum physics is essentially statistical, and the statistical element is by definition rooted in nonlocal space. A number of fundamental fields in nature, such as the weak and strong nuclear forces, have a quantum aspect and are thus connected to nonlocal space. This means that all molecular and submolecular processes are influenced from nonlocal space. And while the underlying causes of these processes are unfathomable to physicists and chemists, their effects can be demonstrated. The gravitational fields possibly and the electromagnetic force fields probably have their basis in nonlocal space. Here too the fields themselves are by definition invisible, unlike their physical effects. The conclusion that most fundamental fields and forces in the universe seem to have their basis in nonlocal space is important for our later discussion and understanding of the nonlocal aspects of consciousness that are experienced during an NDE, and for our understanding of the relationship between consciousness and our physical body. Is it possible to consider this relationship more than just an analogy?

Nonlocal space harbors a hidden reality that exerts a constant influence on our physical world. Everything in our physical world, which is also known as space-time, is subject to time and distance. But ev-

Interference pattern in water.

erything is based on constant interaction between quantum states and this invisible nonlocal space. Everything visible emanates from the invisible. Again this is a matter of complementarity, just like the wave and the particle. The visible particle complements the invisible wave function. The visible, physical world, space-time, complements the invisible and imperceptible nonlocal space. The physical world is influenced at quantum level by nonlocal space, just as our physical body at all levels seems to be influenced by our consciousness. The foundation of our physical universe seems by definition not measurable.

The Quantum Hologram

In a holographic, two-dimensional photo, that is, a photo in a flat plane, a three-dimensional image is conveyed with the help of a coherent laser light. If this photographic plate smashes into a hundred pieces, the total three-dimensional image will, in principle, be present in each shard. The information of the overall image exists in each segment of the plate as an interference pattern.

Interference is what you see when you throw pebbles into a pond (see figure). Waves interact, and the resulting interference creates a pattern of weaker or stronger waves.

Information can be stored in this interference pattern. In a coherent field, where waves are interacting to form a particular pattern, the interference patterns are distributed across the physical medium of the field, for instance, in water or on a photographic plate. The holographic information in a hologram is therefore not stored in the field itself but in the field's physical medium, and the holographic principle means that the stored information in its totality can be retrieved from each location in this physical medium.

Both quantum physics and holography are based on the principle of coherence. Wave functions that cohere, or together form a particular pattern, carry the information in a quantum hologram. A practical example of the principle of the quantum hologram, and a technique that proves the concept of nonlocal information exchange, is functional magnetic resonance imaging (fMRI).[21]

Contrary to classical physics, in quantum physics, a nonlocal connection cannot be described with a field. As described before, nonlocal connections are established instantaneously, that is, faster than the speed of light, which is possible only in a nonlocal space. In a nonlocal space all parts of that space react en masse to all events. The interaction or correlation here does not depend on time and distance and happens at the very smallest subatomic level up to the biggest level of cosmological time and space. A disturbance of space is seen as the carrier of the information that connects or correlates all different parts nonlocally and instantaneously. But what is true for the field also applies to the hologram, namely that a nonlocal connection can never be described with a hologram. The information in a hologram—encoded as an interference pattern in the physical medium of a field with at most the speed of light—can be retrieved from any location in that field. In quantum physics the information is not encoded in a medium but is stored nonlocally as wave functions in nonlocal space, which also means that all information is always and everywhere immediately available. So in both a nonlocal space and in a hologram all information is available from all locations, but the method of information storage and the speed of information retrieval are fundamentally different.

Dutch Nobel laureate Gerard 't Hooft believes that the entire universe might be based on the holographic principle, a view he sees as compatible with string theory.[22] In this theory the strings are one-dimensional oscillating lines (wave functions) floating in space-time. The idea of a holographic universe is based on an as-yet-unknown medium, believed to be strings or branes (this medium used to be known as the ether); in a nonlocal universe everything is encoded as wave functions in nonlocal space. Scientists now know that a vacuum is not empty; at absolute zero, -273.15 degrees Celsius, it is full of energy (a "plenum"), and at the subatomic level it undergoes constant quantum fluctuations that create new quanta "from nothing," which then immediately disappear again. What we see here is a kind of universal process of constant creation and annihilation. These quantum fluctuations are also known as the vacuum's zero-point energy. It can generate virtual particles (with antiparticles) that instantly destroy one another. The same pertains to the appearance and disappearance of virtual energy (waves). *Virtual* means that which is seemingly real or a possibility. There is general agreement about the (extremely short) existence of virtual particles and virtual waves (energy). In two recent and accessible books, *The Connectivity Hypothesis* and *Science and the Akashic Field*, systems theorist Ervin Laszlo uses holographic field theory to argue that the entire universe is a fully interconnected holographic information field. His ideas are based on the theory of a zero-point field in the quantum vacuum or "cosmic plenum."[23]

Consciousness and Nonlocal Space

Perhaps nonlocal space could also be called the absolute vacuum: it lacks structure, has no time, and is an empty space in which quarks (elementary particles and fundamental constituents of matter), electrons, gravity, and electricity have all become one and as such do not exist. This space forms the foundation for an infinite number of possibilities.

This absolute vacuum, this nonlocal space, could be a basis or foundation for consciousness. I support the interpretation of the aforementioned researchers von Neumann, Wigner, Josephson, Wheeler, and Stapp that this nonlocal space is more than a mathematical description; it is also a metaphysical space in which consciousness can exert influence because it has phenomenal properties. *Phenomenal* means based on subjective perception, or literally "subjective perception in the mind." According to this interpretation, consciousness has a primary presence in the universe, and all matter possesses subjective properties or consciousness. In this view, consciousness is nonlocal and the origin or foundation of everything: all matter, or physical reality, is shaped by nonlocal consciousness. There is no longer any distinction between nonlocal space and consciousness. This is not a new insight. As far back as the seventeenth century, Newton held that the omnipresent space might be filled with a "spiritual substance"; he called space the "divine observatory."[24]

The philosopher David Chalmers, who specializes in questions of consciousness, calls this approach monism or panpsychism. He seems to share the belief in the fundamental relationship between consciousness and matter. In this view, physical systems have phenomenal properties at a fundamental or intrinsic level (nonlocal space) and therefore possess subjectivity or a certain degree of consciousness. Phenomenal or subjective properties can be found at the most fundamental level of physical reality and form the basis of physical reality itself. According to this theory, the intrinsic properties of the physical world are themselves phenomenal properties (consciousness). Chalmers thus gives consciousness a clearly causal role in the physical world.[25] Not everybody will be able to endorse this view, but it is certainly worth exploring in more detail. I will come back to this in the next chapter.

The Complementarity of Nonlocal Space

Light behaves either like a particle or like a wave, depending on the environment, but never like both at the same time. Particles and waves are complementary aspects of light; they are incompatible and never visible at the same time, but they are intrinsically linked together. At the speed of light, the speed of a particle equals the phase speed of the particle's corresponding wave function. The particle's speed ranges from zero to the speed of light, and the phase speed of the corresponding wave function ranges from the speed of light to infinity because speed in the quantum-mechanical phase is the opposite of speed in normal space-time, our physical world. The slower the particle, the faster its corresponding phase speed. And when the particle's speed slows to zero, as it does during an observation in a photographic emulsion, its corresponding phase speed is infinite. This results in an instantaneous entanglement (nonlocality) with everything in the universe, including nonlocal aspects of consciousness.

As mentioned, on theoretical grounds physicists cannot carry out observations within this nonlocal space. The gravitational field itself cannot therefore be made visible or measurable. It is possible to exert an external influence by manipulating the wave or localizing the particle. However, as soon as an observation takes place, this multidimensional nonlocal space is once again reduced to our three-dimensional physical world, space-time. An observation reduces the countless possibilities (probability waves) to a single fact, that is, the particle's position at that moment in time. Mathematician and physicist Roger Penrose has called this collapse of the wave function "objective reduction."[26] If no observations take place in the nonlocal space, the phase speed can range from the speed of light to infinity. In other words, not everything in nonlocal space is constantly entangled—only during an observation.

Nonlocal space resembles the "implicate order" of quantum physicist David Bohm. He viewed the implicate order as a basic and multidimensional field of information with holographic principles, in which observation-induced collapse (objective reduction) plays no role. In his vision, "in-formation" is the subtle influence that only affects, or "forms," the phase of a wave, a process in which consciousness plays an essential part. "In-formation" thus has an effect in the physical, visible world without any energy transfer: it "in-forms," or "forms," the physical system receiving the information.[27]

Field Theories in Living Systems

The concept of coherent fields is used not only in physics but also in biology. In the 1920s biologist Paul Weiss drew on limb regeneration in amphibians to formulate the concept of morphogenetic fields, which are organizing fields of formative information that guide the development of the particular form of a living structure or being. And the biologist and medical scientist Alexander Gurwitsch postulated that neither the cell's individual properties nor its relationship with adjacent cells could explain the role of individual cells during embryogenesis (the process by which an embryo is formed and develops), but that a factor external to the embryo appeared to determine overall development. He called this factor a force field or an embryonic field.[28]

Information transfer from fields takes place via resonance, that is, vibration with the same frequency and phase. Resonance is not limited to acoustic resonance in sound or to the electromagnetic resonance we get when tuning in to a radio or television station, but it also exists at the smallest subcellular level as electron spin resonance and nuclear magnetic resonance. Morphogenetic (formative) fields feature nonenergetic information transfer, and these fields, like the probability fields in quantum physics, are based on probability. It is this property that makes these fields difficult to describe. As living systems, all organisms have a rhythmic oscillation, vibration, or periodic motion, each with its specific and characteristic frequency. Each living cell has countless vibrating molecular structures, which in turn have specific oscillations. The reciprocal information transfer between the field and the living cell structures takes place via resonance with these specific frequencies. The English biologist Rupert Sheldrake developed the concept of morphogenetic (formative) fields quite brilliantly in his books *A New Science of Life* and *The Presence of the Past*.[29]

Systems theory and field theories are also making their way into biology and pharmacology mainly because scientists are starting to realize that it is impossible to determine the behavior of an intact and living organism on the basis of its isolated components. A living organism is home to constant information exchange between all of its constituent parts. This is why a living organism is more than just the sum of its parts. In a recent publication in *Nature*, chemist and scientific director of systems biology research Jan van de Greef described his pioneering ideas about systems theory in general, and about systems biology, systems pathology, and systems pharmacology in particular.[30]

Does Quantum Physics Apply to Living Systems?

Quantum theory has been corroborated by countless experiments and refuted by none. It has become a key part of the description of the world around us, but the question remains: Does quantum theory also apply to living systems? Quantum physicists differ on the matter. Schrödinger considered quantum physics to be incomplete, a view shared by Einstein and de Broglie. Schrödinger believed that there ought to be a comprehensive scientific explanation for life and that quantum physics ought to provide the complete biological foundation with which to fathom life's chemical and physical aspects. Current quantum mechanics does not yet allow this; hence his opinion that the discipline is incomplete.

In contrast to Schrödinger, Bohr viewed life as complementary to what can be verified or proven by quantum physics, which only describes processes in "dead" matter. This is his version of the "Copenhagen interpretation" of quantum physics. In Bohr's view, life is "unknowable," and quantum physics can never provide a scientific explanation for life processes because they involve nonstatistical processes of a "higher" order (that is, they defy statistical computation). Bohm too was of the opinion that reality in its deepest sense is unknowable.[31] In living matter, the transition from nonlocal space to the physical world, that is, space-time, is a nonstatistical (chaotic) and nonperiodic (unpredictable) process because this transition is actually possible with only small numbers of atoms or even a single atom. Contemporary quantum physics only describes statistical processes in "dead" matter because the transition from nonlocal space to our physical and measurable world is essentially a statistical, lower-order process. Based on everything I have read, I am (intuitively) drawn to Bohr's interpretation.

Another problem for quantum physics in living systems is the fact that quantum physics applies only to coherent and closed systems. A living system, with heat loss and respiration, exchanges information with its surroundings and thus triggers decoherence (the leaking of information), that is, a loss of coherent and harmonious processes. According to some interpretations, this rules out the possibility of quantum physical processes. However, interference, and hence coherence, has been demonstrated in huge soccer ball–like molecules at 650 degrees Celsius, while in 2000 *Nature* published two articles about quantum superposition in macroscopic states in a superconducting quantum interference device ("squid"), which featured billions of paired-up electrons in a coherent state. These findings have a practical as well as a philosophical significance.[32]

Quantum Theory, Self-Organization, and Consciousness

Despite the aforementioned objections, some scientists, among them quantum physicists, believe in quantum coherence in all living systems at both a cellular and a subcellular level. This could be explained by the self-organizing capacity of living matter, in which unstructured, inert, and chaotic matter from the immediate surroundings is absorbed into a dynamic structure of ordered coherence, as described by Nobel laureate and physical chemist Ilya Prigogine. The physicist Herbert Fröhlich made a convincing case for such processes in living matter, even at body temperature. He described how molecules and cells start to vibrate and form a coherent whole with identical frequencies so that in an ordered state they can be compared to a Bose-Einstein condensate, a system in which the many constituent parts do not just *behave* like a whole but actually *become* a whole. The constituent parts thus lose their identity. This only happens when all properties and all information merge into a coherent whole. One might compare it to the many voices in a choir becoming one harmonic whole, one voice, or an orchestra sounding like one. For many years scientists have been locked in debate about whether the principles of such condensates also apply to macroscopic and living systems.[33]

A living system consists of various subsystems, which actively cohere but which also exhibit various levels of order and disorder, and which can be regular and irregular, stable and unstable at the same time. The end result is organized chaos, constituting what is known as a self-organizing system, with patterns or structures arising from interaction with the outside world without being directly caused by these external factors. A good example of self-organization is a vortex in flowing water, in which the shape of the vortex is influenced by the rate of flow and the volume of water, but the vortex itself is a spontaneous and self-regulating event. Based on the theoretical possibility of living matter's self-organizing capacity, some scientists have sought a quantum-mechanical explanation for the relationship between consciousness and the brain.

Drawing on the principle of coherent systems created through self-organization, neurobiologist Herms Romijn proposed that the constantly changing electric and magnetic fields of neuronal networks (photons or possibly virtual photons), which can be regarded as a biological quantum-coherence phenomenon thanks to their self-organizing aspect, may be the "carriers" or the "product" of consciousness and its memories. His model is akin to neurosurgeon Karl Pribram's idea that memories cannot be stored in small groups of neurons, but only in the coherent patterns formed by the electromagnetic fields of neural networks. In Pribram's view, the brain functions like a hologram. This hologram is capable of storing the human memory's vast quantity of information. He developed his idea in response to the remarkable experiments conducted by Karl Lashley, who already proved in 1920 that memories are not stored in any single part of the brain, but throughout the brain as a whole. These experiments on rats showed that it did not matter which parts and how much of the rats' brains were removed, the animals were still capable of carrying out the complex tasks that they had been taught before the brain operations. The only problem was that at the time nobody could conceive of an explanatory mechanism for memory storage based on a "whole in each part" principle. A vast amount of evidence suggests that our brain draws on the holographic principle to perform its tasks, because Pribram also demonstrated that when he removed 90 percent of the visual cerebral cortex or 98 percent of the optic nerve of a cat the feline was still capable of performing complex visual tasks. These experiments suggested that both memory and visual perception can only be accounted for on the basis of the holographic principle. The same was recently demonstrated for acoustic phenomena (our hearing).[34]

Given the current insights afforded by quantum physics and the theory that consciousness and memories are stored in nonlocal space as wave functions, we should speak no longer of holographic organization but rather, like Romijn, of nonlocal information storage in which memory is nonlocally and instantaneously accessible. If this is the case, visual and auditory information processing also occurs along nonlocal rather than holographic principles. This might explain the possibility of perception during an out-of-body experience as well as a life review with detailed memories and images during an NDE in a dimension without time and distance.

According to anesthesiologist Stuart Hameroff and mathematician and physicist Roger Penrose, microtubules (the tiny structural components of the skeleton of cells that are involved in many cellular processes) inside neurons may initiate information processes via self-organizing patterns that trigger coherent states, and these might explain our ability to experience consciousness. Their suggestion is based in part on the still-speculative theory of quantum gravity, which is the field of theoretical physics that tries to reconcile or unify the theories of quantum mechanics and general relativity. In her book *The Quantum Self,* quantum physicist Danah Zohar also posits biological quantum coherence as an organizing principle, which could explain a "quantum relationship" between consciousness and the body.[35]

Quantum physicist Anton Zeilinger also describes the mind with its thoughts as a quantum process because it is impossible to experience a half-thought, a half-feeling, or a half-yes or half-no, only a complete thought, a complete feeling, and a definite yes or no. The information, the answers our mind receives to our questions, also constitutes a binary system: yes or no, one or zero, on or off.[36] Our consciousness is therefore not a continuum but is split into "quanta" or components, even if we experience it as a continuum. This is something we encounter in everyday life when we watch a film; we see a moving image, even though the film actually consists of twenty-five static projections per second. Rapid processes are perceived as a continuum when perception is slower than the speed of events. The same applies to events at the subatomic level.

Quantum physicist Stapp combines the ideas of psychologist William James, quantum physicist Heisenberg, and mathematician von Neumann into a comprehensive theory drawing together classical physics, quantum physics, quantum chemistry, neuroscience, psycho-pathological experiments, and various fields of psychology. He writes, "The connection between consciousness and the brain is primarily a problem in physics and addressable by physics—but only the correct physics. The causal irrelevance of our thoughts within the classical physics constitutes a serious deficiency of that theory."[37]

He calls quantum laws fundamental "psycho-physical" laws, and with this he explains the causal effect of consciousness on neural processes. According to Stapp, a decision made in the mind of a researcher can have no direct effect on the physical system under investigation (for example, whether the light behaves like a particle or a wave), but because it has an effect on the researcher's neural processes it ultimately also determines the outcome of the research. This explains the mind's effect on the outcome of a study, that is, on creating this outcome or creating reality as we see it. If we carry out a series of successive measurements in a quantum system, the effect of the observation appears to freeze, and the ever-changing system appears to come to a halt (the quantum Zeno effect). Stapp compares this to the mind's effect on the brain: if somebody repeatedly, that is to say with undivided attention, concentrates on an idea or concept, it will bring about a permanent change in brain function. William James called this mindfulness a "holding-attention-in-place" action of volition. In Stapp's view the empirical fact of neuroplasticity, the permanent change in brain function through mindfulness that was discussed earlier, could be an indication of the brain's quantum function. The crux of Stapp's approach is that his quantum description of the brain is essentially holistic: it describes overall brain function rather than a model of the brain based on computer science. And by using the principle of the quantum Zeno effect, Stapp also avoids the criticism that the brain is a macroscopic, warm system, which inherently causes decoherence (the leaking of information) and thereby rules out quantum processes. As von Neumann states, "Consciousness creates reality." Observation is not a passive registration in our consciousness but is rather an active creation by our consciousness. This model by Stapp and von Neumann also retains the possibility of free will. Given the results of the prospective NDE studies, I find the approach of Stapp and von Neumann extremely appealing.[38]

The stream of knowledge is heading toward a non-mechanical
reality; the universe begins to look more like a great thought
than like a great machine.
—SIR JAMES JEANS

As we have read in this chapter, some well-known quantum physicists believe that each observation is determined by our consciousness. Reality, as we experience it, is not a fixed, objective given but is shaped by our consciousness. Similarly, each interpretation of quantum physics is determined by our consciousness. Quantum physics admits a great many interpretations, especially in relation to the theory's application to macroscopic phenomena, living nature, and the role of our consciousness. Everything in quantum physics is still in flux. In fact, sometimes I get the impression that there are almost as many interpretations of quantum theory as there are physicists who specialize in the field. And what's more, during the course of their working lives, most of these physicists also change their mind about the ideas that they once wholeheartedly endorsed.

Not everyone will be able to accept the ideas, concepts, and interpretations of quantum physics, partly out of ignorance and partly because of the many crucial but still unanswered questions. It remains to be seen if and how quantum physics can contribute to finding answers to questions such as: Is quantum physics "complete" (Bohr) or "incomplete" (Schrödinger, Einstein, de Broglie)? Or what exactly are the "dark" matter and "dark" energy that appear to constitute 96 percent of our universe? Other important questions include: What is the origin of life? What is the origin of consciousness? Or is science by definition incapable of answering the latter two questions? I personally believe that quantum theory cannot answer these fundamental questions about the origins of life and consciousness. But I do believe that the foundations of quantum physics, as currently accepted by the majority of quantum physicists, such as nonlocality, wave-particle complementarity, entanglement, and a nonlocal space with probability waves, are crucial to our understanding of the mind-brain relationship. Additionally, the quantum physics idea that the mind determines if and how we experience reality is, in my view, extremely important, but it does not yet enjoy the support of a majority of quantum physicists.

Some prospective and many retrospective studies of near-death experience have shown that various aspects of an NDE correspond with or are analogous to some of the basic principles from quantum theory, such as nonlocality, entanglement or interconnectedness, and instantaneous information exchange in a timeless and placeless dimension. I believe that while quantum physics cannot explain the origins of our consciousness, nonlocal consciousness does have a lot of common ground with widely accepted concepts from quantum theory. So in my opinion, quantum physics could also help us understand the transition from consciousness in nonlocal space to our physical brain. In the next chapter I will try to find answers to the many questions about nonlocal aspects of consciousness and the mind-body relationship, based on principles of quantum theory.

Chapter Twelve

The Brain and Consciousness

*The progress of science is inhibited by imbuing young minds
with an incorrect idea of the nature of reality, and the pernicious
philosophical idea that man is made of classically conceived
matter is not exposed as being incompatible with the empirical
facts.*
—HENRI STAPP

A great deal has been published in recent years about the brain and consciousness. The philosopher David Chalmers, who specializes in questions of consciousness, has written an excellent overview of all the different theories about the brain-consciousness relationship.[1] He starts by describing three materialist and reductionist models, A, B, and C. The first model (A), which he labels "monistic materialism," is based on the premise that everything is matter. Because the brain is made up of neurons undergoing physical and chemical processes, adherents of this theory believe that by explaining these processes in the brain they can also explain consciousness. The most commonly heard interpretation of this model posits that consciousness is merely an illusion. The second materialist model (B) is based on the premise that consciousness must be identical to processes in the brain because in a functioning brain there is a link between certain activities in the brain and certain experiences of consciousness. Adherents of the third model (C) admit that consciousness cannot be reduced to brain function just yet but believe that with scientific progress this will only be a matter of time.

Chalmers presents a detailed case against these three materialist approaches. His first counterargument is that while the structures and functions of the brain can be explained, this in itself is not enough to explain consciousness. His second counterargument involves zombies, imaginary creatures that are physically identical to human beings but lack a human consciousness. If zombies are a theoretical possibility, their brain function must be identical to the human brain, in which case the absence of consciousness in these fictional creatures means that consciousness is immaterial. His third antimaterialist argument invokes known facts about consciousness that cannot be explained on the basis of physiological brain activities. In theory, scientists could know everything there is to know about brain function and still not be familiar with all aspects of consciousness. Even with complete materialist knowledge, Chalmers suggests, we cannot know everything there is to know about consciousness.

Chalmers follows this with a description of three nonreductionist and immaterial models, D, E, and F, furnishing each model with commentaries from proponents and opponents. The fourth model (D) describes the "interactionist-dualism" developed by the Nobel Prize winner and neurophysiologist John Eccles and philosopher of science Karl Popper on the basis of the radical dualism of the sixteenth-century mathematician and philosopher René Descartes. Consciousness and the brain are radically different yet somehow highly interactive. According to Chalmers, this model is seen as incompatible with *classical* physics, whereas concepts from *quantum* physics, such as the collapse of probability waves caused by a deliberate observation or measurement, could actually support this model. As outlined in the previous chapter, not all quantum physicists accept the role of consciousness in quantum physics. Chalmers notes that philosophers usually reject interactionism with arguments from quantum physics, while physicists tend to reject the model on philosophical grounds (dualism).[2]

The fifth explanatory model (E) for the mind-brain relationship is called "epiphenomenalism" or "weak dualism," which posits that certain areas of brain function trigger certain experiences of consciousness, but that consciousness has no effect on brain or bodily function. This concept resembles the materialist vision. Consciousness is said to be the effect of chemical and electrical processes but cannot actually influence these processes. If so, the experience of pain could never cause a physical reaction, and people could never decide to take action. Neuroplasticity also argues against this model given that empirical studies have shown that the mind is capable of permanently changing the anatomy and function of the brain, as discussed earlier.

As his sixth and final model (F), Chalmers cites "phenomenalism" or "immaterial (or neutral) monism." This model is also known as "panpsychism" or "idealism." According to this model, all material, physical systems contain a form of subjective consciousness at an elementary or fundamental level, and all matter has phenomenal properties (that is, properties based on subjective observation). This model posits that consciousness has a primary presence in the universe and that all matter possesses subjective properties or consciousness. In this model consciousness is not only an intrinsic property of all matter, but physical reality is even formed by consciousness. Chalmers thus ascribes consciousness a distinctly causal role in the physical world. He mentions another theoretical possibility in which the intrinsic properties of the physical world *are* not phenomenal properties but rather *possess* phenomenal properties, which makes them protophenomenal properties. For this reason he prefers the term *panprotopsychism*.

Not everybody will agree with the latter model, in which all matter possesses subjective properties or consciousness; most people will favor a materialist model to explain the brain-consciousness relationship. But Chalmers appears to support panprotopsychism and believes that it merits further study.

Empirical research into human consciousness during a loss of all brain function (NDE) also seems to corroborate model F. The content of an NDE suggests that consciousness may be nonlocal. When brain function is impaired, NDErs experience an enhanced consciousness detached from the body followed by a conscious reentry into the body, rendering a materialist explanation of consciousness highly unlikely. When the brain functions normally, an NDE with an out-of-body experience can be triggered by mortal fear or stress while spontaneous out-of-body experiences are not uncommon at a young age. Neurophysiological studies have shown that brain activity cannot account for the content of thoughts and feelings whereas there is incontrovertible evidence for the mind's influence on the brain, given that the anatomical structure of the brain and its associated functions can change in response to experiences in the mind (neuroplasticity). On the basis of all these findings I personally have a strong preference for panprotopsychism, the nonmaterialist model of the fundamental or intrinsic relationship between consciousness and matter. In this chapter I will elaborate on this vision.

The Materialist Approach

A majority of contemporary Western scientists specializing in consciousness research, such as neuroscientists, psychologists, psychiatrists, and philosophers, espouse a materialist and reductionist explanation for consciousness (model A, B, or C). The well-known philosopher Daniel Dennett, for example, adheres to model A, monistic materialism.[3] As I wrote in the introduction to this book, Dennett, like many others, is of the opinion that consciousness is nothing but matter and that our subjective experience of consciousness as something purely personal and distinct from other people's consciousness is merely an illusion. According to Dennett, consciousness is produced by the matter that comprises our brain. This materialist hypothesis is supported by scientific patterns of thought and paradigms that he and many other scientists and philosophers deem absolutely unassailable and are therefore reluctant to challenge. Scientists often struggle to free themselves from prevailing paradigms. And such dogmatic convictions seem to beget prejudice. It prompted Albert Einstein to say, "It is harder to crack a prejudice than an atom."

If the materialist standpoint were correct, everything we experience in our consciousness would be nothing but the expression of a machine controlled by classical physics and chemistry. In Dennett's view, our behavior is the inevitable result of neuronal activity in our brains. The idea that all thoughts and feelings are no more than a consequence of brain activity obviously means that free will is an illusion. In response to this materialist position I cite neurophysiologist John C. Eccles:

> *I maintain that the human mystery is incredibly demeaned by scientific reductionism, with its claim in promissory materialism to account eventually for all of the spiritual world in terms of patterns of neuronal activity. This belief must be classed as a superstition. . . . We have to recognize that we are spiritual beings with souls existing in a spiritual world as well as material beings with bodies and brains existing in a material world.*[4]

The materialist approach, which is based on the premise that consciousness is a product or effect of brain function, is taught at many medical schools in the Western world. The approach is generally not made explicit and simply taken for granted without any kind of debate. Not surprisingly then, nearly all Western doctors believe that consciousness is the result of brain function. I myself am the product of an academic environment and was taught that there is a reductionist and materialist explanation for everything. I always blindly accepted this perspective, not just as a medical student and doctor, but also as the son of a neurologist. According to the materialist approach, the experience of consciousness during a spell of unconsciousness, cardiac arrest, coma, or a period of brain death is of course impossible. If, citing the results of empirical studies of NDE, somebody hints at the possibility of consciousness at a moment when all brain function has ceased, this is usually rejected as unscientific. Such a response from the scientific community is not new. Here is a quote from Dutch author and psychiatrist Frederik van Eeden from 1894:

> *The main concern is that the claims of a few scientists are fiercely disputed by most of their peers, not on the basis of research, but a priori; not even with rational arguments, but with emotional motives. Emotional motives with their aftermath of ridicule, contempt, and insinuation, based solely on an unphilosophical attachment to a closed system. It seems barely credible.*[5]

Sadly, the scientific community has changed little over the past hundred years.

Near-Death Experience, Consciousness, and the Brain

What have we read about the relationship between consciousness and the brain in the previous chapters?

- Many serious and trustworthy people have reported that, to their great surprise, they were able to experience an enhanced consciousness, independently of their body.
- On the basis of a few scientifically sound studies of NDE among cardiac arrest survivors, researchers have come to the conclusion that current scientific knowledge cannot offer an adequate explanation for the cause and content of a near-death experience.
- Some prospective, empirical studies provide conclusive evidence that it is possible to experience an enhanced and lucid consciousness during a cardiac arrest.
- We appear to have scientific proof that the cerebral cortex and brain stem are devoid of measurable activity during a cardiac arrest and that the clinical picture also reflects a loss of all brain function.
- Brain studies have shown that under normal circumstances a functioning, collaborative network of brain centers is a prerequisite for the experience of waking consciousness. This is absent during a cardiac arrest.
- Oxygen deficiency in itself provides no explanation because NDEs can be reported under circumstances that are not life-threatening, such as mortal fear or a serious depression.
- Our mind is capable of altering the anatomy and function of the brain (neuroplasticity).
- In many respects, both consciousness and brain function remain a huge mystery.

Some prospective and many retrospective studies of near-death experience have shown that various aspects of an NDE correspond with or are analogous to some of the basic principles from quantum mechanics, such as nonlocality, entanglement or interconnectedness, and instantaneous information exchange in a timeless and placeless dimension. Past, present, and future are everywhere at once (nonlocally). Earlier I outlined a few generally accepted principles of quantum physics because I am convinced that these are essential to our understanding of the brain-consciousness relationship. In my view, the quantum physics idea that consciousness determines if and how we experience our reality is particularly important for the further theoretical underpinning of this relationship. However, this radical interpretation of quantum physics is not yet commonly accepted.

The Continuity of Consciousness

Science challenges us to devise, test, and discuss new ideas that might explain the reported connection between one's own consciousness and that of other living persons or deceased relatives. The same applies to nonlocal phenomena such as the life review and preview, in which past, present, and future can be experienced simultaneously and which elude our conventional embodied conception of time and space. For me the biggest challenge is to find an explanation for the fact that an enhanced consciousness can be experienced independently of the body during the temporary loss of all cortical and brain-stem function.

A final theoretical possibility, one that has not been mentioned so far, is the theory of transcendence, or rather the continuity hypothesis. It views the NDE as an altered state of consciousness in which memories, self-identity, lucid thought, and emotions can be experienced independently of the unconscious body and in which (extrasensory) perception outside the body remains a possibility. The Dutch study and other empirical studies have shown that NDErs can experience an enhanced consciousness independently of their normal, embodied waking consciousness. I am reluctant to use the word *transcendence* because it suggests something transcending or rising above the body. Transcendence is usually associated with the supernatural or with the concept of transcendental meditation; hence my preference for the term *continuity hypothesis*. Besides, because consciousness is continuous and nonlocal, I do not believe that consciousness rises above the body. It is always present outside and often inside the body. This chapter will shed further light on the concept of the continuity of consciousness.

New Scientific Concepts

As mentioned, current medical and scientific knowledge cannot account for all aspects of the subjective experiences reported by cardiac arrest survivors with an NDE. However, I believe that science means asking questions with an open mind. And science is also about searching for possible explanations for new, initially perplexing problems instead of clinging to old facts and concepts. The problem lies less in accepting the content of new ideas than in rejecting old and familiar conceptions. The history of science tells us that sooner or later—and sometimes very soon—new empirical findings will force us to abandon our acquired knowledge. Quantum physicist David Bohm believed that "fixed ideas which underlie scientific hypotheses are not aids but obstructions to clarity, and that a methodology which combines discipline with openness would be better equipped to keep pace with the truth that is revealed as scientific investigation progresses and deepens."[6]

I do not expect new ideas to find immediate acceptance, but they should at least be examined more closely. As Frederik van Eeden put it back in 1890: "I am more convinced than ever that the a-priori rejection of and refusal to examine unfamiliar and unusual phenomena is the greatest foe of scientific progress."[7]

In the introduction I cited the well-known American philosopher of science Thomas Kuhn, who claimed that contrary to popular belief, most typical scientists are not objective and independent thinkers. This is a generalization, of course, but he believed that scientists tend to be rather "conservative individuals who accept what they have been taught and apply their knowledge to solving the problems that their theories dictate." Most scientists try to reconcile theory and fact within the accepted paradigm, which Kuhn describes as essentially a collection of "articles of faith shared by scientists." All research results that cannot be explained by current scientific theories are labeled "anomalies" because they threaten the existing paradigm and challenge the expectations raised by such paradigms. Needless to say, these findings are initially overlooked, ignored, rejected as aberrations, or even ridiculed.[8]

Near-death experiences are such anomalies because their cause and content cannot be accounted for with current medical and scientific ideas about the various aspects of human consciousness and the mind-brain relationship. I believe that anomalies can make a vital contribution to the critical reassessment and, where necessary, rejection of old concepts in favor of new and better theories that do explain these anomalies. In the past anomalies have always been the key to scientific paradigm shifts, just as the initially inexplicable behavior of heated metal prompted the development of quantum physics.

A New Perspective on Consciousness and the Brain

I developed the following views in response to the commonly reported experiences of an enhanced consciousness during a cardiac arrest. This enhanced consciousness features nonlocal aspects of interconnectedness, such as memories from earliest childhood up until the crisis that caused the NDE and sometimes even visions of the future. It offers the chance of communication with the thoughts and feelings of people who were involved in past events or with the consciousness of deceased friends and relatives. This experience of consciousness can be coupled with a sense of unconditional love and acceptance while people can also have contact with a form of ultimate and universal knowledge and wisdom.

In this new approach, complete and endless consciousness with retrievable memories has its origins in a nonlocal space in the form of indestructible and not directly observable wave functions. These wave functions, which store all aspects of consciousness in the form of information, are always present in and around the body (nonlocally). The brain and the body merely function as a relay station receiving part of the overall consciousness and part of our memories in our waking consciousness in the form of measurable and constantly changing electromagnetic fields. In this view, these electromagnetic fields of the brain are not the cause but rather the effect or consequence of endless consciousness.

According to this concept, our brain can be compared to a television set that receives information from electromagnetic fields and decodes it into sound and vision. Our brain can also be compared to a television camera, which converts sound and vision into electromagnetic waves, or encodes it. These electromagnetic waves contain the essence of all information for a TV program but are available to our senses only through a television camera and set. In this view, brain function can be seen as a transceiver; the brain does not produce but rather facilitates consciousness. And DMT or dimethyltryptamine, which is produced in the pineal gland, could play an important role in disturbing this process, as we saw earlier. Consciousness contains the seeds of all the information that is stored as wave functions in nonlocal space. It transmits information to the brain and via the brain receives information from the body and the senses. That consciousness affects both form and function of the brain and the body has been described in the discussion of neuroplasticity ("The mind can change the brain"). This view corresponds with what David Bohm has written: "Consciousness informs and in-forms."[9]

Nonlocal Consciousness in Nonlocal Space

My term for the wave functions in nonlocal space, which possess both personal and universal information, is *nonlocal consciousness*. According to this approach, consciousness has no material basis. I already outlined this vision in 2004 and 2006, but the terminology I now use is slightly different. In this model nonlocal space is more than a mathematical description; it is a metaphysical space in which consciousness can exert influence because nonlocal space possesses subjective properties of consciousness. In this view consciousness is nonlocal and functions as the origin or basis of everything, including the material world.[10]

As I explained in the previous chapter, observation is, by definition, impossible in nonlocal space because everything is based on probability fields (wave functions); time and distance play no role either. In other words, the physical aspect of our consciousness in the material world, which we experience as waking consciousness and which can be compared to the particle aspect of light, stems from the wave aspect of the "complete" and "endless" consciousness created by collapse of the wave function in nonlocal space. This particle aspect, the physical effect of our waking consciousness, is observable and demonstrable in the brain through EEG, MEG, fMRI, and PET-scan technology whereas consciousness in nonlocal space is not directly demonstrable on (quantum) theoretical grounds: everything that is visible emanates from the invisible. To better understand this immeasurable and invisible nonlocal consciousness, one could think of gravity. While gravity itself is not directly demonstrable or provable, its physical effects certainly are.

Theory of Complementarity

Aspects of consciousness must resonate in different parts of the brain in order to be experienced as waking consciousness. Resonance involves oscillation with the same frequency. As we saw earlier, neurological imaging techniques such as fMRI and PET show that different states of consciousness activate various brain centers. Depression, joy, fear, pain, meditation, decision making, cognitive functions, mindfulness, sleeping, or perception all activate different centers of the brain. But while the imaging techniques can establish the neurological correlations, they do so without accounting for the content of the different aspects of consciousness. They merely point out the place of resonance of the different aspects of consciousness. Our waking consciousness has a biological basis because our body functions as an interface. But there is no biological basis for endless or enhanced consciousness, which is rooted in a multidimensional nonlocal space. So enhanced consciousness is not limited to our brain because it is nonlocal, and under normal circumstances our brain only allows us to experience waking consciousness. Like the particle and wave aspects of light, this perspective on the relationship between nonlocal and waking consciousness constitutes a complementary theory rather than a dualistic one.[11] Conscious subjective experiences and their corresponding objective and visible brain activities, the physical effects of waking consciousness, which can be established with the help of an fMRI or

EEG, are two different manifestations of one and the same underlying reality; they cannot be reduced to one another.

A Comparison with Global Communication

How can this theoretical approach help us understand the possibility of experiencing consciousness, complete with memories and occasional glimpses of the future, during an NDE when the brain has stopped functioning? It may be helpful to compare this continuous, invisible, and instantaneous interaction between the mind and body to modern worldwide communication. Time and distance appear to play no role in the nonstop global exchange of information because of all the electromagnetic information waves for the ubiquitous cell phone, television, radio, and computer technologies that surround and penetrate us at all times. These information waves propagate at the speed of light. We are not aware of the hundreds of thousands of telephone calls, hundreds of television and radio broadcasts, and the billions of Internet connections around us day and night, passing through us and through the walls, including those of the room in which you are reading this book. We are not aware of these electromagnetic information waves until we switch on our cell phone, TV, radio, or laptop. But what we receive is not actually in the appliance. The voice we hear through the receiver is not inside the telephone. The images and music of the TV broadcast are not inside the TV set, and the concert is not inside the radio. We only see and hear the program when we switch on a TV set, and when we switch it off again we stop seeing and hearing it even though the broadcast continues. When we switch on another TV set, we receive the same program again. The connection appears to be nonlocal, and in actual fact all electromagnetic information is disseminated at the speed of light.

We can also compare endless and nonlocal consciousness with the Internet, which does not originate in the computer but is received and made visible to the senses by the computer. Akin to the brain's role in consciousness, a computer has a facilitating function: with the right access codes, a computer allows us to access more than a billion different Web sites. The computer does not produce the Internet any more than the brain produces consciousness. The computer allows us to add information to the Internet just like the brain is capable of adding information from our body and senses to our consciousness. Like a computer, the brain functions as a transceiver. As soon as you switch your computer off, you lose access to all those Web sites. Yet the sites themselves remain available worldwide, in Australia, Africa, Europe, Asia, as well as North and South America. And so it is with consciousness. It is always present. During life, we can experience aspects of consciousness in our body as our waking consciousness. Life allows us to make the transition from nonlocal space to our physical world, space-time. The oxygen deficiency brought on by the stopping of the heart temporarily suspends brain function, causing the electromagnetic fields of our neurons and other cells to disappear and the interface between consciousness and our physical body to be disrupted. This creates the conditions for experiencing the endless and enhanced consciousness outside the body (the wave aspect of consciousness) known as an NDE: the experience of a continuity of consciousness independent of the body. This concept of an enhanced and nonlocal consciousness can account for all elements of an NDE. When the body dies, consciousness can no longer have a particle aspect because all brain function is permanently lost. Endless (nonlocal) consciousness, however, will exist forever as wave functions in nonlocal space.

Scientific Proof of the Nonlocal Entanglement of Consciousness

Experiments appear to provide scientific proof of the nonlocal entanglement or connectedness of consciousness. Pairs of people were placed in two separate Faraday cages, which are rooms shielded from electromagnetic radiation to block out any electromagnetic information transfer. If these two people were strongly connected to each other, such as parent and child or two people who practiced many years of joint meditation, simultaneous changes in their EEG could be registered. In one isolated Faraday chamber, sensory stimulation through random computer-generated flashes of light caused visual evoked potentials in the EEG registration of the stimulated person, and this activity was instantaneously received by the other, unstimulated person in the second Faraday cage.[12] As a result, the registered patterns in the EEG of the unstimulated person changed the moment the lights flashed in the other Faraday cage. This transferred electrical activity, the so-called transferred potentials, the coherence or correlation between the two EEGs, can be ascribed only to nonlocal influence. Because the experiment design excluded electromagnetic information transfer, this correlation cannot be explained with classic scientific models.

Physicist Fred H. Thaheld has outlined a potential scientific basis for this macroscopic and biological nonlocal entanglement. The first studies of this nonlocal entanglement of consciousness were carried out at the University of Mexico by the neurophysiologist Jacobo Grinberg-Zylberbaum. The research initially met with criticism because of its poor design, but scientists at three different laboratories later replicated identical EEG correlations. Two fMRI studies found evidence of nonlocal entanglement between the brains of two isolated individuals while nonlocal influence has also been identified in subjects whose fMRI registration changed significantly when a healer at some distance focused attention on these subjects. And a recent study using laser stimulation and local EEG registration has shown nonlocal biological and macroscopic entanglement between two cultivated specimens of fully isolated human neural networks.[13]

All of these carefully executed and replicated empirical studies confirm the nonlocal properties of consciousness and point to a nonlocal entanglement in biological and macroscopic systems such as the brain. Neither the classical physics model of science nor contemporary biological theories can account for this correlation of biological systems. Whether or not quantum theory is capable of doing so is a question I will try to answer in the next few sections.

The Interface Between Nonlocal Consciousness and the Brain

The human brain is an extremely complex and in many respects mysterious organ with physiological, chemical, and biological properties. But because consciousness is not physiological, chemical, or biological, the brain is much harder to analyze. Mathematician and physicist Roger Penrose has argued that on theoretical grounds consciousness cannot be produced by the brain.[14] He has also demonstrated that computers will never be able to fully replicate or produce consciousness.

I believe that while quantum physics cannot explain the origins of our consciousness, nonlocal consciousness does have a lot of common ground with widely accepted concepts from quantum physics. In my opinion, quantum physics can help us understand the transition from consciousness in nonlocal space to embodied waking consciousness in our physical, visible world. The aforementioned nonlocal entanglement of consciousness in biological and macroscopic systems, which has been demonstrated by the instantaneous information transfer between the brains of two separated subjects resulting in identical EEG and fMRI patterns, can be considered as an initial contribution to explain the transition from aspects of nonlocal consciousness to the brain.

Theories Addressing the Transition from Nonlocal Consciousness to the Physical Brain

The following sections feature technical descriptions of three different interface or place-of-resonance models that may be able to explain the transition from nonlocal consciousness to the physical brain.

All three are complementary models in which subjective conscious experiences and their corresponding objective, physical brain activities are two fundamentally different manifestations of the same underlying nonlocal reality that cannot be reduced to one another. It is important to realize that, in keeping with current interpretations of quantum physics, all three models see the electromagnetic fields of the brain not as the cause but as the effect or consequence of consciousness. The three theories on the interface or place of resonance are, respectively, the link between nonlocal consciousness and (virtual) photons; the influence of nonlocal consciousness on the brain via the quantum Zeno effect; and the nonlocal information transfer from consciousness via quantum spin correlation. All three models require future elaboration by the scientific community. It is possible to skip these technical sections and finish this chapter with the conclusion.

How the exact transition ("place of resonance") from nonlocal space to the physical world comes about is not known. In fact, the process will probably never be fully knowable or verifiable. The potential role of DMT in establishing or disrupting this transition or interface will be equally difficult to prove.[15] This means that we will probably never have any experimental evidence for the actual transition or interface between consciousness and the brain. Quantum physics allows for several theoretical possibilities, which are all speculative to a certain degree—fundamentally difficult to prove or disprove. In the previous chapter I discussed a few quantum-mechanical concepts for the transition from consciousness to the brain. Of the following three theories, my personal preference goes to the third, although I believe that all three models are a genuine possibility and in some way complement one another. In the near future, these three models will have to be researched and developed in greater detail.

The Link Between Consciousness and (Virtual) Photons

Consciousness is nonlocal, that is, everywhere in nonlocal space and intrinsically entangled with all potential information stored in wave functions. Consciousness triggers collapse of the wave function and is thus the source of embodied waking consciousness. There is a theoretical possibility that consciousness in nonlocal space is linked to—or serves as the basis for—the electromagnetic field connected to the nervous system and the brain. In that case consciousness would be hitchhiking, as it were, on the electromagnetic field that probably originates, like consciousness, in nonlocal space. As we saw earlier, neurobiologist Herms Romijn developed this hypothesis, which is based on coherent systems.[16] In physics, coherence is used as a measure of the possible interference of waves. Two waves are coherent when they are capable of forming an interference pattern and storing information. On the basis of the principle of coherent systems created through self-organization, Romijn posits that the constantly changing electric and magnetic fields of the neuronal networks can be seen as a biological quantum-coherence phenomenon. It creates the conditions for complementary systems. This would make the electromagnetic fields, which Romijn believes may be based on "virtual" photons, meaning seemingly or possibly real photons, the carriers or the product of nonlocal consciousness.[17] There is general agreement about the (extremely short) existence of virtual particles, being in a constant process of creation and annihilation. By viewing the electromagnetic fields as a biological quantum phenomenon, Romijn avoids the criticism that the brain is a macroscopic, warm system that naturally causes decoherence (the leak of information) and therefore rules out quantum processes. Given the nonperiodic (unpredictable) nature of consciousness, he proposes a complementary theory with a hitchhiking consciousness capable of translating the physical periodicity (regular recurrence) of dead matter into the nonperiodic processes of living matter in nonlocal space.

The process shows a certain analogy with the double-slit experiment, in which as soon as the intensity of the light dwindles from a massive bombardment to the transmission of individual photons there is a shift from an electromagnetic wave to a probability wave. In the case of a single photon, no electromagnetic wave can be measured, but the (immeasurable) probability wave is used to statistically predict where the photon will hit the photographic plate. Perhaps we could apply this to the brain, with brain activity measured through the registration of the electromagnetic field (EEG). In the event of a cardiac arrest this electromagnetic activity will slow to individual pulses with extremely low electromagnetic energy so that these minimal energy packets (pulses) come to resemble individual photons. These minimal energy packets must then be described with the probability waves from quantum physics instead of the electromagnetic waves from classical physics. When the electromagnetic activity can no longer be measured, it does not mean that there are no more probability waves. In fact, this is where the probability wave becomes a useful descriptor. In theory, the complete loss of brain function is still accompanied by (immeasurable) probability waves. Any potential influence on the minimal processes occurring in the brain at that moment cannot be ruled out (the neurons' pilot-light state). NDE studies suggest that during the loss of all measurable brain function people continue to experience nonlocal consciousness; this nonlocal consciousness is theoretically based on probability waves.

The Influence of Consciousness via the Quantum Zeno Effect

We turn now to the position of quantum physicist Henry Stapp. As mentioned in the previous chapter, he calls quantum theories "psychophysical" laws, and with this he explains the causal effect of consciousness on neural processes. In this model consciousness can exert an influence only on neural processes and not on physical reality. If we make a series of successive observations in a quantum system, the effect of the observation appears to freeze, the constantly changing system appears to slow to a halt, and this is known as the quantum Zeno effect.[18] Stapp compares this quantum effect with the mind's effect on the brain: constant attention to an idea or concept produces a permanent change in brain function. He believes that the empirical fact of neuroplasticity could be an indicator of the brain's quantum function. The crux of Stapp's approach is that his quantum description of the brain is essentially holistic: it describes overall brain function rather than a model of the brain based on computer science. Using the principle of the quantum Zeno effect, Stapp, like Romijn, avoids the criticism that the brain is a macroscopic, warm system that rules out quantum processes. From the studies cited we know that quantum processes really do take place in the brain. But like Romijn's theory, Stapp's fails to explain why certain consciousness processes activate certain brain centers.

Information Transfer via Quantum Spin Correlation in the Brain

Nonlocal information transfer could happen via nuclear spin resonance, which is also known as quantum spin correlation or quantum spin coherence. To recap, a spin is a fundamental property of nature, akin to electric charge or mass. All the smallest particles, such as protons, neutrons, and electrons, have a spin that is either positive or negative and that is always a multiple of one-half. Individual, unpaired particles have a spin of ½. This unpaired nuclear spin plays an important role in MRI. A positive and a negative spin together make the visible manifestation of the spin disappear. In a magnetic field a particle with spin can absorb a photon or a wave function with a particular frequency (as noted before, information is encoded as differences in frequencies). The nuclei of all molecules in all of our body's cells, including neurons and neurons' cell membranes, consist of a number of positively charged protons and uncharged neutrons, collectively known as nucleons, usually with a neutralized spin. Like all molecules in our body, the same applies to the DNA in each cell.

In information transfer via quantum spin correlation (nuclear spin resonance), overall brain function should be likened to a quantum hologram and the brain to a parallel-linked quantum processing unit capable of decoding information that enters nonlocally. Nuclear spin resonance is the principle underlying magnetic resonance imaging (MRI), for which, as for a quantum hologram, there is evidence of nonlocal information exchange on the basis of the coherence of the phase relationship of wave functions.[19] For an MRI the hydrogen nuclei in the water and the fats of the neurons must align themselves with the magnetic field on the basis of quantum-mechanical laws, with the protons of the hydrogen nuclei aligning themselves parallel or antiparallel to the inducing magnetic field. The nuclear spin resonance of MRI is induced by a magnetic field that is approximately 100,000 times more powerful than the Earth's magnetic field, but this is not to say that a far less powerful field cannot have the same effect on a smaller scale. This is suggested by some recent studies, which will be quoted later. Alain Aspect's experiment, which provided definitive proof of nonlocal entanglement, also drew on magnetic influence and measurement of the spin direction of a "first" particle, instantaneously (faster than the speed of light) revealing the spin direction of the "second," the remote particle.

The reciprocal information transfer between nonlocal consciousness in nonlocal space and the brain (the interface) could also rest on quantum spin coherence, which is established under the influence of (virtual) photons. The sole explanation for this information transfer is the fact that the brain is capable of performing a Fourier transform. A Fourier transform is a mathematical operation used to convert a linear system (time) into wave functions or frequencies and vice versa. This analysis or transformation is frequently applied to harmony in music, the forecast of tides (high and low tide), digital signal processing, and systems analysis. In quantum mechanics the Fourier transform also applies to Schrödinger's probability waves. We have proof of the brain being capable of performing a Fourier transform because this Fourier transform also forms the basis for the effect of nuclear spin resonance during an MRI brain scan. But what is the effect of a Fourier transform in the brain? It may be compared to a musician with perfect pitch who hears a tone and knows which note (frequency) is being played. A reverse Fourier transform may be compared to a musician who, when reading sheet music, can tell from a note (frequency) which pitch it has or what the note sounds like. The musician can do this only because his or her brain is capable of performing this transformation. The same principle informs the imaging process of an MRI scan.

I have a strong preference for the third model of (reciprocal) information transfer between nonlocal consciousness and the brain via quantum spin coherence, with a possible role for (virtual) photons. My preference is informed by the many recent publications that appear to prove this form of information transfer. A recent article in *Nature* provided evidence of quantum coherence in photo synthesis in living systems, whereby solar energy (photons) was converted into chemical energy by wavelike energy transfer through quantum coherence of coherent electronic oscillations in both the donor and acceptor molecules.[20] This link between electronic and molecular oscillating states is a result of resonance triggered by the superposition of interference patterns of wave functions of energy (photons). In other words, what we are seeing here is nonlocal energy transfer in living systems on the basis of the quantum coherence of photons, which is akin to the process of nonlocal information transfer in the brain through (virtual?) photons.

Photons (waves or particles) are intrinsic quantum objects and natural long-distance carriers of information both in classical communication via radio, TV, mobile phones, and wireless Internet and in quantum communication. In *Science* and *Nature* recently the results were published of research carried out under laboratory conditions that proved information transfer between matter and light through electron spin and nuclear spin resonance on the basis of nonlocal quantum entanglement. This form of information transfer between light and matter is comparable to reciprocal information transfer between nonlocal consciousness and the brain via the model of nuclear spin correlation or nuclear spin coherence.[21] DNA may play an important role in this form of information transfer, and this will be discussed at length in the next chapter.

Recent studies among volunteers have found strong indications of a nonlocal therapeutic effect of certain drugs such as morphine, when the substance was placed between a pulsating magnetic source and the brain. The subjective therapeutic effect in these volunteers was identical to the effect of receiving this drug directly into the body. And the same subjective therapeutic effect was achieved when the subjects drank water that had been exposed to a pulsating magnetic source, to laser light, microwaves, or even to a flashlight, with the drug placed between the photon source and the water.[22] The authors ascribe this empirically proven positive effect to quantum entanglement between nuclear spin and/or electron spin in the water and nuclear spin and/or electron spin in the brain. The nonlocal information transfer is made possible by, respectively, the magnetic, laser, or flashlight source or the microwaves.

The search for truth is more precious than its possession.
—ALBERT EINSTEIN

In conclusion, these three possible models of an interface between nonlocal consciousness and the brain will have to be elaborated though future research because the questions continue to outnumber the answers. As mentioned, nonlocal and reciprocal information exchange between consciousness and the brain will never be fully knowable or verifiable, rendering any theories on the subject by definition difficult to prove or disprove. Perhaps a combination of data from empirical and theoretical scientific research could contribute to more definitive answers. As I said, I personally prefer the model of nuclear spin or quantum spin resonance.

On the strength of the prospective studies of near-death experience and recent data from neurophysiological research and concepts from quantum theory, I strongly believe that consciousness cannot be localized in any particular place—not even in the brain. It is nonlocal (that is, everywhere) in the form of probability waves. For this reason it cannot be demonstrated or measured in the physical world. There is, independent of the body, a continuity of consciousness that is intrinsically connected to or entangled in nonlocal space, though not identical to this space. The different aspects of consciousness are all nonlocal and accessible, although there is probably some kind of hierarchy. The essence or foundation of consciousness (protoconsciousness) probably lies in the vacuum or plenum of the universe, from where it has a nonlocal connection with consciousness in nonlocal space (panprotopsychism). In this view, the vacuum is the source both of the physical world and of consciousness. Perhaps nonlocal space could be called the absolute or true vacuum because the vacuum and nonlocal space are either identical or nonlocally connected and therefore indistinguishable. Everything is a form of space. Consciousness encompasses nonlocal space, and both my consciousness and yours encompass all space. In fact, each part of our consciousness encompasses all space because each part of infinite is infinite itself. This is exactly what the concept of nonlocality means.

Nonlocal consciousness is the source of our waking consciousness. The two are complementary aspects of consciousness. Under normal, everyday circumstances people experience waking consciousness (the "particle" aspect), which is just one small part of overall and endless nonlocal consciousness (the "wave function" aspect). During life people perceive with the senses while the brain functions as interface. Under abnormal circumstances, people can experience the endless aspect of nonlocal consciousness independent of the body, which is called the continuity of consciousness, and perceive directly via consciousness in space. This is known as a near-death experience. DMT from the pineal gland, of which the release seems to be triggered or stimulated by events in our consciousness, could play a key role in establishing and disrupting the interface between the brain and nonlocal consciousness. As mentioned, this interface may be based on quantum spin coherence (nuclear spin resonance).

Nonlocal consciousness is endless, just as each part of consciousness is endless. But our body is not endless. Every day, fifty billion cells are broken down and regenerated in our body. And yet we experience our body as continuous. Where does the continuity of the constantly changing body come from? How can we explain long-term memory if the molecular composition of the neurons' cell membrane is completely renewed every two weeks? And how can we have a long-term memory if the millions of synapses in the brain undergo a process of constant adaptation (neuroplasticity)? The next chapter will consider these questions in more detail.

Chapter Thirteen

The Continuity of the Changing Body

*The important thing in science is not so much to obtain new
facts as to discover new ways of thinking about them.*
—SIR WILLIAM LAWRENCE BRAGG

How is it possible to have constant interaction between nonlocal consciousness and the brain in an ever-changing body? How can there be continuity if the physical constitution of the interface is subject to constant change? The substance changes all the time, but the pattern remains the same. Our cells may be seen as our body's physical building blocks, yet every day some fifty billion cells in our body are broken down and regenerated. This is the equivalent of 500,000 cells per second. Every two weeks all of the molecules and atoms in our body's cells are replaced. How can we account for long-term memory if the molecular makeup of the cell membrane of neurons is completely renewed every two weeks and the millions of synapses in the brain undergo a process of constant adaptation? At a subnuclear level, the quarks and gluons that constitute the neutrons and protons of our body's cells are destroyed and regenerated within the time frame of a staggeringly small 10^{-23} seconds. So in fact throughout our lives our bodies are destroyed and reconstructed once every 10^{-23} seconds. And yet we experience our body as a continuity. How can we explain this experience of continuity of the ever-changing body?

DNA

It is possible that DNA (deoxyribonucleic acid) plays a key role in the exchange of constantly changing information between the body and nonlocal consciousness. Because DNA plays a central role in the formation and function of all cells, including the neurons, it also forms the basis of the constantly changing electromagnetic fields of these cells. DNA is person specific: each human being—in fact, each living being here on this earth—has his or her individual DNA, and DNA is the only permanent aspect of each bodily cell from cradle to grave. All approximately one hundred quintillion cells in our body, with their virtually infinite differentiation and specialization of functions, stem from the one unique DNA molecule created at conception. The first cell division follows about thirty-six hours after conception while all of the eight embryonic cells created after two more cell divisions could serve as a stem cell for an entire organism. The fourth division, however, creates sixteen cells with the differentiation potential for future functions, for instance, for a cardiac cell, neuron, skin cell, or liver cell. From here on, the cells' future is fixed.

But what is responsible for the change in DNA function after the fourth cell division? How and why does this differentiation occur after the fourth cell division? Because the molecular structure of DNA remains identical in all cells, its function has to change from that moment on. If you cut down a tree, each single cell from its bark has the potential to produce a new tree with branches, leaves, flowers, and fruit. In other words, the DNA of each cell contains the potential for a whole tree. The DNA structure is contained in this one cell, but the act of cutting down the tree completely changes its function. How do we account for this change in function? And how can we explain the transformation of a larval caterpillar into a pupa and eventually a butterfly when its DNA structure stays the same?

Nontechnical Synopsis

What follows is a brief synopsis for those readers who would rather not explore the likely role of DNA in the exchange of nonlocal information with the ever-changing body. This synopsis does not contain any references to scientific literature

It is clear that as the only person-specific and permanent component of each cell in our body, DNA must play an essential role as interface for the creation and continuity of all bodily functions as well as for the interaction between nonlocal consciousness and the body. Human DNA is a molecule with a double-helix structure, which consists of 23 pairs of chromosomes and contains approximately 30,000 genes made up of more than 3 billion base pairs. Approximately 5 percent of DNA encodes protein while the remaining 95 percent has an as-yet-unidentified function and is therefore called "junk" DNA. This portion of the DNA sequence is sometimes described as the biggest surprise of the human genome. The more complex an organism is and the more advanced its development, the higher its percentage of junk DNA.

The precise function of DNA is still far from clear. Some scientists believe that junk DNA could be used for identification purposes, comparable to a kind of bar code. In this view DNA receives instructions via nonlocal information exchange. This idea ties in with the theory of epigenetics. Epigenetics is the study of reversible changes in gene function brought about by factors outside of the DNA without changing its underlying structure.

Living cells emit coherent light in the form of biophotons, a pulsating stream of tens of thousands of photons per sec/cm^2, which is about a hundred million times weaker than daylight. This coherent, very low intensity light ("biological laser"), the source of which is DNA, is involved in intracellular communication that is responsible for directing biological functions such as cell growth, cell differentiation, and cell division. This is known as bioinformation. DNA appears to be the direct and indirect personal coordinator of all information needed for the optimum function of our body. Our individual DNA receives the necessary information for this from nonlocal space. According to this theory, a life's worth of immunological information is stored in nonlocal space too and directly accessible via the DNA in each cell.

The differentiation of cell functions during the embryonic phase cannot be explained by the genetic code recorded in DNA structure alone but also rests on nonlocal information. According to this hypothesis, DNA does not contain the hereditary material itself, but it has a nonlocal capacity for receiving hereditary and morphogenetic information. (*Morphogenetic* relates to the biological development and shape of the organism). All information about the genesis and morphogenesis of the body with its different cell systems and specialized functions is stored nonlocally, and this information is necessary for the continuity of all bodily functions because of the constant breakdown and regeneration of molecules and cells. Each cell must therefore be in contact with "morphogenetic consciousness" via the DNA in the cell nucleus.

Nonlocal consciousness contains all past experience (memories). In all probability individual DNA provides the different forms of our consciousness, such as waking consciousness and our individual subconscious, with different places of resonance, both in our brain and in other cell systems in the body. In fact, there is evidence that remote cells are capable of communicating and responding to the thoughts and feelings (consciousness) of the cells' "owner." We have proof of instantaneous and nonlocal communication between the consciousness of a subject and his isolated white blood cells in a growth medium at a considerable distance away. This shows that even at a distance each cell is capable of responding, via the individual DNA, to the cell owner's state of mind. This remote communication between discrete cells supports the possibility of nonlocal information exchange via DNA and also explains another phenomenon. The donor organ in a heart transplant contains the donor's person-specific DNA. The organ recipient can sometimes sense snippets of feelings and ideas that are later found to match the deceased donor's personality and consciousness. This is known as "transplanted memory." This phenomenon of "cellular memory" can also be explained by the view that DNA continues to be the place of resonance for the deceased donor's individual nonlocal consciousness, which the transplant patient can receive through the donor organ's DNA.

In short: DNA has an interface function in each cell via the process of nuclear spin resonance, as mentioned earlier. This makes it possible to explain the continuity of our ever-changing body because this

The double-helix structure of DNA.

process enables the exchange with all hereditary information from nonlocal space and with nonlocal consciousness. DNA also has a coordinating function in the reciprocal information exchange between cells, cell systems, and organs.

What Exactly Is DNA?

DNA is a molecule with a double-helix structure and is made up of nucleotides. (Nucleotides are molecules that make up the structural units of DNA and RNA and also play central roles in the metabolism and function of the cell: they serve as sources of energy, participate in cellular signaling, and are cofactors of enzymatic reactions.) Human DNA consists of 23 pairs of chromosomes and contains approximately 30,000 genes made up of over 3 billion base pairs.[1] So each gene has about 100,000 base pairs, which consist of combinations of adenine, guanine, thymine, and cytosine (A, G, T, and C). The DNA in each human cell is about three meters in length (imagine that!) and is coiled up within a cell nucleus of one-thousandth of a millimeter. DNA is found not only in each cell nucleus but also in the "power stations" of each cell, the mitochondria. Most research into DNA is done in vitro and looks at the structure of severed strands of DNA; very few laboratories carry out in-vivo research into the function of living DNA.

DNA function is different for each cell type because certain genes are active while others have been deactivated. The question why certain genes are switched on or off cannot be answered by research into DNA structure alone. The function of DNA, with its different combinations of A, G, T, and C, is sometimes compared to a piano keyboard, which has several octaves consisting of seven white and five black keys, tones and semitones, with each octave a replication of the previous one. Yet each composer and musician is capable of creating unique music from a limited palette of tones and rhythms. All well-known melodies and compositions are based on these few octaves. Just as a musical instrument (structure) does not equal the music being played (function), DNA structure is not the same as DNA function. It takes a musician to play the musical instrument, but how does DNA function? Everything that happens in the human body originates from the virtually unlimited possibilities arising from this one unique DNA molecule. How can a single human cell with a diameter of less than one-thousandth of a centimeter contain so many instructions in its DNA that it would take a thousand six-hundred-page books to record them all? The more we learn about our body, the more we realize what an immensely intelligent system it is and how little we actually know.

Approximately 5 percent of DNA encodes protein and is called exon. All proteins in our body are small DNA copies. To produce cell protein, the exon part of DNA constantly interacts with messenger RNA (mRNA), transfer RNA (tRNA), and ribosomal RNA (rRNA). Interference RNA (RNAi), which is thought to block and possibly also release individual genes and groups of genes, has received a great deal of attention recently. However, this chapter will not dwell on new insights into the possible functions of the various kinds of RNA.

The remaining 95 percent of our DNA has an as-yet-unknown function and is therefore called junk DNA, noncoding DNA, or intron.[2] This section of DNA is sometimes described as the greatest surprise of the human genome. The more complex an organism is and the more advanced its development, the higher its percentage of introns. The difference between the human genome (human DNA) and the DNA of a mouse is only 300 genes (1 percent), but the real difference is that mice have 10 percent fewer base pairs and far less junk DNA. Because the difference between human and chimpanzee DNA is 2 percent, one might argue that humans bear a closer resemblance to mice than to apes, but this is nonsense of course. Some single-cell animals (amoebas) have about thirty times as much DNA as humans but virtually no junk DNA whereas most plants have far more DNA (50,000 genes) than mammals (25,000 genes).[3] It is not the number of genes that determines the complexity of an organism but the number of collaborative patterns among the genes that determines an organism's DNA function. Of course the question remains which mechanism determines these collaborative patterns among the genes.

Epigenetics

At this point I should mention the recent development of epigenetics, which seeks to explain the functional differences between genes and especially why some genes are active where others are not. According to Nobel laureate and molecule biologist Joshua Lederberg, the functional differences are certainly not always a consequence of the hereditary structure of DNA but result from the environmental factors around the DNA. Epigenetics is the study of reversible changes in gene function that are brought about without altering the DNA sequence in the cell nucleus. This means that the function of the DNA changes but its structure does not. Research has shown that while identical twins have the same DNA, their epigenetic material can vary. This supports the view that DNA function is determined by information outside of the DNA itself and that nonlocal information exchange via resonance might play an essential role. Recent research supports the possibility of remote information exchange: pieces of intact DNA can recognize matching DNA strands at a distance without direct physical contact and without the presence of proteins. Somehow these pieces of DNA are able to identify one another at a distance, and the tiny bits of genetic material tend to congregate with similar DNA without help from any other molecules or chemical signals. From a current theoretical standpoint this feat should not be chemically possible because classical science has no explanation for this process of remote recognition. Recognition of matching DNA sequences forms the basis of recombination, a process that plays an important role in the evolution of an organism and in the repair of damaged DNA.[4]

Epigenetics also involves the study of all the processes involved in an organism's development. Its approach to these processes is fundamentally different from Darwin's evolutionism, in which, independent of environmental factors, only natural selection and random variation play a decisive role. Could the origins of life on earth, along with the origins of the many different species of plants, animals, and eventually humans, as well as the origins of an incredibly complex molecule such as DNA, be explained through coincidence alone? To quote nineteenth-century biologist and zoologist Edwin Grant Conklin, "The probability of life originating from accident is comparable to the probability of the unabridged dictionary resulting from an explosion in a printing shop."

In the next few sections I will come back to the basic principle of epigenetics that DNA function is determined by information outside of the DNA because it forms the foundation of my view on the role of DNA as interface between nonlocal consciousness and the ever-changing body.

The Possible Function of Junk DNA

What exactly is the function of DNA? The euphoria that greeted the discovery of the complete structure of human DNA has since subsided a little because the precise mechanism of DNA has become increasingly unclear. The function of junk DNA in particular has continued to perplex the scientific community. Scientists had originally hoped to discover the genes underlying certain illnesses and to use them to find cures. They have indeed identified some genes that play a part in certain malignant or hereditary diseases, but the precise mechanism remains a mystery.

Is the information stored *inside* a gene because of a particular sequence of A, G, T, and C, or does the gene merely provide access to information? Can DNA receive instructions via (nonlocal) information exchange? As mentioned, this is the theoretical premise of epigenetics. Computer expert Simon Berkovich posits that 95 percent of DNA with an as-yet-unidentified function, junk DNA or intron, might be used for identification purposes, comparable to the bar codes that contain all important information about products that are bought and scanned at the supermarket checkout. Needless to say, with three billion base pairs, DNA has a vast number of information-coding options. Berkovich sees DNA as the individual access code to individual information, shared information (of the species), as well as morphogenetic (formative) information stored in nonlocal space. Similarly, physicist Rosario N. Mantegna suggests that the noncoding regions of DNA (junk DNA) might play an important role in biological information exchange.[5]

The development of DNA in living organisms is a nonlocal affair and, because of this nonlocality, therefore a quantum process, which makes it—in keeping with the orthodox, "Copenhagen interpretation" of quantum physics—a fundamentally unknowable process. According to the quantum physicist Niels Bohr, life is unknowable, and quantum physics, given its incompleteness, can never provide a scientific explanation for life processes. In his "Copenhagen interpretation" of quantum physics, life is viewed as complementary to the physically verifiable, just like waves and particles are complementary aspects of all elementary processes. Not the development itself but only the result of the development of macromolecules such as DNA or protein can be demonstrated. Most quantum physicists believe that all molecular and submolecular processes are influenced and coordinated from nonlocal space and therefore immeasurable. Quantum physicist Erwin Schrödinger posited as early as 1944 that DNA might be a nonstatistical macromolecule; according to his theory, DNA could function as a "quantum antenna" for nonlocal communication. *Nonstatistical* (unpredictable, chaotic) processes are quantum mechanical processes, which originate in nonlocal space for living organisms, as opposed to *statistical* (predictable, orderly, regular) processes, which originate in the nonlocal space of dead matter. Schrödinger speculates that the DNA of living organisms has "receptor" or "resonance" potential for receiving and decoding information from nonlocal space. Physicist and computer scientist Peter Marcer too believes that a living organism can obtain its nonlocal and holographic information only through DNA.

All information is present and stored as wave functions in nonlocal space and is thus nonlocally accessible and essential for the complex organization and evolution of living organisms.[6]

Anesthesiologist and consciousness researcher Stuart Hameroff, by contrast, sees DNA as a potential quantum computer, in which the three billion base pairs function as qubits (quantum bits), with bits in a quantum superposition of *both* 1 and 0. In a normal computer this is *either* 1 *or* 0. Quantum superposition occurs during the coherent organization of photons and the self-organization of biological processes.[7]

Biophotons

In the 1920s biologist and medical scientist Alexander Gurwitsch discovered that ultraviolet radiation plays an essential role in cell division. Since 1972 researchers have been studying the release of photons in living organisms, known as biological luminescence. Much of this research has been carried out by biophysicist Fritz-Albert Popp and bioscientist Marco Bischof, who developed the concept of biophotons. The research has been comprehensively documented in Bischof's book *Biophotonen*.[8]

Living cells emit coherent light, a pulsating stream of tens of thousands of photons per sec/cm², which is about one hundred million times weaker than daylight but can still be registered with special cameras. The light spectrum of these biophotons has a frequency ranging from 200 to 800 nanometers (in the range of ultraviolet to infrared light). This extremely low intensity coherent light ("biological laser") appears to be involved in intracellular communication, which has prompted the hypothesis that electromagnetic and other coherent fields are responsible for directing biological functions such as cell growth, cell differentiation, and cell division. This is known as bioinformation. Because the release of biophotons has been demonstrated only in the DNA and cell nuclei of living, fractioned (divided into pieces) cells of plants and mammals, Popp and Bischof view the DNA molecule as the probable source for the creation of a coherent field of photons, for which reason it could function as interface between nonlocal space and the living organism. A review article by Dutch cell biologist Roel van Wijk details the current state of scientific research into the origins and effects of information exchange by biophotons. But even though no specifics of the biophoton theory have been refuted so far, and researchers have actually corroborated many aspects, the theory still lacks definitive proof.[9]

What is the source or the origin of the photons that originate in living systems? In the previous chapter I explained how, in current interpretations of quantum physics, the electromagnetic fields of the brain are considered to be the effect rather than the cause of consciousness. According to neurobiologist Herms Romijn, as we saw, the electromagnetic fields, possibly based on "virtual" photons, may be the carriers or the product of consciousness. The electromagnetic fields of the brain and of the heart are of course much more powerful than those of biophotons, but by viewing electromagnetic fields (photons) as a biological quantum phenomenon, we can view the latter, far weaker fields also as an effect of information from nonlocal space that cells receive through their DNA. Both the emission of single photons with informative properties and the nonlinear, hyperbolic reaction of biophotons in living cells to sunlight seem indicative of a biological quantum phenomenon. A nonlinear, hyperbolic reaction is specific to a quantum process. This hyperbolic reaction to sunlight is identical to the process of photosynthesis, which is now also seen as a biological quantum process.[10] Without sunlight, life on earth would be impossible.

DNA as a Source of Information for Each Cell

As the only person-specific and permanent cell component in our body, DNA plays an essential role as interface for the body's design, the continuity of all bodily functions, and the interaction between nonlocal consciousness (and memories) and the body. This approach retains the already-discussed interface model based on nuclear spin resonance (quantum spin correlation).

All matter, including all of our body's cells, molecules, and atoms, is made up of 99.999 percent emptiness or vacuum, and this vacuum is filled with energy and information that originates in nonlocal space, just as the universe around us is saturated with information and energy.

As a result, our DNA is always in contact with all possible forms of information from nonlocal space. DNA transmits information directly and nonlocally via coherent systems in remote molecules, cells, and organ systems. But information is also communicated indirectly via signal proteins, messenger proteins, and antibodies produced by DNA and transported in the bloodstream to the cell membrane. Information is also exchanged via the DNA-induced electromagnetic fields in neurons, which find their way to the body via the autonomous nervous system (the sympathetic and the parasympathetic systems) and the central nervous system. Finally, there is indirect information exchange from the brain (hypophysis, epiphysis or pineal gland, brain stem) via hormones and neuropeptides, which are also produced by DNA in certain cells. DNA appears to be the direct and indirect personal coordinator of all information required for the optimum function of our body. And for this our individual DNA receives the necessary information from nonlocal space.

This conclusion is reinforced by our immune system, which protects our individual organism from foreign invaders such as viruses, bacteria, and foreign cells that enter the body with blood transfusions and tissue and organ donations. Governed by DNA, the immune system must identify self and nonself antigens, coordinate the generation of necessary antibodies from an unprecedented number of options, and provide an immunological memory storage that remains directly accessible and up-to-date throughout life. People enjoy lifelong immunity from the infectious diseases they suffered in childhood.

Where in the body could this immunological memory be stored while the body's composition changes every second? And how could this ever-changing immunological information be stored in DNA? In my opinion, this immunological information could also be stored in nonlocal space and directly accessible to the individual DNA in each cell via nonlocal information exchange. This view seems to be corroborated by an article in *Nature* that provides evidence of resistance to certain antibiotics among strains of bacteria in animals living in the wild in extremely remote areas, thus ruling out any contact with the antibiotic in question.[11] We can only assume that the bacterial DNA received information via nonlocal space from strains that developed resistance as a result of irresponsible and careless use of antibiotics elsewhere in the world.

Nonlocal Information Transfer via DNA

In view of the these facts and arguments, DNA probably plays a central role in the reciprocal information transfer between nonlocal space and the field of resonant and coherent cell structures. I would compare DNA to the high-speed processor in my computer. This processor, which consists of a tiny oscillating quartz crystal and a couple of million transistors on a few dozen square millimeters, is constantly moving, switching, and copying data at a speed of four hundred million bits per second. The processor with its oscillating crystal does not contain any information itself, but it transmits information that enters encoded in the form of electromagnetic waves with certain frequencies. For living systems too the fact that all organisms possess rhythmic oscillation, vibration, or periodic movement is essential for (nonlocal) information exchange. Each living cell is composed of countless vibrating molecular structures. All molecules (including DNA) and atoms in a human cell are part of a living organism with an oscillating activity, each with its own characteristic frequency between 100 and 1,000 Gigahertz.

The oscillating activity of cells and the propagation of waves in and between cells are nonlinear processes (quantum processes), which have been demonstrated in calcium ions in genes, in proteins, and in cellular networks of neurons and cardiac muscle cells. There is evidence for transitions from simple to complex oscillating behavior, for instance during the development of action potentials (in the heart or neurons) or of chaotic patterns and complex feedback mechanisms in living systems. This means that in living organisms many processes, like the development of the electrical signals in the heart or brain, or feedback mechanisms in and between cells, are considered to be typical for quantum processes. Scientists have even found evidence of this oscillating property in activated white blood cells. The oscillating activity triggers resonance between molecules with identical frequencies, thus producing a cohesive whole of vibrating molecules. *Resonance*.

refers to vibration with the same frequency. The coherence of vibrating molecules creates powerful interference patterns that, in an ordered state, not only behave like a whole but actually become a whole. The parts lose their individual identity. Recent experiments with epithelial cells in the intestine appear to prove this. When a group of cells was damaged by toxic substances and significantly changed as a result, a group of identical detector cells, which are mechanically separated and which cannot communicate via chemical or electrical mechanisms, underwent an identical significant change despite a lack of exposure to any toxic substances. Their synchronous (coherent) response without direct contact suggests nonchemical and nonelectric remote communication between these cells. The authors of the research paper do not exclude biophotons as a possible source of this remote information exchange.[12]

According to developmental biologist Brian Goodwin, the differentiation of cell functions during the embryonic phase cannot be accounted for solely by the genetic code contained in DNA structure. Goodwin too proposes self-organizing fields in and among cells to explain the differentiation and coordination of cells and cell systems.[13] What else could explain the hundreds of thousands of well-coordinated chemical reactions in each cell every second, coupled with a reciprocal feedback mechanism in cells, organs, and entire organisms (systems biology)? Besides, chemical processes sometimes happen a million times faster in living organisms than under the most auspicious laboratory conditions. How can the living cell, governed by DNA, achieve this? It is highly likely that nonlocal information exchange between cells and cell systems plays a role. Another possible explanation is the fact that all cells are interconnected because they originate from a single source, namely the fertilized egg cell with the person-specific DNA. As we saw, Alain Aspect's experiment, which provided definitive proof of nonlocal information exchange, also made use of two particles originating from one and the same source.

The electrocardiogram (ECG), which shows the heart's electrical activity, can be registered on the skin of people's arms, legs, and chest because this electrical activity can be found in each of the body's cells. Presumably all of the body's quintillion cells are interconnected via the heart's rhythmically changing electromagnetic field. Also the registration of the brain's electrical activity, the EEG, reveals the heart's electrical activity. It is possible that self-organization enables the heart, with its intensive electromagnetic fields with coherent patterns, to create reception potential (an interface) for certain aspects of our consciousness and to transmit this information via its electromagnetic fields to the body as a whole. However, this supposition requires a great deal of additional research.

To understand correctly the evident effect of nonlocal information I would like to mention the effective functioning of groups of thousands and sometimes millions of living organisms, such as bees, wasps, ants, and termites. These colonies are examples of living and self-organizing systems composed of animals with different tasks but with a collective consciousness coordinated by the queen. If the queen is isolated from her colony but alive, everything continues as normal, but if the queen is killed away from her colony, chaos ensues and all work stops. The queen coordinates at a distance (nonlocally)—and probably on the basis of her DNA function—all of the colony's activities by creating and maintaining a collective consciousness.[14]

A flock of starlings reacting to a falcon attack.
Photo by Manuel Presti, Italy

Collective consciousness also appears to play a role in the extremely rapid coordination displayed by schools of fish or by migrating birds (see picture). A flock of birds recorded on film was found to have a reaction speed of 38 milliseconds, which is far too quick for normal communication between hundreds and thousands of birds that are often dozens of yards apart.[15] The birds behave no longer like separate beings but like a cohesive and coherent entity. Because the collective coordination of these species is innate, it makes sense to assume that their DNA plays an important part in this kind of remote communication.

DNA, Heredity, and Consciousness

According to the theory set out in this chapter, DNA does not contain hereditary material itself but is capable of receiving hereditary, morphogenetic (formative), and person-specific information from nonlocal consciousness. As mentioned earlier, morphogenetic fields involve a nonenergetic form of information transfer, which makes them comparable to probability fields in quantum physics. The reciprocal information transfer between the field and the living cell structures takes place via resonance with specific frequencies, even at the smallest subcellular level of electron spin resonance and nuclear magnetic resonance (quantum spin correlation). As mentioned before, the concept of morphogenetic fields was conceived independently by the biologists Weiss and Gurwitsch and brilliantly developed by the English biologist Rupert Sheldrake. These morphogenetic fields store information about the development and design of the body with all its different cell systems with specialized functions. This information is also essential for the continuity of all bodily functions, because of the constant breakdown and regeneration of molecules and cells.[16]

Each cell must therefore be in contact with the morphogenetic consciousness via the DNA in the cell nucleus. The process is best illustrated by the stem cell, which has been much written about in recent times. These stem cells are called omnipotent because they have the potential to grow into any cell type, such as a neuron, heart cell, or muscle cell, depending on the environment in which stem cells originate or are cultivated. This environment appears to contain the information (the morphogenetic consciousness) that the cells need to develop and specialize. When stem cells are cultivated among neurons, they will be able to grow only into neurons and never into heart or muscle cells. All cells in the body communicate with each other and with consciousness fields via resonance, electromagnetic fields, hormones, and messenger proteins. The cell collaboration that enables the continuity of function of the various cell systems must occur within nanoseconds given the aforementioned replacement of 500,000 cells per second. Because of the distance between the various cell systems in the body, the speed of this information exchange must approach the speed of light, which is obviously much faster than signal proteins alone could manage.

What is heredity other than the capacity to remember what was and what may be? Heredity is the preservation of physical possibilities and unconscious properties. But heredity is also the preservation of conscious properties, "being conscious," which results in our capacity for conscious recollection, conscious expression of will, and conscious decision making guided by our conscious (and unconscious) experiences, which in turn leads to knowledge and insight. Heredity is therefore another word for memory. Whether we call memory a property of consciousness or a biological principle is irrelevant, because these words indicate the different levels at which the same kind of information functions and manifests itself nonlocally. Both heredity and consciousness with memories form, via our DNA, the connecting link between past and future.

Consciousness is the nonlocal repository of all past experience. The reception of information from nonlocal space rests on our free will (intention), attention, and the state of our (waking) consciousness. In all probability, our person-specific DNA has given the different manifestations of our consciousness, such as waking consciousness and the individual subconscious, their different places of resonance, both in the brain and in other cell systems, each with an individual access code comparable to a private phone number. There is also a universal or collective human consciousness that links each individual human being with everything in existence or everything that ever was or will be, and this happens via the universal-human DNA with a shared access code. This is comparable to an international access code. This universal-human access code (DNA) differs from that of animals and plants. According to psychologist C. G. Jung, who himself experienced an NDE during a cardiac arrest in 1944, the shared human consciousness is similar to the collective unconscious.[17] Alongside the ego, as waking consciousness, Jung recognizes the self: a higher or broader aspect around the ego, which encompasses both the conscious and unconscious components of personality. Individuality, therefore, is distinct from the embodied ego. The unconscious individual component of consciousness is in contact with other aspects of the collective human unconscious, of which the individual unconscious essentially forms a part. Each part is linked nonlocally with the whole. The next chapter will elaborate on the various aspects of consciousness.

Communication with Remote Cells

There is evidence that remote cells are capable of communicating and responding to the thoughts and feelings of the owner of these cells. The inventor of the lie detector, Cleve Backster, not only tested his equipment on people but also conducted experiments with plants and white blood cells. A lie detector registers minute changes in skin conductivity via hypersensitive electrodes. Backster also registered changes in the conductivity of the leaf surface of plants responding to emotional or negative thoughts, even after those leaves had been ground and their remains distributed across the electrodes. Finally, he measured white blood cells, leukocytes, taken from an oral swab and kept alive in a culture medium. This technique is sometimes applied in dental surgery. As mentioned earlier in this chapter, activated white blood cells exhibit an oscillatory activity while the distribution of waves in and between cells is nonlinear, that is, a quantum process. Even when the white blood cells in Backster's experiments were transported twelve to twenty kilometers from their owner, or when the cells were placed in an environment shielded from electromagnetic radiation (in a Faraday cage), the cells were found to react when the subject was shown horrific or sexually arousing images (see figure). At the same time, he registered anomalies in the subject's skin conductivity and found proof

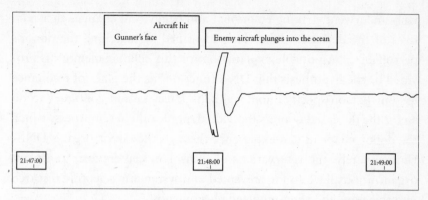

The response of the white blood cells of a Pearl Harbor veteran who, twelve kilometers from his cells, is shown a filmed close-up of the face of a gunner shooting a Japanese aircraft that subsequently crashes into the ocean.[18]

of instantaneous and nonlocal communication between the subject's consciousness and his white blood cells. In other words, each cell is capable of responding, via its person-specific DNA, to the state of mind of its owner. This communication between separate cells at a great distance ties in with the possibility of nonlocal information exchange via individual DNA.[19]

Transplanted Memory

The donor organ in a heart transplant contains the donor's person-specific DNA. Because this material is foreign to the recipient, it will in principle trigger a rejection response, which requires suppression with strong medication. But sometimes the DNA in the transplanted organ continues to function as a place of resonance or interface of the donor's consciousness, allowing the organ recipient to sense snippets of feelings and ideas that turn out to match the personality and consciousness of the deceased donor. This phenomenon was the subject of a 2003 Discovery Channel documentary called *Transplanting Memories*. It is an appropriate label for the experiences described by Claire Sylvia in her book *Change of Heart* and by physician Paul Pearsall in his book *The Heart's Code*. A recent article features ten well-documented cases of heart transplant patients, some of them still extremely young, who underwent striking emotional and behavioral changes that relatives of the deceased donor later identified as matching the donor's personality.[20] A possible explanation for this cellular memory is provided by the hypothesis that DNA functions as the place of resonance for our person-specific, nonlocal consciousness and continues to do so for the deceased donor's individual, nonlocal consciousness, which the donor recipient can experience through the donor organ's DNA. Unfortunately, the reservations of transplant centers and transplant organizations have so far prevented any systematic scientific research into this now-and-then-reported phenomenon.

In summary, DNA seems to be more than just a complex molecule that encodes proteins on the basis of hereditary information. Approximately 95 percent of DNA is called junk DNA and has an as-yet-unknown function. However, there is increasing evidence that this part of our DNA plays a significant role as interface between nonlocal consciousness and the body, thus enabling each individual cell to function as interface. I see heredity as a form of memory. Heredity is the preservation, via DNA, of physical possibilities and conscious and unconscious properties (individuality). In addition, DNA could play a coordinating role in the collaboration of cells, cell systems, organs, and the living organism as a whole through the reciprocal information exchange regulated via DNA (biocommunication).

The fact that the DNA in each cell has an interface function and facilitates the exchange with the hereditary information from nonlocal space and with nonlocal consciousness may explain the continuity of our ever-changing body. Additional analysis of living DNA is needed to elaborate and confirm this hypothesis.

Chapter Fourteen

Endless Consciousness

If we have learned one thing from the history of invention and discovery, it is that, in the long run—and often in the short one—the most daring prophecies seem laughably conservative.
—SIR ARTHUR C. CLARKE

This book is about consciousness: about endless consciousness; about experiences of an exceptionally lucid and enhanced consciousness that may follow the loss of all brain function; about the brain and consciousness; about quantum mechanics and consciousness; about nonlocal consciousness; about being conscious.

But what is consciousness? The term is extremely difficult to define, because it is often used to describe many different forms of consciousness. Someone in a state of deep and dreamless sleep usually experiences no consciousness, whereas somebody who is awake can be said to be conscious. This is known as waking consciousness. Waking consciousness requires an observing subject, a person who is aware. People can be aware of thoughts, feelings, emotions, and memories, all of which are known as objects of consciousness. A person's ability to perceive or experience an object in waking consciousness depends on selective intention and attention. People can be so lost in thought that they are barely aware of themselves or their surroundings. The fact that they are not aware of "being aware" does not mean that there is no consciousness at that moment. The awareness that we exist, the experience of a sense of subjectivity (our self-awareness), is another aspect of consciousness. But in my opinion consciousness encompasses more, and this is what I want to look at in this chapter.

As mentioned, consciousness is subjective and not scientifically verifiable. The ability to experience consciousness is different from the nature or intensity of any other subjective experience. Physicist and psychologist Peter Russell compares the ability to experience consciousness with the light of a film projector.[1] As the projector throws light onto a screen, the projected images change constantly. All of these projected images, such as perceptions, feelings, memories, dreams, thoughts, and emotions, form the content of consciousness. Without the projector's light there would be no images, which is why the light can be compared to our ability to experience consciousness. But the images do not constitute consciousness itself. When all the images are gone and only the projector's light remains, we are left with the pure source of consciousness. This pure consciousness without content is called *samadhi* by Indian philosophers and initiates and can be experienced after many years of meditation. It is said to bring enlightenment.

During an NDE, the encounter with "the light" is felt to be the most intense and most essential part of the experience. This encounter is always accompanied by an overwhelming sense of unconditional love and acceptance. At this point NDErs feel completely enveloped by the enlightening and all-encompassing consciousness.

Our Consciousness and Reality

Normally we derive our sense of self from our body, our appearance, our family, our history, our nationality, our roles (father, mother, child, grandfather, grandmother, lover), our jobs, our social and financial status, our possessions, other people's opinions of us, and so forth. We also derive our identity from our thoughts and feelings, our belief system, our values, our creative and intellectual capacities, our character, and what we call our personality. All this, and much more, determines our sense of who we are.[2]

But who are we in essence, independent of our thoughts, feelings, and ideas? What shapes our consciousness, and what determines how we experience this consciousness? Is consciousness a primary property of the universe, which has always had a nonlocal presence, or is consciousness the product or consequence of something else?

Why and where consciousness originated will probably remain a mystery forever because I think the answer to this question is unknowable. Consciousness is not visible, not tangible, not perceptible, not measurable, and not verifiable. And yet consciousness is what each living being draws on to give form and meaning to life. Without consciousness, there is no living body. Down to every last cell, life appears to be an expression of the will of (unconscious aspects of) consciousness. Without consciousness, there is no perception, no thought, no feeling, no knowledge, and no memory. Consciousness is all-encompassing; reality as we experience it exists only in our consciousness. In fact, it is influenced and ultimately determined by our consciousness.[3] Because the body restricts our ability to perceive "true reality," this true reality is by definition unknowable. We cannot perceive anything without consciousness. During waking consciousness we can perceive the world around us via our senses. This is not an objective, passive registration in our consciousness, but an active creation by our consciousness: it depends on our intention and attention. Everything exists only inside our consciousness, and everything outside it, such as "true objective reality," is unknowable. Electricity, for example, is not immediately perceptible, as the Dutch physician and author Frederik van Eeden wrote more than a century ago, but its physical manifestations are: light to the eye, pain to the skin, sound to the ear, taste to the tongue. Similarly, forces in themselves are not knowable; only their physical effects, such as the motion of objects or the experience of weight, are visible or measurable. Our consciousness is not visible either. Only the physical aspects of our waking consciousness can be seen and registered. Reality around us is unknowable, unlike the physical and visible aspects of reality that we can experience in our consciousness. This view was shared by the philosopher Immanuel Kant, who argued that we can only know reality as it appears to us and not reality as it is in itself. Perception is possible thanks to our power of reason (an aspect of consciousness) because our consciousness shapes reality as it appears to us. But true reality (*Das Ding an Sich* or "the thing-in-itself"), according to Kant, is unknowable.[4]

Personal and Shared Aspects of Consciousness

All aspects of our consciousness are interconnected. Our waking consciousness, which we experience on a daily basis, is an individual aspect of the all-encompassing consciousness, as is the awareness that we exist (self-awareness), that we perceive, and that we have memories. Our ability to reflect on what we think and our awareness that we dream (in the case of lucid dreams) are other aspects of our individual consciousness. In Latin the word *individual* literally means "indivisible." But there is also a universal or collective human consciousness that connects each individual with everything that is, was, or ever will be. As mentioned earlier, there is good reason to believe that this shared human consciousness could be compared with the concept of the collective unconscious as defined by psychiatrist and psychologist Carl G. Jung. He writes that the human collective unconscious manifests itself in images that recur in our dreams as well as in fairy tales and myths. These images are also the source of all religions. Individuality, which includes Jung's concept of the "self," is thus distinct from the embodied ego or waking consciousness because the "self" is the center of the total personality, including both conscious and unconscious aspects and the ego. The unconscious, individual component of consciousness communicates with other aspects of the collective human unconscious, of which it essentially forms a part. Everything is nonlocally connected to everything else. Under normal circumstances, the ability to receive information from nonlocal space (such as memories, knowledge, and associative thoughts) rests on our free will, our attention, and our (waking) consciousness. Some aspects of the personal unconscious, however, can only be experienced in dreams, meditation, regression therapy, or hypnosis. The collective unconscious is, in principle, unlimited; according to Jung, its top and bottom layers will never be accessible to our waking consciousness.[5]

Transpersonal Aspects of Consciousness

I do not know what I am, and I am no longer what I know.

—ANGELUS SILESIUS (JOHANNES SCHEFFLER)

Because, in Jung's view, a collective aspect of consciousness is not experienced as personal consciousness, it is also known as the transpersonal aspect of consciousness. *Transpersonal* refers to those aspects of consciousness that transcend the personal or the ego. In his recent book, psychologist Jorge Ferrer takes stock of the many different versions of transpersonal theory.[6] He offers a comprehensive and systematic overview of the transpersonal aspects of consciousness, which he describes as the spiritual dimension of humanity. In his view, transpersonal consciousness is a basic principle that enables human spiritual growth.

The term *transpersonal psychology* originated in the work of the clinical psychologist and founder of humanistic psychology Abraham H. Maslow and the work of psychologist Stanislav Grof, who is often described as the founder of transpersonal psychology. In the 1970s Grof stressed that certain aspects of consciousness, which may be experienced during the (therapeutic) use of LSD, cannot be accounted for with current scientific ideas about consciousness and the brain. In doing so, he also referred to other experiences of an altered state of consciousness, including the experience of consciousness during a medical crisis (now known as NDE) or during isolation. He wrote that these experiences must be located in a dimension without time and space because states of mind change as soon as inner attention shifts. To me these are aspects of nonlocal consciousness in nonlocal space. Consciousness can be experienced as the sole reality when the distinction between the ego and its surroundings dissolves; people experience themselves as simultaneously part and whole (experience of unity). Sometimes they even relive, with photographic precision, the first few days or weeks of their life or remember specific details of their birth. At times people report successive experiences of birth and death. Wondering how all information can be encoded in each sperm or egg cell,

Grof proposes a form of cellular consciousness, which is compatible with my view of DNA function as described in the previous chapter.[7]

Holistic philosopher and author Ken Wilber elaborates on the concept of transpersonal consciousness. In his book *No Boundary* Wilber describes endless consciousness, or unitary or eternal consciousness. The transpersonal self, or "the witness," is experienced as one with everything it witnesses. But I am not identical to my experiences. I have a body, but I am not my body. I have desires, but I am not my desires. I have emotions, but I am not my emotions. I have thoughts, but I am not my thoughts. All that remains is pure and undivided consciousness.[8] This consciousness transcends the individual and connects humans with a world beyond time and space. This unitary consciousness differs from all other levels of consciousness because it encompasses all levels or aspects of consciousness. To Wilber this boundless, endless consciousness is the spiritual aspect of humanity, and this consciousness encompasses the present ("now") at every moment in time. There is no past and no future here, no beginning and no end, no division between self and nonself. Everything is connected. There is a boundless unity. The "eternal now" or the "timeless moment" *is* consciousness. His ideas bear a striking resemblance to the concept of nonlocal consciousness.

According to Wilber it is extremely difficult to express the ineffable experience of this unitary consciousness because endless consciousness cannot be captured in words and thoughts.[9] This echoes the sentiments of people trying to describe their NDE. More than two thousand years ago, Plato wrote that our language is too limited to describe the essence of things: "Words conceal rather than reveal the inner natures of things."[10]

Psychologist Harald Walach also holds the view that spiritual and religious experiences can be conceptualized as the alignment of the individual with some kind of transpersonal or transcendent reality. According to him this interconnectedness can be explained by the effects of generalized entanglement, predicted by a theoretical model analogous to quantum theory. Generalized entanglement is a formal and scientific way of explaining spirituality as alignment of an individual with a whole, which according to this model inevitably leads to nonlocal correlations.[11]

Experiences of an Altered State of Consciousness

Over a century ago psychologists William James and Frederic W. H. Myers studied extraordinary experiences of consciousness. Both called into question the—still-dominant—view that consciousness is a product of the brain. In their recent and well-documented book *Irreducible Mind,* psychologists Edward and Emily Williams Kelly outline Myers's and James's major contribution to the study and description of special states of consciousness, such as mystical or religious experiences, unconscious aspects of humans ("subliminal consciousness"), hypnosis, trance, premonitions, and physical changes caused by (self) suggestion. The Kellys' book reintroduces us to long-lost knowledge that is now regaining currency thanks to recent scientific treatises on extraordinary aspects of human consciousness. More than a century ago James's and Myers's research into special forms of consciousness asked the same questions that I raise in this book. In a posthumously published book Myers even wrote about the human personality surviving physical death while James too was convinced that human consciousness lends us an aspect of immortality. In 1898 James wrote that the brain's role in the experience of consciousness is not a productive but is instead a permissive or transmissive role; that is, it admits or transmits information. In his view consciousness does not originate in this physical world but exists already in another, transcendental sphere; access to aspects of consciousness depends on the personal "threshold of consciousness," which for some people is lower than for others and which allows them to experience various aspects of enhanced consciousness. James draws on abnormal experiences of consciousness to support his theory. He speaks of the continuity of consciousness, and both his approach and his terminology are remarkably similar to my concept of nonlocal consciousness.[12]

Nonlocal Consciousness

As I wrote earlier, our consciousness is intrinsically connected with nonlocal space. In this view, nonlocal space or the vacuum is the source of both the physical world and consciousness while in turn nonlocal consciousness is the source of both waking consciousness and all other aspects of consciousness. Just as consciousness is infinite and endless, so is each part of consciousness. This is exactly what is meant by nonlocality.

Again I would like to stress that this notion of nonlocal consciousness does not explain the origins of consciousness. It is merely a description that helps us understand the various experiences of consciousness discussed in this chapter. As stated before, the origins of consciousness are and will probably remain a mystery forever.

An NDE Is an Aspect of Endless Consciousness

All-encompassing consciousness is known by many different names. I call it endless or nonlocal consciousness.[13] But it has also been called the higher or supreme consciousness, cosmic consciousness, divine consciousness, or the pure source or essence of our consciousness. Others prefer terms such as boundless consciousness, transpersonal consciousness, ultimate consciousness, unitary consciousness, or eternal consciousness, but all these terms refer to the same all-encompassing principle. Systems philosopher Ervin Laszlo calls this supreme form of consciousness the Akasha field because it stores all knowledge and an infinite amount of information.[14] All of these different names refer to one and the same thing: there is an ultimate source of consciousness in a multidimensional space, and virtually every part of this endless and nonlocal consciousness is accessible to humans.

Forms of enhanced consciousness, coupled with a sense of disembodiment, are experienced under a range of different circumstances. The experience of this special state of enhanced or endless consciousness is totally different from normal waking consciousness, which is merely one element of nonlocal consciousness. Each aspect of our consciousness can be described as an aspect of endless or nonlocal consciousness, and the primary distinction between them is the intensity of the experience.

This endless consciousness can be experienced under different circumstances. In life-threatening situations we speak of a near-death experience. But this term is far from ideal because enhanced consciousness can also be reported under circumstances that are not life-threatening. *Experience of insight* and *enlightenment experience* may be suitable terms as well as *religious* or *mystical experience*. But perhaps *experience of nonlocal or endless consciousness* is even better.

Surprisingly, these kinds of experiences seem to be quite common. In response to a separate question in a new poll by the Pew Research Center's Forum on Religion and Public Life in 2009, half of Americans (49 percent) say they have had "a religious or mystical experience—that is, a moment of religious or spiritual awakening." This represents a sharp increase over the past four decades because in 1962 only 22 percent of Americans reported having had such an experience.[15] Here is an overview of the circumstances (listed in chapter 6) under which an enhanced consciousness can be experienced:

1. *Near-death experiences* can occur during a medical crisis, for example during a cardiac arrest, coma, or near drowning, in which brain function is usually seriously impaired.

2. *Fear-death experiences* are brought on by sudden and intense mortal fear and are reported after a seemingly inevitable death caused by, for example, almost having a traffic or mountaineering accident.

3. *Identical experiences* are reported under circumstances that involve feelings of despair, loneliness, or isolation. For example, Charles Lindbergh reported one such experience during his first solo flight across the Atlantic Ocean, and they are also known to affect astronauts during space flight. Possible circumstances include isolation (shipwrecked people, astronauts), extreme dehydration or hypothermia, and depression or existential crisis.

4. *Experiences of an enhanced consciousness* are also reported in the absence of mortal fear, despair, or loneliness, such as during meditation or total relaxation (enlightenment or unitary experience), during regression therapy (sometimes done under hypnosis), and during the use of mind-expanding drugs such as LSD or DMT.[16]

Other Forms of Nonlocal Consciousness

This chapter will also look at some additional experiences of enhanced consciousness that can be explained with the theory of nonlocal consciousness. Some of these examples are less familiar because people tend not to talk about what they feel are inexplicable and often overwhelming, emotional, and extremely personal experiences. The social and scientific taboo on them is often even greater than, until recently, the one on near-death experiences. I hope that my detailed and careful descriptions, where possible in combination with reference to scientific literature and the concept of nonlocal consciousness, will contribute to breaking this taboo. Perhaps this will encourage more people to come forward and raise awareness of these intimate and often emotional experiences.

Deathbed Visions

During the end of life people sometimes report encounters with deceased loved ones (usually a partner or parent), the sight of a beautiful, unearthly environment and a bright light, or a sense of unconditional love. The content of such deathbed visions (or end-of-life experiences) can bear a striking resemblance to an NDE except that the dying person experiences waking consciousness during the deathbed vision and can discuss the enhanced and nonlocal consciousness at about the same time as they experience it. Those who have a near-death experience, by contrast, do not experience waking consciousness: they are unconscious and uncommunicative. Like an NDE, a deathbed vision takes away the fear of death. I know the story of a family doctor who made daily visits to a dying man who was terrified of his imminent demise. But one day he was surprised to find the dying patient sitting up in bed, beaming. In response to the doctor's inquiry about what had happened, the patient explained that he had had a visit from Jan, a good friend and neighbor who had died three years earlier and who had reassured him about death. Two days later the patient died peacefully.

Many accounts of deathbed visions are either not recognized as such or are interpreted as hallucinations, terminal confusion, or as the side effects of medication. But since nurses and volunteers, especially in hospices and other palliative care institutions, have become more open to this kind of experience, deathbed visions are now acknowledged more frequently. Deathbed visions are described by Elisabeth Kübler-Ross, the Swiss psychiatrist whose books broke the taboo on death and dying in the medical world; by Marie de Hennezel, a French psychologist with many years' experience of working with terminal patients; by the American doctor John Lerma; and by two American nurses, Maggie Callanan and Patricia Kelley, all with a great deal of practical experience in terminal and palliative care institutions. A recent book on end-of-life experiences has been published by neuroscientist and NDE researcher Peter Fenwick. Earlier I mentioned the empathetic NDE, in which those attending a deathbed are included in the dying patient's experience.[17]

There are few retrospective scientific studies of deathbed visions. Because most people die shortly after or even during their deathbed vision, most of the reports are from nurses, doctors, and family and not from patients themselves. The best-known study is the one by the psychologists Karlis Osis and Erlendur Haraldson. In a more recent scientific paper, psychologist Emily Williams Kelly writes that 41 percent of dying patients in her study reported a deathbed vision. The fact that only 4 percent of respondents to a general survey in Germany reported experiencing a deathbed vision during the terminal phase of a beloved family member suggests that these experiences often go unreported or unrecognized. Better, more systematic, and prospective research is needed to improve our understanding of the incidence, content, and effect of deathbed visions.[18]

Contact with the (Nonlocal) Consciousness of Deceased Persons

The concept of a nonlocal, endless consciousness also explains visions of remote, dying people around the moment of their death and of people who are already dead, usually during the first few days, weeks, or months after their funeral. These experiences or visions, being a form of nonlocal communication, are mostly reported by relatives of the deceased person, and are called perimortem and postmortem experiences, or after-death communication. The term *perimortem* refers to the moment of death and *postmortem* to the time after death. I shall return to this in the next few sections. The most comprehensive book on the subject, featuring anecdotal evidence, is by Bill and Judy Guggenheim.[19]

However, there are no scientific articles or monographs on this topic because the taboo on the possibility of contact with the (nonlocal) consciousness of deceased persons is such that these experiences are usually not mentioned, not even to next of kin. However, the European values survey of 1980–1983 carried out by researchers at Tilburg University in the Netherlands asked the question anonymously whether or not people had ever had a sense of contact with somebody who had died. In Europe 25 percent of people (125 million) answered yes, compared to 30 percent (100 million) in the United States. Although the positive reports in the Netherlands amounted to only 12 percent, this still means that nearly 2 million Dutch people have experienced some form of contact with (the consciousness of) deceased persons. Research into postmortem experiences among widows and widowers revealed a percentage of contact with (the consciousness of) the deceased partner of roughly 50 percent. Among parents who had lost a child, the chances of some form of contact with (the consciousness of) their deceased child was as high as 75 percent. As mentioned earlier, contact with deceased relatives is quite common during an NDE.[20]

Perimortem Experiences

When somebody's presence is sensed at a moment when this person's death is still unknown, we speak of a perimortem experience. This involves sensing the moment and sometimes even the manner of death of a loved one, often at a great distance, when none of this information could have been known. It often does not emerge until later that the perimortem experience took place at the exact same moment of death elsewhere. The experience typically involves an inner awareness that the person in question has died. Occasionally it is accompanied by a proper farewell, with direct, silent contact or with a word of thanks for a shared life. This type of experience can take place during the day, with witnesses or with shared experiences, but more often than not these events take place at night, in the form of what is usually called a lucid dream, which leaves an exceptionally deep impression and feels much more real than a normal dream. What follows is an account of a perimortem experience from the Guggenheims' book:

> Tom and I grew up together. We were next-door neighbors, but I hadn't seen him since he entered the priesthood. I lost complete contact with him and his family after I moved to Texas.
>
> One night over ten years later, I woke out of a sound sleep. I saw Tom standing at the bottom of my bed in a Navy uniform! When I saw his uniform, I couldn't believe it because I thought he was a Catholic priest! He said, "Good-bye, Melinda. I'm leaving now." And he disappeared.
>
> My husband woke up, and I told him what had happened. But he said I was just dreaming.
>
> Three days later, I got a letter from my mother stating that Tom had just been killed in action. I also found out he had been a chaplain in the Navy![21]

Postmortem Experiences

Another big taboo is talking about the sensation of contact with (the consciousness of) a deceased loved one during the weeks, months, or years after his or her death. This contact can consist of sensing a presence, feeling touched, or seeing the deceased person and is sometimes accompanied by communication, certain fragrances, or unexpected, "chance" incidents that are intuitively linked with the dead person. This postmortem experience is often so intensely emotional and coupled with an inner conviction that the experience or message derives from (the consciousness of) the deceased loved one that people initially struggle to accept that the experience actually took place. Any kind of systematic scientific research into postmortem experiences has so far proved impossible.

However, there is a great deal of anecdotal evidence; there are books from the United States as well as from the Netherlands and Belgium, and dozens of people have confided in me their experiences, often with great reticence and with the request for confidentiality. Tsjitske Waanders shares one of those postmortem experiences in her book: "I was aware, without seeing, hearing or touching, of my child next to me. . . . This awareness had nothing to do with reason. It was an intuitive awareness."[22]

People share these experiences only when they feel a great deal of trust and know that they will not be at the receiving end of prejudices or negative comments ("It's just wishful thinking"; "It's your grief talking"). When they do share them, they tend to do so warily for fear of being misunderstood or declared insane. Most encounters with the deceased loved one occur during sleep in a lucid dream and sometimes even involve communication. But this is no dream. Like its perimortem equivalent, this experience feels much more intense and real than any ordinary dream. Sleep seems to enhance the potential for receiving nonlocal consciousness because waking consciousness normally blocks the reception of the interface. A postmortem experience, that is, contact with the consciousness of a deceased loved one, is often extremely comforting, and the inner awareness that all is well with the deceased person aids the mourning process. Contact with (the consciousness of) the deceased person also tends to alter one's perception of death and brings an inner conviction that there is a form of life after physical death. In many cases fear of death disappears. This effect is comparable to the consequences of an NDE or a death-bed vision, which can also involve contact with the consciousness of deceased loved ones.

One of the best-known examples of a postmortem experience is found in Shakespeare's *Hamlet,* where Hamlet sees the ghost of his dead father in his mind's eye and is told by this ghost that his brother Claudius murdered him to marry his widow, Hamlet's mother, and become king of Denmark. The ghost also shares details of his death by poisoning. I was reminded of Hamlet when a woman from the United States, both of whose parents had been murdered, e-mailed me an unmistakable example of a double postmortem experience (although the first is more like an empathetic NDE):

I have not had a near-death experience, but I have had a "dream" that was as though I had one.

My mom was murdered in 1992. After suffering in intensive care for fifty days she passed away. I was very distraught and upset because my mother always had a fear of dying. This preyed on my mind. One night, about a month after her death, I went to bed and said a prayer and asked in that prayer to please let me know my mother was all right. That night I had a dream. I had the same experience that so many people have described in their near-death experience. I went through a darkness towards the light. And the light was an overwhelming feeling of Love. It was a blinding, brilliant light that is indescribable. It was so inviting, it felt so good, and I didn't want to leave. I was told my mother was fine, but I had to go back. I awoke knowing my mom was in a better place, and I had such a sense of peace. I know in my heart that there is something beyond this life.

I have never been an overly religious person. I am reluctant to tell many people this incident but was compelled to write to you after reading this article. Three years ago also my father was murdered. After three weeks the police came to a standstill and put out a call for help in the newspaper. I dreamed of my dad three nights in a row. Each night he told me to look in the files and gave me specific instructions. After the third night I called the head of the ATF who was working on our case. He must have thought I was a real crackpot. But I had looked in my dad's files. In my dream he had given me a date and a name. Sure enough, the name was there. The ATF agents contacted that person, and he gave the police the names of the people who were involved in my father's murder. I really can't give you any more details on this—we haven't gone to trial yet and there is a gag order issued. I don't claim to be psychic. I don't have any idea why these things have happened to me. But it makes me wonder and curious.

Another example, this one of a shared or collective postmortem experience, was sent to me by a young man from Hungary:

I would like to tell you about my mother, who had a massive brain hemorrhage three years ago. It left her paralyzed and unable to speak, and she died about six months later. Three days after her funeral, the following happened: I was sleeping when suddenly a strange cold feeling woke me. I was sleeping on my right side, and when I woke up I rolled onto my left, sensing something there. And to my big surprise, I saw my mother! She was dressed in white, surrounded by a radiant white light, and smiling; she was beautiful. She touched me on the shoulder and told me, though not with words: "Everything is all right now, and there is nothing for you to worry about." I wanted to respond, but somehow or other I fell back asleep.

I didn't wake up again until the following morning, and it wouldn't have been anything other than a strange dream had it not been for the following incident: From the moment I woke up, I kept thinking about what had happened that night, and in the afternoon I went to my father's room to talk to him about it. But to my big surprise, my father said, "You'll never guess what happened last night!" And my father told me, "In the middle of the night, a cold feeling woke me, and when I turned around and sat up I saw your mother at the other end of the bed. She was radiating light, she was dressed in white, she looked happy, and she touched me and said that I shouldn't be worried about her and that she would take good care of us." And after that my father had fallen asleep again!

Neither of us had ever experienced anything like it; neither of us had ever heard anything about contact with the dead. My father is a rational doctor and never mentioned it again. I never dreamed about my mother again. But I'm convinced that it wasn't a dream. I'm convinced of this, because my father and I had the same experience, during the same night, without realizing it.

Belief in a Form of Personal Afterlife

A recent, major European survey on social trends in European countries, conducted by researchers at Tilburg University, not only featured questions about work, leisure, politics, religion, and social issues, but also questioned respondents about belief in a personal afterlife. It emerged that, on average, between 48 and 59 percent of Europeans believe in a life after death, with people in the former East Germany forming the exception at a worldwide low of 15 percent. The same survey found that 22 percent of people in Western Europe believe in an existence before this life, that is, in reincarnation. The European study put the figure of belief in a personal afterlife in the Netherlands at 50 percent of the population (approximately 8 million people). In the United States between 72 percent and 74 percent of the population believes in life after death, and in the United Kingdom about 58 percent believe in an afterlife. In the Netherlands 21 percent (nearly 3.5 million people), in the United States 24 to 27 percent, and in the United Kingdom 29 percent believe in reincarnation. As we know, official Christian doctrine denies reincarnation.[23]

These figures are interesting given that two articles in *Nature* suggest that the majority of today's scientists in the Western world reject these ideas. They generally struggle to accept the possibility of a personal afterlife. But how else can we explain these experiences, which have been reported the world over by different people under different circumstances, other than by assuming that consciousness can be experienced independently of the body in a dimension where time and space play no role and everything is connected nonlocally? Mainstream scientific opinion is in sharp contrast with the findings of the social survey, which suggests that more than 250 million people in Europe believe in some form of continuity after physical death and more than 100 million Europeans believe in the possibility of reincarnation.[24]

The Continuity of Consciousness After Physical Death

The more we become accustomed to this idea of a consciousness which overflows the organ we call the brain, then the more natural and probable we find the hypothesis that the soul survives the body.

—HENRI BERGSON

The findings of NDE research suggest the possibility that (nonlocal) consciousness is present at all times and will therefore last forever.

The content of a near-death experience suggests a continuity of consciousness that can be experienced independently of the body. But as
I outlined above, identical experiences of an enhanced and nonlocal
consciousness (sometimes coupled with contact with the consciousness of deceased persons) are reported during episodes of mortal fear,
despair, isolation, meditation, on a deathbed, and during perimortem
and postmortem experiences.

The questions still outnumber the answers, but in view of all the
reported experiences of consciousness, we ought to seriously consider
the possibility that death, like birth, may be a mere passing from one
state of consciousness into another. Our vision of death is completely
transformed by the near-inevitable conclusion that after physical death
nonlocal consciousness can continue in another dimension in an invisible, immaterial world that encompasses past, present, and future.

Consciousness is not confined to the brain because consciousness is
nonlocal, and our brain facilitates rather than produces our experience
of consciousness. Whereas our waking consciousness has a biological
basis, because our body functions as an interface, there is no biological
basis for our endless and nonlocal consciousness, which has its roots in
nonlocal space. Waking consciousness is experienced via the body, but
endless consciousness does not reside in our brain.

A death notice I came across recently featured the following words:
"What you have perishes; what you are survives beyond time and
space." Death merely marks the end of our physical aspect. In other
words: we *have* a body, but we *are* consciousness. Free from our body,
we are still capable of having conscious experiences, we are still sen-
tient beings. Recently somebody with an NDE wrote to me: "I can live
without my body, but apparently my body cannot live without me."
Once the body has died, following a terminal phase that can last anywhere between hours and days, we are in contact with or, rather, we
become a conscious part of this endless and nonlocal consciousness.

Research into NDEs does not provide us irrefutable scientific evidence for this conclusion because NDErs never actually died. They
regained consciousness. But they all had a very close brush with death,
with temporary total loss of all brain function during a (reversible)
process of dying. Besides, scientific research has shown that conscious-

ness can indeed be experienced independently of the body without brain function. This conclusion alters our view of humankind and has consequences for current medical and ethical issues. Knowledge of near-death experiences can be of great practical significance to both health care practitioners and dying patients and their families. They would all benefit from knowing about the extraordinary experiences that may occur during a period of clinical death or coma, on a death-bed, or after death. I will write in more detail about these medical issues in the appendix.

Other Forms of Nonlocal Information Exchange

After an NDE, most people are affected by an enhanced intuitive sensitivity or nonlocal information exchange. They become more sensitive to parts of the nonlocal consciousness of which they were previously unaware. By the same token, they have an increased sensitivity to aspects of other people's consciousness. The concept of nonlocal consciousness appears to explain not just the NDE, but also enhanced intuitive sensitivity, remote viewing or nonlocal perception, geniality, and the mind's influence on matter (nonlocal perturbation). All of these phenomena can also be reported during and after an NDE, as we have seen.

Enhanced Intuitive Sensitivity

I take enhanced intuitive sensitivity to mean the experience of verifiably correct information about people or events, including future events, that could not have been obtained via the senses or from memory. It refers to (often involuntary) contact with aspects of one's own consciousness, with that of other people and deceased loved ones, and perhaps even with animals and nature. We are mostly unaware of this continuous relationship with our surroundings. Much to their surprise and confusion, NDErs often experience enhanced intuitive sensitivity, such as clairvoyance, clairsentience, clairaudience, or prognostic dreams about events that have yet to take place, as we have seen. The NDE seems to permanently enhance their reception capacity. We could compare this with a TV set, which normally only shows us channel 1, the broadcast of our own personal consciousness, but after an NDE also gives us channels 2, 3, and 4 with aspects of other people's consciousness. As I wrote earlier, there is scientific proof for this nonlocal "entanglement" of consciousness.[25]

An unconscious experience near death and dying is an ineffable premonition, an inner sense that somebody is about to die, in most cases a sudden death caused by an accident or a cardiac arrest. A sudden death is never entirely unforeseen. During the final few hours, days, or weeks before the sudden parting, people often ask general questions about death, express their funeral wishes "for no apparent reason," or make a will.[26] I know of a six-year-old girl who asked her mother what death really meant, whether you disappear completely or live on in some way. Two hours later she died when she crossed the street and was hit by a car. People with enhanced intuitive sensitivity often know or dream that somebody is about to die.

Enhanced intuitive sensitivity is common across Europe and the United States. Both the European values survey and the Gallup Poll in the United States have asked people (anonymously) whether they ever experienced telepathy (contact with somebody at a great distance) or clairvoyance (the sensation of knowing what is happening at a great distance). The question elicited positive responses from 46 percent (230 million people) in Europe and 60 percent (180 million people) in the United States. In May 2006 Reader's Digest published a survey with questions about aspects of heightened intuition among more than a thousand adults in the United Kingdom. The sense that others were secretly staring at them was reported by 68 percent of people while 62 percent of respondents reported that they knew who was calling before they answered the phone. Premonitions were reported by 52 percent while 26 percent sensed when a loved one was ill or in trouble, and 19 percent reported an encounter with a deceased relative.[27]

In summary, we can say that this heightened intuition does not just affect NDErs, but is also quite common among the general public. However, these kinds of experiences are generally hushed up, because today's society, and especially today's scientific community, cannot really accommodate it. The concept of nonlocal consciousness, however, provides an explanation for these experiences.

Remote Viewing (Nonlocal Perception)

Since 1972 physicist Hal Puthoff and his colleague Russell Targ at the Stanford Research Institute have been carrying out research into the accuracy and reliability of remote viewing. Also known as nonlocal perception or nonlocal awareness, this involves the reception of information from or about an object, such as a building, appliance, or location, that is not accessible or available to the senses. It means that people are capable of remote (nonlocal) perception of objects that are either randomly selected or merely indicated by coordinates on a map.[28]

Puthoff was rather skeptical at first, but the results obtained by intuitively gifted subjects were so impressive that during the Cold War the CIA expressed an interest in the method in the hope of obtaining intelligence about secret projects in the Soviet Union. The CIA financed this research into remote viewing for dozens of years and drew on it frequently for intelligence about the enemy. It was not until 1996 that President Clinton declassified some of the research results.

Research at the Princeton Engineering Anomalies Research (PEAR) institute has shown that "normal" people are equally capable of obtaining positive results. The chances that the results of remote viewing were mere coincidence were found to be statistically less than 1 in 1 billion. It later emerged that people were also capable of accurate perceptions of buildings or objects if they were isolated in a Faraday cage blocking all electromagnetic radiation or in submarines at a depth of up to 170 meters under sea, excluding even extremely low frequency transfer. The concept of nonlocal consciousness provides the only explanation for this. The reported nonlocal perceptions are conveyed as if made from a great height but are coupled with the power to zoom in on the minutest details. People could even describe the interior of buildings and the contents of locked filing cabinets or state secrets. Perceptions at extreme distances in space include that of a ring around the planet Jupiter that had never been seen from earth.

Its presence was confirmed only later by images taken by the passing NASA satellite *Pioneer 10*. One of the most recent remote viewing successes was the discovery of Saddam Hussein's hiding place thanks to detailed descriptions of the suspected location.[29]

Researcher Stephan Schwartz has drawn on nonlocal perception that goes back in time for the rediscovery of archaeological sites, including Cleopatra's palace, Mark Antony's palace in Alexandria, and the ruins of the lighthouse of Pharos near Alexandria.[30] Researchers at PEAR have also carried out research into nonlocal perception in the future. Researchers asked people to travel and photograph places like train stations and airports. These photographs bore a striking resemblance to previously documented images that had been remotely viewed by other subjects several hours before the photographed events took place.

This method of perception, with its 360-degree range of vision and simultaneous bird's-eye and detailed view, is reminiscent of people's descriptions of out-of-body experiences during an NDE. The fact of verifiable perceptions from both past and future is also consistent with what happens during an NDE, when clear images from the past or future may be seen during a life review or preview. Schwartz too puts forward a theory of nonlocal perception that corresponds exactly with the concept of nonlocal consciousness.[31]

Genius Insight

Where does sudden scientific insight come from? How do radically new insights enter consciousness? We know that Einstein's theory of relativity came to him in an epiphany. A sudden brain wave inspired Mendeleyev, the Russian chemist, to draw up the periodic table, listing the chemical elements according to atomic mass. What are the origins of inspiration in writers, painters, and other artists? How could someone like Mozart write his beautiful compositions at such a young age? Mozart said, as did Brahms, that he heard the music in his head and that all he had to do was transcribe it, which allowed him to put his brilliant music on paper in near-perfect notation within a very short space of time. Inspiration, creativity, and sudden scientific insight may be explained by (unconscious) contact with aspects of nonlocal consciousness. In the same way, an NDE can give people the feeling of being in contact with a tremendous source of wisdom, although they usually have no recollection of it later. Those who are interested in this subject should have a look at the chapter on the genius insight in Edward and Emily Williams Kelly's book.[32]

The Mind's Influence on Matter: Psychokinesis, Telekinesis, and Teleportation

We earlier looked at the power of the mind to exert an active influence on the anatomy and function of the brain (neuroplasticity). We know that the mind can have an effect on the body from the fact that fear or sexual arousal can trigger clear physical reactions. Then is it also possible that consciousness has an effect on "dead" matter by influencing processes at quantum level?

Psychokinesis or telekinesis involves transforming the visible, outward form of objects through mindfulness, which has been repeatedly demonstrated under controlled circumstances, even inside the U.S. Capitol Building in the presence of an official military delegation. Teleportation involves moving material objects, sometimes at a great distance, by unknown physical forces summoned by the mind. Psychokinesis and telekinesis are now also known as nonlocal perturbation because they involve the mind's direct influence on matter without the intervention of any known physical energy. The research that was done to prove that mindfulness can indeed alter the course of processes of chance has been covered at length in a recent book.[33]

It may look like science fiction, but the U.S. intelligence services and the U.S. Army are spending a lot of money and research resources on these phenomena, as shown by a U.S. Air Force report published under the title *Teleportation Physics Study*. The author of this report, a physicist, cites the many scientific studies that suggest that these phenomena are possible at macroscopic as well as at quantum level. He is referring here to the work of people like quantum physicist Anton Zeilinger, who has written that there is definitive proof of quantum teleportation.[34]

The U.S. Air Force report features an extensive overview of studies on nonlocal perception and teleportation and offers a possible scientific explanation for this kind of phenomenon on the basis of quantum physics, including the concept of zero-point fluctuations, otherwise known as the information stored in wave functions in nonlocal space. Most fascinating are the Chinese studies described here, which were controlled, blind, and double-blind and carried out in the Aerospace Medicine Engineering Institute in Beijing. The articles were all translated from Chinese into English by the Defense Intelligence Agency. The experiments were carried out with intuitively gifted children and young adults and proved teleportation across dozens of yards for small radios, photosensitive paper, mechanical watches, and insects; the tests were later repeated and the results of teleportation recorded on video and with extremely high speed photography.[35] The findings sound spectacular indeed: objects such as nuts, matches, pills, sponges, live insects, and so forth were moved through sealed envelopes and sealed glass bottles and from canisters with sealed caps without any of these storage containers breaking or ripping. Video recordings of the instantaneous transportation showed the object suddenly vanishing from its storage container and reappearing in another location. Sometimes it looked as if the object merged with the side of the container or storage box. Insects remained alive, but the amplitude and frequency of the radio signals of the small radio transmitters that were moved changed during the teleportation process; in fact, during teleportation they were momentarily imperceptible, as if the transmitter was temporarily in another dimension. The steady signal reappeared as soon as the radio transmitter had been moved. The subjects were always blindfolded and had no idea what object they were supposed to move with their mind. Objective spectators and military observers were in attendance at all times to rule out deception. To the author of the American military review article, it is clear that consciousness, in conjunction with theories from quantum physics, plays a key role in the explanation of this kind of phenomenon.

In his recent book *Entangled Minds,* Dean Radin also uses a quantum-mechanical model to explain many carefully researched and well-documented phenomena of nonlocally connected consciousness, both between people and between consciousness and matter. Likewise, quantum physicist Amit Goswami has written extensively on the causal influence of consciousness on matter. Here too the concept of nonlocal consciousness seems to be essential in explaining these kinds of extraordinary phenomena.[36]

He who has never changed his mind has never learnt anything.
—ADVERTISING CAMPAIGN FOR DUTCH NEWSPAPER *NRC Handelsblad*

In this chapter we looked at various aspects of nonlocal consciousness for which there is scientific and often well-founded evidence. Research into near-death experience helped me develop the concept of nonlocal and endless consciousness, which explains many and perhaps all aspects of the extraordinary experiences of consciousness discussed in this chapter. It is hard to avoid the conclusion that the essence of our endless consciousness predates our birth and our body and will survive death independently of our body in a nonlocal space where time and distance play no role. There is no beginning, and there will never be an end to our consciousness. In view of this, we should seriously consider the possibility that death, like birth, may be a mere passing from one state of consciousness into another. During life, the body functions like an interface and facilitates the reception of some aspects of our enhanced consciousness; junk DNA and DMT may play a role in this process.

Chapter Fifteen

Some Implications of NDE Studies

*I spoke out, because it would not do the dignity of science any
good if it could be accused of silence on certain issues, because it
does not know how to interpret them.*
—FREDERIK VAN EEDEN

Following the more theoretical aspects about consciousness and the
brain in the last chapters, with a more scientific view at the various
aspects of nonlocal consciousness, I want to reconsider some of the
implications of NDE and nonlocal consciousness in relation to ethical,
medical, and social issues in our predominantly materialist Western
society. If it really is true that the essence of our endless consciousness
predates our birth and our body and that it will survive death independently of our body in a nonlocal space where time and distance play no
role, there will be no beginning or end to our consciousness.

This idea has raised in me over the last twenty years many mind-
boggling questions. Therefore I would like to devote the next sections
to a few of the profound and sometimes emotional questions about
coma, dying, and death that I have asked myself and also received in
recent years since the publication of our Dutch study in *The Lancet*
in 2001 and of my Dutch book, *Eindeloos bewustzijn* (Endless Consciousness), in 2007 as well as during question-and-answer sessions after the many lectures on NDE that I have given in the last twenty years.

The widespread reports of an enhanced and lucid consciousness during a spell of unconsciousness brought on by the loss of brain function can inspire us not only to change our perception of the relationship between consciousness and brain function but also to change our ideas about life and death. Perhaps nowadays most people still believe that death is the end of everything because they never have heard or read little about NDE. That death is the end used to be my own belief. But after many years of critical research into the stories of NDErs, and after a careful exploration of current knowledge about brain function, consciousness, and some basic principles of quantum physics, my views have undergone a complete transformation. As a doctor and researcher, I found the most significant finding to be the conclusion of one NDEr: "Dead turned out to be not dead." I now see the continuity of our consciousness after the death of our physical body as a very real possibility.

Questions I receive are usually prompted by personal experiences or by the experiences and stories of friends and family, and they tend to be about the content or consequences of an NDE and the greatly enhanced intuitive sensitivity that is often reported afterward. Other frequent questions cover deathbed visions and contact with deceased relatives, what we called postmortem experiences. People also mention what happened when they tried to discuss their NDE with doctors, nurses, or family. They are greatly relieved when they can finally discuss their often profound experiences and receive answers to questions that they never dared to ask for fear of being rejected or derided. Likewise, fellow doctors and scientists frequently ask me, either in person or in writing, why a serious oxygen shortage cannot explain an NDE and how we can be so certain of the complete loss of all brain function during a cardiac arrest. Most of these questions have already been dealt with at length in previous chapters, so I will not revisit them here.

In this chapter I want to focus on the following three questions: (1) Why does the medical and scientific community harbor so much opposition to research into the cause and content of an NDE? (2) If there is such a thing as continuity of consciousness, does this endless and nonlocal consciousness return in another body? (3) An organ transplant cannot go ahead unless the donor has been declared brain-dead.

What is the difference between coma and brain death, and does brain death really equal death?

Scientific Opposition to NDE

I see the learned man in what you say!
What you don't touch, for you lies miles away;
What you don't grasp, is wholly lost to you;
What you don't reckon, you believe not true;
What you don't weigh, that has for you no weight;
What you don't coin, you're sure is counterfeit.
—GOETHE, *Faust II*

I am frequently asked why the scientific and medical community harbors so much opposition to research into the cause and content of an NDE and why the topic receives so little attention in scientific journals. Occasionally the inquirers themselves are extremely critical.

Research into NDE and other manifestations of nonlocal consciousness that cannot be accounted for by current Western science often evokes ridicule or rejection as well as emotional responses or prejudices. The response of doctors and other scientists toward near-death experiences is for the most part shaped by their ideas about life and death, and these in turn are shaped by their religious or spiritual background or lack thereof.

It is useful therefore to reflect on what we know about scientists and their religious beliefs. Surveys have shown that, in contrast to the general public, most scientists are not very interested in religion or the possibility of immortality. Whereas 91 percent of the U.S. population believes in God or a form of personal afterlife, a recent survey among more than a thousand American doctors put the number at 76 percent (belief in God) and 59 percent (belief in a personal afterlife). An article in *Nature,* however, suggests that only 39 percent of scientists describe themselves as religious and that 61 percent are either nonreligious or agnostic. A comparison with an identical survey conducted in 1914 shows that this percentage of nonbelievers has remained stable over more than eighty years. More remarkable still is that another article in *Nature* claims that only 7 percent of the most prominent and influential scientists, who are all members of the National Academy of Sciences in the United States, describe themselves as religious or spiritual. This means that 93 percent of today's leading scientists reject any form of religion or spirituality whereas in 1914 the percentage of religious leading scientists was still 28, while 35 percent believed in a personal afterlife.[1]

If our consciousness denies the possibility of a god or immortality, our (preconceived) ideas about life and death will be informed by that very same consciousness. It looks likely therefore that opinion on these matters will remain divided. The influence of well-known leading scientists must certainly not be underestimated because they often occupy important positions in national and international scientific advisory committees and they have seats on the boards of major scientific journals, where they determine whether or not an article will be published. But surveys suggest that their opinion is not representative of all scientists, especially not of doctors.

The personal opinions of scientists tend to determine whether new ideas gain currency in science and whether articles about new or groundbreaking insights are published. Scientific NDE studies highlight the limitations of our current medical and neurophysiological ideas about the various aspects of human consciousness and the relationship between consciousness and memories and the brain. The view that consciousness is the product of purely neurological processes in the brain remains the most widely held hypothesis. When new ideas do not fit the generally accepted (materialist) paradigm, many scientists perceive them as a threat. It is hardly surprising therefore that when empirical studies reveal new phenomena or facts that are inconsistent with the prevailing scientific paradigm, they are usually denied, suppressed, or even ridiculed. The history of science tells us a similar story. New ideas rarely received an enthusiastic response; they always evoked resistance.

A good example of the reluctance by materialistic scientists to accept new ideas are the following statements by the renowned Dutch neurobiologist Dick Swaab in a recent interview. He calls himself an atheist, stating that "we are our brain" and "everything, like wrong eating habits, sexuality or the conception of God, is a product of our brain." He is also convinced that "consciousness is a product of the brain." And he continues:

I don't believe in a soul. . . . The soul is just a big mistake. . . . I am a person with a huge machine in my skull, which at the same time has its own limitations and also mainly functions automatically. . . . It is said that if one person has a delusion it should be seen as a psychiatric problem. But when a lot of people have the same delusion it is called religion. I can not guarantee the absolute truth of my ideas, but as a scientist I am used to working with an uncertainty to be expressed in percentages. The chance that I am right is very high compared with the people who are convinced of their religion. I am sure they will be wrong.[2]

His comments on the Dutch edition of this book are in a similar vein:

All those aspects of an NDE can be imitated by lack of oxygen in the brain or by stimulation of certain areas in the brain that cause the same characteristics. It is just a disturbance of the information processing in the brain. But Van Lommel is not interested in this theory. His ideas are solely based on spirituality and theology. His science is in fact pseudoscience. He seems to be fundamentally converted because he keeps his own ideas despite clear opposition.

True science does not restrict itself to narrow materialist assumptions but is open to new and initially inexplicable findings and welcomes the challenge of finding explanatory theories. Science equals asking questions with an open mind and ought to be based on curiosity. Abnormal findings offer the chance of modifying existing scientific theories or replacing them with new insights that do offer an explanation; we have historical precedents for this. Contemporary science remains rooted in a picture of reality based solely on physically observable data. In my opinion contemporary science ought to review its implicit assumptions about the nature of reality because they have led to the neglect or denial of important and as-yet-unanswered questions about consciousness.

Reincarnation

Common sense is the collection of prejudices acquired by the age of eighteen.
—Albert Einstein

After a lecture on NDE and the possible continuity of consciousness after physical death, the question often arises whether consciousness can return in a new body (reincarnation).

Reincarnation is a generally accepted principle in Hinduism and Buddhism. Encompassing the law of karma, which is also known as the law of cause and effect, reincarnation involves a new life in which people are given the chance to make up for what they neglected to do, did not do well, or failed to learn in a previous life. Reincarnation, or the transmigration of the soul, has been common throughout history and in many cultures, among them the ancient Egyptians, Romans, Greeks (Plato), Celts, and the Cathars. The same was true for the Indian tribes in North America and the Tlingits in Alaska, the Aztecs, Mayas, and Incas of Central and South America, nations and tribes in Africa, the Australian Aboriginals, and the Druze in Lebanon. The anthroposophist Rudolf Steiner wrote, "Everything . . . is subject to the law of reincarnation."[3]

The psychiatrist Ian Stevenson carried out systematic research into reincarnation and published widely on the subject; his accessible book *Where Reincarnation and Biology Intersect* contains many well-documented cases of apparent reincarnation. When a child starts talking about a previous life, it usually happens spontaneously between the ages of two to four. Such children are very emotional when they talk about their previous life and remember details of a previous marriage, including the names of their spouse and offspring and often those of neighbors and family as well. In their recollections they do not distinguish between past and present; their memory appears to spring from a timeless source. In more than half of these cases, the children's stories feature a violent death, ranging from 29 percent of children in Alaska to 74 percent in Turkey. Some of these children exhibit unusual behavior, such as identity confusion or phobias related to the premature, unexpected, and violent end to their previous life.[4]

For his research Stevenson spoke not only with the child and next of kin but also with the family of the deceased and supposedly reincarnated person. This often resulted in surprisingly detailed similarities between the child's story and the family's information, even though the two sets of families had never met or spoken before. Stevenson paid special attention to unusual birthmarks or congenital problems in the children in places where the fatal wounds were sustained in the previous life, and he drew on autopsy reports and other documents for verification. He studied most of his cases in Burma, Alaska, Sri Lanka, Lebanon, India, and Turkey. While reluctant to claim that reincarnation definitely exists, he considers the evidence convincing. Reincarnation may never be scientifically proven, but studies such as Stevenson's make a reasonable case for it.[5]

The question remains, How can we explain young children's detailed memories from a previous life, in conjunction with physical characteristics that are consistent with the violent death in that life, without the idea of reincarnation? And how can it be possible for some Buddhist lamas, such as the Karmapa or the Dalai Lama, to write down, before their deaths, exactly where and when they will be reborn? And what about so-called annunciation dreams telling a mother-to-be, sometimes even before conception, the gender and character of her unborn child?[6]

Some NDEs involve the experience of what appear to be previous lives cut short by a violent death, as we saw in an earlier chapter, while regression therapy under hypnosis, which lifts the inhibiting function of our waking consciousness, also frequently elicits reports of experiences from a previous life.[7] These experiences have left many NDErs open to the idea of reincarnation.

Memories of a previous life can be explained with the idea of a nonlocal consciousness because it posits a nonlocal connection with the consciousness of somebody who has died, that is, somebody who was previously alive. The same concept explains encounters with deceased persons during an NDE and perimortem and postmortem experiences. While some see the possibility of contact as proof of reincarnation, I regard it as an indication of the nonlocal aspect of endless consciousness. This does not exclude the possibility of reincarnation of one or more aspects of a nonlocal consciousness, resulting in memories of a previous life. I am disinclined to believe that people return in a body with their entire personality, which is associated with their current sense of self or ego. However, I have come to believe that memories of a previous life are possible.

Organ Donation: What Is the Debate About?

After my NDE lectures, and not just in the Netherlands, I frequently receive pressing questions about brain death and organ transplantation: Does brain death really equal death? How can somebody be declared dead when nearly 100 percent of the body remains warm, intact, and seemingly functional? What is the difference between coma and brain death? Does brain death mark the start of the process of dying, which normally takes hours or days to complete, and how is this process affected by the removal of organs such as the heart and lungs? What is the state of consciousness during brain death and the process of dying? And what to make of the stories about changed thoughts and feelings following a heart transplant, so-called transplanted memory? I want to explore these questions in considerable detail because brain death and organ transplantation constitute a clear and practical example of some of the ethical and medical questions raised by contemporary health care.

Let me start by saying that I am not in principle opposed to organ transplantation, on the condition that the decision to donate an organ was made with due consideration and with loving intentions, with the full understanding that the operative removal of organs has an impact on and accelerates the process of dying. Inadequate and often one-sided information hampers any well-considered choice, especially when people are expected to fill out a donor registration form while lining up to renew their driver's license. When they fill out their form, few people recognize the true significance of the brain-death diagnosis and its practical consequences for the family of a potential donor who has been certified brain-dead.[8] Ignorance about the true meaning of organ donation only becomes apparent when a loved one is on a ventilator and the doctor informs you that your husband, wife, brother, sister, or child is actually dead and asks permission for organ donation. At this emotional moment, 70 percent of next of kin, uncertain about their brain-dead relative's views on the issue, refuse permission for donation. But even when patients are registered donors, the family often withholds permission. The sight of a relative attached to a ventilator and fluids, unconscious but with the body still warm, often triggers an intuitive uncertainty about whether brain death really equals death.

Pressing Questions

Most of the questions I receive are also formulated in the following two letters, which were written in response to a Dutch government debate on the introduction of a new system for organ donation under which all Dutch residents would automatically be registered as donors unless they express their objections and opt out:

Isn't it about time for us to shift the focus of this recurring debate to the question why people don't "simply" make this choice? Where is the scientific evidence that organ transplantation doesn't interfere with the process of dying? The word *process* indicates phases, and the organs are removed at the start of this process. Why is "the dead person" put under general anesthetic before the organs are removed? Physicians speak of patient reflexes, but is this what they really are? Why do criteria for brain death differ the world over? What is the relationship between the body, soul, and mind? Who studies the psychological consequences for a parent forced to say good-bye to a warm and breathing child? What about the many publications on near-death experience in situations when brain death was diagnosed and yet the patient regained consciousness and lived to tell the tale? There are many other questions that may well be too big for our limited human understanding and that get in the way of a simple yes or no. An aggressive government campaign will only fuel distrust. Everybody should be free to choose in their own time.

Another letter reads:

For fifteen years I carried an organ donor card, never thought twice about it. After hearing the following story from one of my students, I ripped it up.

While she was thought to be in a deep coma, without any apparent brain activity, her specialist and husband were having a conversation by her bedside. The specialist predicted that his patient would be a "vegetable" for the rest of her life and asked the husband to consider taking her off the equipment that was keeping her alive. The husband was still hopeful of a recovery, so she was kept on the ventilator. Several months later the woman woke up. It emerged that she had been able to hear throughout most of her coma and had overheard the conversation between her doctor and husband about passive euthanasia! She said how awful this had been and that while she had been trying to shout that she was still there, that she wanted to live, be with her husband and children, they were discussing her possible demise. In principle, I still support organ donation, but I haven't yet filled in my donor card. Why not? Because there are too many unanswered questions. The government campaign has focused primarily on the donor shortage and the need for people to register as donors.

Inadequate and One-Sided Information

One problem with current donor policies is that they are often publicized with inadequate and one-sided information. This is what it says on the Dutch donor registration form:

Waiting times for donor organs are long. People who could have been saved by an organ transplant die unnecessarily. That is why we need donor organs. You can help! Join the donor register. Let us know your wishes and help save lives.

Under current Dutch law, donor registration is possible from the age of twelve, and there are plans to give young people of fourteen and over a donor registration form when they collect their identity cards. The government campaign focuses only on recruiting more donors and not on providing objective information about the background to "postmortem" organ donation, that is, donation after somebody has been declared dead. The debate centers almost exclusively on the organ shortage, but this shortage will always exist, irrespective of the number of donors and the number of organ transplants.[9] In other words, the emphasis is the positive aspects of donorship (noble, heroic, life-saving) at the expense of the negative ones.

A few examples of the inadequate information:

1. Despite the many information campaigns, most people remain ignorant of the major difference between organ donation and tissue donation. They do not know that tissue donation is still possible when the dead body has been at the mortuary for twenty-four hours. Tissue donation covers skin, heart valves, bone and muscle tissue, and the eye's cornea. Postmortem organ donation involves the removal of organs from so-called brain-dead patients, whose warm bodies are still in a deep coma on a ventilator. These organs include the kidneys, liver, heart, lungs, pancreas, and parts of the intestine. Any contraindications for organ donation, such as malignant diseases, arteriosclerosis, chronic infections, HIV, and recent piercings, are rarely mentioned.

2. The receipt of a new organ does not guarantee a normal life expectancy. Organ recipients require intensive medical checkups for the rest of their lives because of the risk of rejection and the side effects of immunosuppressant medication, and they are at a greater risk of malignant diseases, high blood pressure, diabetes, and serious infections.

3. There is no mention of the potential physical and psychological side effects of an organ transplant. The "tyranny of the gift" puts immense psychological pressure on recipients: they often suppress any negative feelings for fear of being labeled ungrateful.

4. The information lacks any kind of reference to the necessary measures to keep organs in a fit state for donation even before the donor is diagnosed as brain-dead and permission for organ donation has been obtained.

When Is Somebody Brain-Dead?

With the technical expertise to transplant organs, such as the first kidney transplant in 1965 and the first heart transplant in 1967, came the problem of obtaining suitable organs. In 1968 an ad hoc committee at Harvard Medical School decided after lengthy deliberation that henceforth *Coma dépassé* (long-term, irreversible coma) would be called death, thus creating the possibility to obtain transplant organs from "dead" patients. There is a difference between brain death and coma. During a coma, also known as "apparent death," electrical activity can still be registered in the brain. The heart beats normally, blood pressure is regulated by medication, the patient receives artificial respiration, and there is usually some brain function left. A "permanent coma," by contrast, is a state in which a large part of the cerebral cortex and brain stem have sustained serious damage, and if there is any indication of irreparable damage the diagnosis will be "brain death." A coma brought on by loss of cerebral cortical activity, but with a functioning brain stem, can last for years in exceptional cases because brain-stem reflexes such as breathing and swallowing are still possible; this is known as a "vegetative state" or sometimes also as a "locked-in" syndrome.

According to international organ transplantation guidelines (the Netherlands, Europe, United States), any confusion between permanent coma and brain death is out of the question because in the latter case the brain is devoid of any measurable electrical activity and the brain stem is irreparably damaged.[10] But I am often asked how accurate such measurements are. And what to make of the many reports of consciousness during a period of coma with demonstrable loss of brain function? The Dutch organ transplantation guidelines, like other international guidelines (United States, United Kingdom), have this to say about brain death:

The brain needs a constant supply of oxygenated blood. If the brain . . . is deprived of oxygenated blood for more than a few minutes, it is irreparably damaged. The result is irreversible loss of brain function. Continued treatment is futile. The brain-dead patient has died.[11]

This raises a question about the process of dying, which can last hours or even days and which the guidelines ignore. What is more, when brain death has been diagnosed, 96 percent of the body is alive and is being kept alive whereas the patient is legally dead. When, according to the Dutch and other international guidelines, is somebody brain-dead?

Somebody is brain-dead in the case of irreparable and complete loss of brain, brainstem, and medulla oblongata function. The person in question can no longer breathe independently. All brain function has ceased and the body is incapable of regulating blood pressure and body temperature.[12]

About artificial respiration, the Dutch (and other international) guidelines say:

Donor organs need oxygenated blood. That is why the brain-dead donor receives artificial respiration until the organs are operatively removed. This can only be done at the hospital's Intensive Care Unit (ICU). Because of the artificial respiration, the brain-dead donor does not look dead. He appears to be asleep, has normal skin tone, and feels warm. The monitor shows his heartbeat. And yet he is dead![13]

This is why the law speaks of "the respiration of the dead body" even though every doctor and every layperson knows that it is impossible to give artificial respiration to a "real" dead body from the mortuary.

About brain activity, the protocol says:

An electro-encephalogram (EEG) is the registration of the electri-
cal activity of the cerebral cortex. A "flat" EEG, i.e., a straight line,
shows that any electrical activity in the cerebral cortex is absent.[14]

A flat EEG plays an important role in the diagnosis of brain death although a flat EEG during a cardiac arrest does not rule out any immeasurable activity in the brain. The EEG provides no information about the brain stem. "When an EEG is impossible (in the case of a crushed skull, for example) . . . doctors inject a contrast fluid into the blood vessels in the brain to see whether there is any blood circulation in the brain."

Following serious head trauma or a massive brain hemorrhage, swelling in the brain usually causes high tissue pressure, which makes it difficult to demonstrate blood circulation in the brain, whether with contrast fluid or with isotope analysis. These cases often prompt the erroneous conclusion that there is no more blood circulation in the brain. Cooling the brain (hypothermia therapy) reduces the cerebral swelling and offers the patient a chance of recovery. This type of treatment is still rare, however.

Does Brain Death Equal Death?

The Dutch and other international organ transplantation guidelines claim that brain death equals death. The predominant image of death holds that life and death can never overlap; a person is either dead or alive but never both at once. However, it is scientifically impossible to determine exactly when all life has left the body. The process of dying lasts between hours and days, takes a different course for everybody, and takes places at organ level down to cellular and subcellular level, with different processes and rates of disintegration for each system. Besides, when brain death is diagnosed, nearly 100 percent of the body is still alive. The criteria and diagnostic methods for brain death vary from country to country, and the more experts learn about diagnostic problems, the more uncertain they become. Most people are unaware that the removal of organs from "dead" patients usually requires general anesthesia because of the so-called Lazarus syndrome: violent reflexes by the certified dead organ donor. Would a corpse need general anesthesia? Patients who have been certified brain-dead also exhibit significant changes in blood pressure, vascular resistance, and heartbeat during the operative removal of organs, which is possible only if parts of the brain and the spinal cord reflexes remain intact.[15]

The fact that "dead" patients can bear living children also calls for reflection. Dozens of pregnant women in a deep coma, who were diagnosed as brain-dead, have been kept on a ventilator with intravenous medication and food for weeks or even months until their child was born, after which the machinery was switched off. Can a corpse give birth to a living child? The matter is further complicated by the assertion that a "clinically brain-dead pregnant woman is personally, but not biologically, dead"! There are medical guidelines and even a book on the subject of sustaining pregnancies in brain-dead women, called *Management of Post-mortem Pregnancy: Legal and Philosophical Aspects.*[16]

A Drop in the Number of Brain-Dead Donors

The number of potential brain-dead patients whose organs can be harvested drops every year thanks to advances in the treatment of serious brain trauma or cerebral hemorrhage. Doctors in the Netherlands are highly skilled in "donor prevention," that is, the prevention of brain death. In fact, the Netherlands' donor potential is the lowest in Europe, partly thanks to improved treatment techniques for brain injury, but above all because of strict legislation on compulsory safety belts in cars, compulsory helmets for moped and motorbike riders, the introduction and enforcement of speed limits, and the ban on the use of alcohol while driving. In other countries the number of (young) road traffic victims is much higher. If legislation were less strict, the Netherlands would have far more organ donors.

In 2002 there were in the Netherlands 1,131 potential donors, but due to contraindications for donation, such as age and chronic illness, permission for organ donation not always requested, refusal by family (in approximately 35 to 45 percent of all cases), and other circumstances, only 232 donors were registered, resulting in 202 actual donors. Over the past five years the number of actual donors has remained stable at around 200 a year. However, the number of patients waiting for organs, especially for kidneys, increases every year because of broader indications for hemodialysis.[17]

Also in the United States the number of people needing a transplant continues to rise faster than the number of donors. About 3,700 transplant candidates are added to the national waiting list each month, and each day about 77 people receive organ transplants. However, 18 people die each day waiting for transplants that can't take place because of the shortage of donated organs.[18] In a recent review article in the *New York Times* by pediatric cardiologist Darshak Sanghavi, titled "When Does Death Start?" some causes of the lack of organs for donation in the United States are explained by the less frequent occurrence of the diagnosis brain death, and he also describes the theoretical possibility of letting "dying" patients in coma, but not brain-dead, die in the operation room to be able to use their organs for donation from about five minutes after the heart has stopped beating (nonbeating heart donation). In the United Kingdom 3,513 organ transplants were carried out in 2008 thanks to the generosity of 1,854 donors, but despite the fact that more people than before were living donors, there are still many patients waiting for a transplant.[19]

Only new practices, such as nonbeating heart donation (the removal of organs when the heart has stopped beating), encouraging donors to donate a kidney or a (small part of the) liver during life, and perhaps also ongoing research into the possibility of xenotransplantation (genetic manipulation of pigs for the purpose of creating organs, especially the heart, fit for human transplantation) could theoretically reduce waiting lists for organs. But the problem of long waiting lists could also be eased by paying more attention to the nonphysical aspects of organ donation and by addressing questions about the meaning of illness and fear of death. Questions of meaning ought to be part and parcel of the debate on organ transplantation.[20] Not everything that is medically possible is automatically meaningful or necessary.

In conclusion, as a cardiologist I am frequently asked questions about ethical and medical aspects of the Dutch health care system, especially about the extent to which more knowledge about NDE can contribute to a more humane treatment of patients and their families. And this happens also in the United States and the United Kingdom. Fear of death and of the process of dying often informs decisions on ethical and medical issues on the part of doctors, patients, and families. Knowledge of near-death experience can be of great practical significance to health care practitioners and to dying patients and their families. In the appendix I will look into more detail at what it would mean for patients and doctors if more people were aware of the extraordinary experiences that may occur during a period of clinical death or coma, on a deathbed, or after death.

Chapter Sixteen

Epilogue

*The reasons cited against the acceptance of inexplicable
phenomena were always emotional reasons, such as the fear that
the beauty or efficiency of the scientific system would suffer.
These are entirely unscientific and irrational grounds resulting
from inadequate reflection.*
—FREDERIK VAN EEDEN

I sincerely hope that I have succeeded in explaining the concept of
nonlocal consciousness and its consequences for science, health care,
and our image of humankind. I am aware that this book has not always been easy and can be no more than a springboard for further
study and debate because we still lack definitive answers to many important questions about consciousness and its relationship with the
body.

Near-Death Experience and Science

The roles of DMT, junk DNA, and nuclear spin resonance, in particular, require further analysis. Given the various forms of nonlocal consciousness that cannot be explained by current Western science, this
book had to challenge a purely materialist scientific paradigm. This
paradigm is to blame for the scientific and social taboo on near-death
and other inexplicable experiences. I hope that by making a reasonable case for a nonlocal and therefore ubiquitous consciousness, this

book can help engender new views on consciousness. The acceptance
of new scientific ideas in general and ideas about endless consciousness
in particular requires an open mind unhindered by dogma. Sometimes
it only takes a single anomalous finding that defies explanation with
commonly accepted concepts and ideas to transform science.

Research into near-death experience has helped me to develop the
concept of nonlocal and endless consciousness, which can explain
many and perhaps all aspects of the extraordinary experiences of
consciousness discussed in this book. These include near-death experiences, fear-death experiences, identical experiences triggered by
despair, depression, isolation, meditation (religious and mystical experiences), and total relaxation (experiences of enlightenment or unity),
as well as experiences prompted by regression therapy and the use of
mind-expanding substances such as LSD or DMT. Deathbed visions,
perimortem and postmortem experiences, enhanced intuitive sensitivity or nonlocal information exchange, nonlocal perception, and the
influence of mind on matter (nonlocal perturbation) can also be seen
as manifestations of nonlocal consciousness.

It is hard to avoid the conclusion that our endless consciousness
preceded birth and will survive death independently of the body and
in a nonlocal space where time and place play no role. According to the
theory of nonlocal consciousness, there is no beginning and no end to
our consciousness.

Near-Death Experience and Health Care

NDEs are much more common than previously assumed, and the personal consequences of such an experience are far more profound than
doctors, nurses, and relatives ever imagined. All health care practitioners, dying patients, and their families ought to be aware of the
extraordinary experiences that may occur during a period of clinical
death or coma, on a deathbed, or after death. These experiences often
result in significant life changes, including the loss of the fear of death.
By accommodating rather than judging these experiences, patients and
their families are given a chance to integrate them into the rest of their
lives. Because I realize that knowledge of near-death experience can be

of great practical significance to health care practitioners and to dying patients and their families, I will write in the appendix in more detail about some ethical and medical consequences of NDE for the health care sector in the West.

Near-Death Experience and Our Image of Humankind

An NDE is both an existential crisis and an intense learning experience. People are transformed by the glimpse of a dimension where time and space play no role, where past and future can be viewed, where they feel complete and healed, and where infinite wisdom and unconditional love can be experienced. These transformations are primarily fueled by the insight that love and compassion for oneself, others, and nature are essential. After an NDE, people realize that everything and everybody are connected, that every thought has an impact on oneself and others, and that our consciousness survives physical death. The realization that everything is nonlocally connected changes both scientific theories and our image of mankind and the world.

In his book *Markings* Dag Hammarskjöld wrote, "Our ideas about death define how we live our life." When people think that death is the end of everything, they choose to invest in the ephemeral, the material, and the outward and are less inclined to respect the environment, the future world of our children and grandchildren. Our mind shapes our perception of the world. When we are in love, the world is beautiful; when we are depressed, the world is terrible; and when we are anxious (when we allow ourselves to be frightened by the press and politicians), our world becomes a fearful place. "The mind is its own place, and in itself, can make heaven of Hell," John Milton wrote in *Paradise Lost*.[1]

We have to change our mind to change our way of life and our world, as Ervin Laszlo explains in his book *You Can Change the World*.[2] All change in the world starts with ourselves. As an American woman with an NDE wrote to me in an e-mail: "When the power of love becomes stronger than our love for power, our world will change."

This calls for another consciousness. This is the insight acquired by being open to the meaning of an NDE and by really listening to people who want to share their NDE with us. These insights are age-old and timeless, but near-death experiences have brought them back within our reach, giving us the chance to learn to listen with our heart.

It often takes an NDE to get people to think about the possibility of experiencing consciousness independently of the body and to realize that consciousness has probably always been and always will be, that everything and everybody are connected, that all of our thoughts will exist forever and have an impact on both ourselves and our surroundings, and that death as such does not exist. An NDE provides an opportunity to reconsider our relationship with ourselves, others, and nature, but only if we continue to ask open questions and abandon preconceptions. I hope that this book has contributed to this process.

The Practical Significance of NDE in Health Care

It is nice to be important but it is more important to be nice.
—ACTOR WILL ROGERS

Knowledge of near-death experience can be of great practical significance to health care practitioners and to dying patients and their families. All parties ought to be aware of the extraordinary conscious experiences that may occur during a period of clinical death or coma, around the deathbed and the dying, or even after death. These experiences often result in major life changes, including the loss of the fear of death. By accommodating rather than judging these experiences, patients and their families are given a chance to integrate them into the rest of their lives.

In this appendix we look at the role health workers can play in the process of coming to terms with NDEs. Igor Corbeau's survey measured the psychopathology of 84 NDErs on the basis of an existing and widely used list of symptoms, the Symptom Checklist 90 (SCL–90).[1] These symptoms in people with an NDE include the interpersonal problems by the negative interaction with others and the intrapersonal problems like depression due to the difficult acceptance of the forced return to life. Compared to the general population (score: 118), the average NDEr scores higher (138), but still noticeably lower than the average client who is referred to a psychologist (178), and much lower than the average psychiatric patient (205). The survey found no significant differences between the types of problems suffered by people who had gone through an NDE as a child and those who had experienced one as an adult. Their only-slightly raised score on the Symptom Checklist 90 might suggest that NDErs have relatively few mental problems. The opposite is true. It emerges that 19 percent of NDErs have a higher psychopathological score (more severe psychological problems) than the average psychiatric patient.

More than half of the NDErs questioned as part of Corbeau's survey indicated that they felt or continued to feel the need for support. The services of a family doctor were most frequently called upon, followed by a psychologist, a paranormal therapist, a pastor, a psychiatrist, or a psychotherapist. The consultations with psychologists, psychiatrists, family doctors, and social workers were generally rated as bad to very bad. Approximately half of the NDErs who had received treatment from a family doctor or psychologist said that the intervention had had an adverse rather than a beneficial effect. Many felt that regular health care providers did not take them seriously and that the therapist's level of knowledge about the various aspects of NDE was highly inadequate. Spiritual therapies and transpersonal psychotherapists appeared to yield the best results, but people also appreciated peer support from fellow NDErs, for instance through Merkawah, the Dutch chapter of IANDS, the International Association of Near-Death Studies.[2] The survey identified no link between successful treatment and the passage of time since the NDE.

NDE in the Hospital

To provide NDErs with better help and support than they currently receive, all health workers ought to consider the possibility of an NDE after a patient suffers a life-threatening crisis. In the event of an NDE, the doctor or therapist should not reject the experience as a pathological or anomalous incident but should regard it as an existential crisis with all the disorientation and psychological problems such a crisis entails. They must try to help NDErs distinguish between the experience and its consequences and, where possible, involve the partner in this process. NDErs and their families should also be made aware of information in books and on the Internet and of the activities of Merkawah and/or IANDS.

The best support comes from health workers who not only are open to such an extraordinary experience but who are also familiar with the scientific literature on the subject. Unfortunately, not all of them are. Doctors and nurses at cardiac care units ought to be aware that patients who are recovering from a cardiac arrest and who express their disappointment at successful resuscitation may have had an NDE. In fact, all cardiac arrest survivors should be routinely asked whether they have any recollection of the period of their cardiac arrest, that is, their spell of unconsciousness. It is vital for patients to be given the opportunity to talk about their experience without being told it was a hallucination or merely a side effect of medication or lack of oxygen in the brain. A negative response from friends and family can also increase confusion and doubt. Tell patients that such an experience is known as a near-death experience, and reassure them by saying that they are quite common after a cardiac arrest.

Patients in a coma after a serious traffic accident or after a cerebral hemorrhage or stroke, and who are in an ambulance, emergency department, or intensive care unit, can be aware of themselves and their surroundings. People ought to listen carefully to patients regaining consciousness and ask whether they have any recollection of their period of coma. At the same time doctors and nurses should be careful when talking about a comatose patient and realize that patients may be able to see and hear everything during their coma. Their condition prevents them from signaling that they are aware of what is happening around them ("locked in" or perhaps even better, "locked out"), but after regaining consciousness some of these patients indicate that they experienced a lucid consciousness. Sometimes a coma involves observation from a position outside and above the body, from where patients can see relatives, nurses, and doctors and hear what is being said. Communicating with the comatose patient, playing their favorite music, explaining what is happening, and a positive approach can speed up their recovery. Even before it became widely known that the experience of consciousness (an NDE) was possible during a coma,

articles in medical journals suggested that communication with comatose patients by doctors, nurses, and relatives could have a positive
effect on the clinical outcome and that memories of the period of coma
after a traffic accident were reported by more than 50 percent of recovered patients, especially if they were comatose for ten days or more.[3]

Staff at surgery, neurology, and internal medicine wards ought to
consider the possibility of an NDE in patients recovering from a coma
caused by a traffic accident, a stroke, diabetes, or severe blood loss
during major complex surgery (for instance when an aorta ruptures).
Obstetricians should bear in mind that NDEs can occur after complicated deliveries, given that severe blood loss during or after the birth
of a baby is a frequent NDE trigger in young women. At pediatric
wards, meanwhile, young patients sometimes report NDEs after a
near-drowning, asphyxiation, or a coma caused by acute encephalitis
or a traffic accident.

Failed Suicide Attempts and Suicidal Patients

Research has shown that at least 20 percent of failed suicides report an
NDE, which can have a profound and sometimes positive impact on
the future lives of these often seriously depressed patients. The NDE
teaches them that taking their own life does not solve any of the problems they tried to flee; they simply take these problems with them,
and once they are without a body, a solution proves much harder to
find. Besides facing up to the fact that suicide offers no solution to
their problems, most people experience their NDE precipitated by a
failed suicide attempt as largely positive, given that it is accompanied
by a sense of acceptance, love, and understanding.[4] Staff at psychiatric wards and health visitors should routinely ask all failed suicides
whether they experienced an NDE.

Thanks to these newly obtained insights, renewed suicide attempts
after an NDE are extremely rare. Some studies have suggested that if
suicidal and severely depressed patients in psychiatric clinics are given
information about NDE and its consequences, it actually reduces the
risk of suicide.[5] The confrontation with an NDE can have a positive
therapeutic effect on suicidal patients.

Terminal and Palliative Care Units

Talking about post-NDE views on death and dying and about the experience of the continuity of consciousness ("dead turned out to be not dead") can be extremely reassuring for patients and nursing staff in hospices and at hospital wards where patients receive terminal and palliative care. Several researchers have shown that talking about the content and consequences of an NDE helps reduce fear of death among terminal patients. Videos and books about NDE can offer support during the process of dying and reduce the fear of death felt by both patients and next of kin. When young children are dying, parents, grandparents, and other relatives can derive a lot of comfort from books about deathbed visions or childhood NDEs.[6] After an NDE, many people volunteer for terminal care at home or in a hospice, where their newly acquired insight can be of great comfort to dying patients and their families.

Nurses, doctors, and the families of terminal patients ought to be open to deathbed visions or end-of-life experiences, which sometimes are comprised only of vague, intuitive images and an inner sense that the moment of transition is near. This is true for adults as well as for children. Shortly before the moment of death, a look of intense peace and calm will suddenly appear on the dying person's face, sometimes coupled with a far-off gaze, a blissful smile, or words such as "the light." And people who have been suffering from dementia for years can suddenly become extremely lucid in the final moments before death, recognize family members, and say good-bye to them. This is called "terminal lucidity."[7]

Terminal heart or lung patients, who are extremely short of breath during the final stages of their cardiac or respiratory disease, can also experience deathbed visions in which they talk about encountering a dead partner or seeing beautiful landscapes.[8] These dying patients ought to be encouraged to talk about their experiences without nurses or family expressing their doubts. Open questions from nurses are of great comfort to both patient and family because a deathbed vision can ease their fear of imminent death.

If all the family does is cry and complain that they cannot do without their dying relative, the moment of death is postponed and the suffering prolonged. The process of letting go becomes easier if one is prepared and able to thank the dying person for all the good times together, thus allowing him or her to depart in love and confidence. The process of dying also benefits from friends and family helping the dying person to let go of sorrow or guilt. There is still time to reestablish lost contact or address unresolved issues with children. In the words of Elisabeth Kübler-Ross: "One should diminish the emotional burden of harboring unfinished business."[9]

Experiences After Death

The death of a parent, partner, or child is followed by a dark period of sorrow and mourning. During the first few days, weeks, and months, there is a great chance of contact with (the consciousness of) the dead person, often in a lucid dream. As mentioned, after-death communication is quite common but is rarely reported for fear of disbelief and rejection. Talking about these experiences is taboo in our society, even though about 125 million Europeans, 100 million Americans, and nearly 2 million Dutch people have had a sense or an actual experience of contact with a dead relative. The chances of some form of contact with a dead partner or child can be as high as 50 to 75 percent.

Health workers and family members should not dismiss this experience of contact with a dead person as wishful thinking or a hallucination triggered by the overwhelming loss but should instead listen to the story and explain that these kinds of experiences are common. Encounters with dead relatives and loved ones are usually very comforting, and they have a positive impact on the mourning process. Health workers can also refer people to books about perimortem and postmortem experiences.[10]

Views on Death in the Health Care Sector

It goes without saying that more knowledge about NDE research and the possibility of a personal afterlife could have a significant impact on the practice of medicine. This knowledge informs ideas about the

treatment of patients in a coma or in the final stages of a terminal illness, as well as views on subjects such as euthanasia, assisted suicide,
and abortion. Our approach to these medical and ethical problems is
shaped in part by our belief in a possible continuity of consciousness
after physical death or, in contrast, by our conviction that death is the
end of everything. These views are usually based on religious beliefs
or the lack thereof.

As already mentioned, research in the United States has shown that
doctors' religious beliefs play a significant role in their practical approach to these kinds of issues. A recent survey among nearly 1,150 US
doctors revealed that 76 percent of them believe in God and 59 percent
believe in a personal afterlife. Twenty percent describe themselves as
spiritual rather than religious. Of the doctors surveyed, 55 percent admitted that their faith had an influence on their medical practice. The
percentage of family doctors (70 percent) who tried to live and work
according to their religious beliefs was higher than the percentage of
specialists (48 to 60 percent). This obviously results in differences in
medical practice regarding procedures such as euthanasia, assisted suicide, the issue of "do not resuscitate" forms, initiating or ending life-
prolonging treatment, birth control, and abortion.[11]

Euthanasia and Assisted Suicide

Needless to say, a doctor's principles are not the sole determining factor in ethical and medical questions, such as the issue of a "do not
resuscitate" form or a request for euthanasia. Patients' views on death
play an important role in their desire for a voluntary and early death.
A request for euthanasia or assisted suicide is probably based not just
on a desire for an end to suffering and for a more humane and digni-
fied death, but also on the perception of death as the end. The patient believes that after physical death nothing personal will be left.
As someone wrote to me, "When I'm dead, all thoughts, feelings and
memories are gone, and my suffering is over. I'll be free."

Would patients' desire for euthanasia or assisted suicide change if
they realized that consciousness survives death because it has no beginning and no end?

The public at large remains confused about the precise meaning of euthanasia. Political debates on the subject have certainly added to this confusion. Active euthanasia involves the administration of a lethal injection, usually in the final stages of a malignant disease or AIDS, precipitating an accelerated and unnatural death. In the Netherlands doctors are legally allowed to carry out euthanasia only if the patient's suffering is unbearable, without any prospect of improvement, and when a second doctor has been consulted. Doctors who refuse to perform active euthanasia on principle are obliged to refer their patient to a doctor who is prepared to go ahead with the procedure. Assisted suicide usually involves the prescription of lethal drugs.

The term *euthanasia* does not cover the practice of withholding medication to prevent further suffering. For example, the decision to withhold antibiotics from a ninety-year-old demented patient will, in combination with the patient's refusal to eat and drink, probably result in death due to an untreated infection. Some people call this passive euthanasia. But the deliberate decision not to treat a complication results in a natural death. Our current medical techniques are capable of keeping people alive until a very old age, sometimes with a very poor quality of life, and often expressly against the wishes of patients ("I'm tired of life") and families. How useful and desirable is it to keep treating malignant diseases at an old age, with disfiguring operations and intensive radiation treatment? To what extent are all these interventions informed by fear of death?

A request for euthanasia is often a request for support and assistance during the final stages of an intractable illness. If the doctor provides proper care and attention, as well as adequate medication for pain, shortness of breath, and fear, patients will often withdraw their request for euthanasia. As the work of Elisabeth Kübler-Ross in the United States and Marie de Hennezel in France demonstrates, this can be achieved through good twenty-four-hour home care or by placing terminal patients in a palliative care institution (a hospice).[12]

The strict national guidelines for active euthanasia appear to be closely observed in the Netherlands. Euthanasia is a difficult and emotional intervention for patient, family, and practicing doctor alike; no doctor embarks on it lightly. In 2005 doctors in the Netherlands performed active euthanasia on 2,325 patients, which represented a 33 percent drop from 2001. About 96 percent of these patients who underwent active euthanasia were in the end stage of a malignant disease or HIV with a mean life expectancy of one to maximal two weeks and with untreatable pain or extreme shortness of breath. Euthanasia is not allowed in patients who are not able to express their wish for euthanasia, as in the end stage of dementia. Palliative sedation, the prescription of powerful medication for pain or shortness of breath, usually slightly reduces the terminal phase of dying patients. In 2005, 9,700 patients received palliative sedation in the Netherlands while 66,000 patients were denied life-prolonging treatment or given (in the case of mostly very old patients) intensive pain medication.[13]

The Netherlands appears to be the only country in the world with so much openness about euthanasia and such careful monitoring of the various treatments available to terminal patients. There is certainly no sign of the downward spiral so feared abroad. In the many countries where abortion, assisted suicide, and euthanasia are officially banned, these procedures still take place but illegally and thus without proper medical guidelines, resulting in medical and psychological risks for patient, family, and doctor.

When asked about my own opinion on active euthanasia, my usual response is that I judge neither the patients or families who request euthanasia nor the performing doctor. Each case is unique. But one must not forget that families sometimes put pressure on doctors to perform euthanasia because the adult children "cannot bear it any longer" even though the elderly patient no longer exhibits any signs of conscious or visible suffering. Euthanasia also cuts short the time for patients and their families to settle matters or talk things through. Every procedure that curtails the natural life span reduces the available time for a peaceful, meaningful, and loving good-bye.

I have been told of people who used to be opposed to euthanasia on religious or political grounds and who changed their minds as soon as someone in their own family was affected by a terminal illness and suffered a great deal of pain, shortness of breath, anguish, and fear. I myself am not an advocate of active euthanasia, nor do I condemn it. My views on the issue have become more nuanced thanks to the research into the content and consequences of an NDE, in particular the stories about the continuity of consciousness and the loss of the fear of death. Familiarity with NDE research and the possibility of a personal afterlife can reduce fear of death and help inform people's opinions and decisions on ethical and medical issues.

> *You'll ne'er attain it, save you know the feeling.*
> —GOETHE, *Faust I*

In conclusion, knowledge about near-death experience can be of great practical significance to health care practitioners and to dying patients and their families. NDEs are much more common than previously assumed, and the personal consequences of such an experience are much more profound than doctors, nurses, and relatives ever imagined. Openness, sympathy, and proper support help NDErs accept and integrate this experience.[14] Fear of death and of the process of dying often informs decisions on ethical and medical issues on the part of doctors, patients, and families. A new perspective on death, which conceives of a continuity of consciousness after physical death, will have consequences for the way in which health care providers deal with coma patients and resuscitated, seriously ill, or dying patients and with stories about contact with the consciousness of dead relatives. Continuing improvement of the quality of health care depends not just on technical and medical advances but also on compassion for individual patients and their families.

Acknowledgments

This book would not have been possible without the assistance and support of a great many people. First of all, I would like to thank everybody who shared their NDE with me, either in person or in writing, and who helped lay the foundation for the ideas in this book. Specifically I would like to thank Mickey Broekhuysen, Machteld Blickman, Monique Hennequin, Cees de Kort, Elly Moerman, Bert van Schuijlenburg, Joke Crone-Ravestein, and the late Evert ter Beek, Willem Witteveen, and Ben Blondé. Through them I express my thanks to all the other NDErs I cannot mention by name.

The reports of near-death experiences and the scientific approach of George Ritchie, Raymond Moody, Michael Sabom, Melvin Morse, Kenneth Ring, Bruce Greyson, P. M. H. Atwater, Sam Parnia, Peter Fenwick, and Mario Beauregard have been a source of inspiration to me, and I feel privileged to count most of them among my personal friends. Ken Ring's help in getting the Dutch study published in *The Lancet* deserves special mention.

When Ina Vonk, Nico Vissel, Ruud van Wees, Vincent Meijers, and I decided to set up Merkawah (http://www.merkawah.nl), the Dutch chapter of IANDS (The International Association of Near-Death Studies; http://www.iands.org) in 1988, we never expected it to mark the start of a study of near-death experience that would last about ten years and that would be published in *The Lancet* in 2001 and receive worldwide attention. Merkawah was set up to raise awareness about NDE, organize meetings for NDErs (peer support), and promote scientific research into the subject. The prospective study that Ruud, Vincent, and I initiated at the cardiac care units of ten Dutch hospitals in 1988 was brought to a successful conclusion with the help of heart patients, nursing staff, and many volunteers; Ingrid Elfferich was responsible for coordinating and analyzing the interviews eight years later.

The ideas for this book were first conceived in the 1990s and gradually developed in subsequent years. I am indebted to Paul de Haas for our many conversations and for his advice on quantum physics. When I finally started writing this book in January 2007 I received a great deal of support from my readers Rudolf Smit, Tsjitske Waanders, my daughter, Sabien, and my wife, Niekje. Monique Hennequin deserves special mention for contributing her remarkable story and for her invaluable advice on the manuscript. The critical comments from Régine Dugardyn, my editor at Uitgeverij Ten Have, helped me express my ideas more clearly. Peter van Huizen was indispensable in preparing the definitive Dutch version of this book. Additionally I would like to thank Laura Vroomen for her excellent translation and Eric Brandt for his invaluable help and positive support for this American edition.

Finally, without the support, sympathy, and space that Niekje has given me over the years, I would never have been able to write this book. She has always been and remains my indispensable and loving sounding board.

Velp, The Netherlands, December 2009.

Art Credits

(Page 28) Jeroen Bosch, *Visions of the Afterlife: Earthly Paradise and Ascension to the Empireum*. Palazzo Ducale, Venice, Italy. Reprinted by permission of Scala / Art Resource, NY

(Page 75) The out-of-body experience of a six-year-old girl during her near death experience. From the collection of Dr. Melvin Morse.

(Page 84) Robert Fludd, *The Cabalistic Analysis of the Mind and the Senses*. 1617. Oxford Science Archive, Oxford, Great, Britain. Reprinted with permission of HIP / Art Resource, NY

(Page 107) Doctor examining a near-death experience. © Steve Michiels / ComicHouse.nl

(Page 179) The anatomical structures in the brain. Illustration by Maura Zimmer.

(Page 180) MRI of the brain. Image © Larry Mulvehill/Corbis.

(Page 187) A neuron with dendrites and synapses. Illustration by Maura Zimmer.

(Page 189) Transcranial Magnetic Stimulation (TMS). Illustration by Maura Zimmer.

(Page 197) fMRI scan of a three-year-old girl, following removal of the left part of her brain (right on picture). Reprinted from *The Lancet*, Vol 359, Issue 9305, Johannes Borgstein and Caroline Grootendorst, "Half a Brain," 2002, with permission from Elsevier.

(Page 230) Matter as a complex field of standing waves. Photograph by Dr. Erwin W. Mueller.

(Page 234) Double-slit experiment, with interference creating bright and dark bands. Illustration by Maura Zimmer.

(Page 243) Interference pattern in water. Photo by Martin Dohrn / Photo Researchers, Inc.

(Page 285) The double-helix structure of DNA. Reprinted by permission of SPL / Photo Researchers, Inc.

(Page 295) A flock of starlings reacting to a falcon attack. Photo by Manuel Presti, Italy.

(Page 299) The response of the white blood cells of a Pearl Harbor veteran who, twelve kilometers from his cells, is shown a filmed close-up of the face of a gunner shooting a Japanese aircraft that subsequently crashes into the ocean.

Notes

Introduction

1. G. G. Ritchie, *Return from Tomorrow* (Grand Rapids, MI: Zondervan, 1978.

2. G. G. Ritchie, *Return from Tomorrow* (Grand Rapids, MI: Zondervan, 1978); R. A. Moody Jr., *Life After Life* (Covington, GA.: Mockingbird Books, 1975).

3. D. Kennedy and C. Norman, "What We Don't Know," *Science* 309, no. 5731 (2005): 75.

4. A. M. Owen et al., "Detecting Awareness in the Vegetative State," *Science* 313 (2006): 1402.

5. A. Korthals Altes, *Uit coma* [Out of Coma] (The Hague, the Netherlands: Mirananda, 2002).

6. Kerkhoffs, J. *Droomvlucht in coma* [Dream Flight in Coma] (Melick, the Netherlands: Marga Genot Melick, 1994).

7. For the Dutch population, see J. Becker and J. de Hart, *Godsdienstige veranderingen in Nederland* [Religious Change in the Netherlands], Werkdocument 128, Sociaal Cultureel Planbureau (2006), and the Association of Religion Data Archives, http://www.thearda.com. For the American studies, see F. A. Curlin et al., "Religious Characteristics of U.S. Physicians," *Journal of General Internal Medicine* 20, no. 7 (2005): 629–34, and the Pew Forum on Religion and Public Life, *U.S. Religion Landscape Survey* (2007), http://www.religions.pewforum.org. For the United Kingdom, see L. Halman, *The European Values Study: A Third Wave. Sourcebook of the 1999–2000 European Values Study Surveys,* question 30-B (Tilburg: EVS, WORC, Tilburg University, 2001).

8. I. Maso, "Argumenten voor een inclusieve wetenschap" [Arguments in Favor of an Inclusive Science], paper presented at the conference Weten-

364 NOTES

schap, wereldbeeld en wij [Science, Worldview, and Us], Brussels, Belgium, June 2003.
9. A. H. Maslow, *The Psychology of Science* (New York: Harper & Row,
1966), chap. 8.
10. T. S. Kuhn, *The Structure of Scientific Revolutions* (Chicago: University
of Chicago Press, 1962).
11. D. Dennett, *Consciousness Explained* (Boston and London: Little,
Brown, 1991).
12. D. J. Chalmers, "Facing Up to the Problem of Consciousness," *Journal
of Consciousness Studies* 3, no. 1 (1995): 200. Chalmers's review essay is
"Consciousness and Its Place in Nature," in *Philosophy of Mind: Classical and Contemporary Readings,* ed. D. J. Chalmers (Oxford: Oxford
University Press, 2002). Also at: http://consc.net/papers/nature.html.
13. W. Heisenberg, "The Representation of Nature in Contemporary Physics," *Symbolism in Religion and Literature* (1960), 231, cited in R. D.
Boisvert, "Heteronomous Freedom," in *Philosophy and the Reconstruc-
tion of Culture,* ed. John. J. Stuhr (Albany: State University of New York
Press, 1993), 139.
14. P. van Lommel et al., "Near-Death Experience in Survivors of Cardiac
Arrest: A Prospective Study in the Netherlands," *Lancet* 358 (2001):
2039–45; B. Greyson, "Incidence and Correlates of Near-Death Experiences in a Cardiac Care Unit," *General Hospital Psychiatry* 25 (2003):
269–76; S. Parnia et al., (2001). "A Qualitative and Quantitative Study
of the Incidence, Features, and Aetiology of Near Death Experiences in
Cardiac Arrest Survivors," *Resuscitation* 48 (2003): 149–56; P. Sartori, P.
Badham, and P. Fenwick, "A Prospectively Studied Near-Death Experience with Corroborated Out-of-Body Perception and Unexplained Healing," *Journal of Near-Death Studies* 25, no. 2 (2006): 69–84.

Chapter 2: What Is a Near-Death Experience?

1. C. Zaleski, *Otherworld Journeys: Accounts of Near-Death Experience in Medieval and Modern Times* (Oxford: Oxford University Press,
1987).
2. B. Greyson, "Near-Death Experiences," in *Varieties of Anomalous Experiences: Examining the Scientific Evidence,* ed. E. Cardena, S. J. Lynn,
S. Krippner (Washington, DC: American Psychological Association,
2000), 316.
3. IANDS USA, http://www.iands.org.
4. Zaleski, *Otherworld Journeys.*
5. For the American figures, see G. Gallup and W. Proctor, *Adventures
in Immortality: A Look Beyond the Threshold of Death* (New York:
McGraw-Hill, 1982). For the German incidence, see I. Schmied, H.
Knoblaub, and B. Schnettler, "Todesnäheerfahrungen in Ost- und Westdeutschland: Eine empirische Untersuchung" [Near-Death Experiences

in East and West Germany: An Empirical Study], in *Todesnähe: Interdisziplinäre Zugänge zu einem außergewöhnlichen Phänomen* [Near-Death: Interdisciplinary Approaches to an Extraordinary Phenomenon], ed. H. Knoblaub and H. G. Soeffner, 65–99 (Konstanz, Germany: Universitätsverlag, 1999).

6. R. M. Hoffman, "Disclosure Needs and Motives After Near-Death Experiences: Influences, Obstacles, and Listener Selection," *Journal of Near-Death Studies* 14 (1995): 29–48.

7. R. A. Moody Jr., *Life After Life* (Covington, GA: Mockingbird Books, 1977); G. K. Athappilly, B. Greyson, and I. Stevenson, "Do Prevailing Society Models Influence Reports of Near-Death Experiences: A Comparison of Accounts Reported Before and After 1975," *Journal of Nervous and Mental Disease* 194, no. 3 (2006): 218–33. For culture-specific differences, see Zaleski, *Otherworld Journeys*. For the results of studies with indigenous peoples, see A. Kellehear, "Culture, Biology, and the Near-Death Experience: A Reappraisal," *Journal of Nervous and Mental Disease* 181 (1993): 148–56.

8. K. Ring, *Life at Death: A Scientific Investigation of the Near-Death Experience* (New York: Coward, McCann & Geoghegan, 1980); M. B. Sabom, *Recollections of Death: A Medical Investigation* (New York: Harper & Row, 1982); B. Greyson, "The Near-Death Experience Scale: Construction, Reliability and Validity," *Journal of Nervous and Mental Disease* 171 (1983): 369–75.

9. Greyson, "Near-Death Experience Scale." I think the terms *paranormal* and *transcendental* are confusing, and I choose not to use them myself; they are Greyson's terms.

10. Ring, *Life at Death*.

11. Greyson, "Near-Death Experience Scale." Greyson also calculated that both scoring systems are accurate and highly correlated, with a correlation coefficient of 90.

12. J. M. Holden, "Veridical perception in near-death experiences." In J. M. Holden, B. Greyson, and D. James, eds., *The Handbook of Near-Death Experiences* (Santa Barbara, CA: Praeger/ABC-CLIO, 2009), pp. 185–211.

13. P. van Lommel et al., "Near-Death Experience in Survivors of Cardiac Arrest: A Prospective Study in the Netherlands," *Lancet* 358 (2001): 2039–45; R. Smit, "Corroboration of the Dentures Anecdote Involving Veridical Perception in a Near-Death Experience," *Journal of Near-Death Studies* 27, no. 1 (2008): 47–61; T. Rivas, "Een gesprek met TG over de man met het gebit" [A Conversation with T.G. About the Man with the Dentures], *Terugkeer* 19, no. 3 (2008): 12–20.

14. C. G. Jung, *Memories, Dreams, Reflections,* ed. Aniela Jaffé, trans. R. and C. Winston (New York: Random House, 1961), 289–90.

15. British Broadcasting Company, *The Day I Died: The Mind, the Brain, and Near-Death Experiences* (2002); K. Ring and S. Cooper, *Mindsight:*

Near-Death and Out-of-Body Experiences in the Blind (Palo Alto, CA: William James Center/Institute of Transpersonal Psychology, 1999).

16. Ring and Cooper, *Mindsight,* 26.

17. BBC, *Day I Died.*

18. N. Evans Bush, "Afterward: Making Meaning After a Frightening Near-Death Experience," *Journal of Near-Death Studies* 21, no. 2 (2002): 99–133.

19. G. G. Ritchie, *Return from Tomorrow* (Grand Rapids, MI: Zondervan, 1978), 63–66.

Chapter 3: Changed by a Near-Death Experience

1. G. Groth-Marnat and R. Summers, "Altered Beliefs, Attitudes and Behaviors Following Near-Death Experiences," *Journal of Human Psychology* 38 (1998): 110–25.

2. K. Ring, *Heading Toward Omega: In Search of the Meaning of the Near-Death Experience* (New York: Quill William Morrow, 1984); M. Grey, *Return from Death: An Exploration of the Near-Death Experience* (London: Arkana, 1985); P. M. H. Atwater, *Coming Back to Life: The Aftereffects of the Near-Death Experience,* rev. ed. (New York: Citadel, 2001); C. Sutherland, *Transformed by the Light: Life After Near-Death Experiences* (Sydney, Australia: Bantam Books, 1992); M. Morse, *Transformed by the Light* (New York: Villard Books, 1990); P. Fenwick and E. Fenwick, *The Truth in the Light: An Investigation of Over 300 Near-Death Experiences* (New York: Berkley Books, 1997); K. Ring and E. Elsaesser-Valarino, *Lessons from the Light: What We Can Learn from the Near-Death Experience* (New York and London: Insight Books/Plenum, 1998); A. Opdebeeck, *Bijna dood: Leven met bijna-doodervaringen* [Nearly Dead: Living with Near-Death Experiences] (Tielt, Belgium: Uitgeverij Terra-Lannoo, 2001).

3. Sutherland, *Transformed by the Light.*

4. Sutherland, *Transformed by the Light;* Groth-Marnat and Summers, "Altered Beliefs"; B. Greyson, "Near-Death Experiences and Personal Values," *American Journal of Psychiatry* 140 (1983): 618–20; B. Greyson, "Reduced Death Threat in Near-Death Experiences," *Death Studies* 16 (1992): 533–46; Ring, *Heading Toward Omega.*

5. Atwater, *Coming Back to Life;* Sutherland, *Transformed by the Light;* Opdebeeck, *Bijna dood* [Nearly Dead]; Greyson, "Near-Death Experiences and Personal Values."

6. Schmied, Knoblaub, and Schnettler, "Todesnäheerfahrungen" [Near-Death Experiences].

7. Opdebeeck, *Bijna dood* [Nearly Dead].

8. Opdebeeck, *Bijna dood* [Nearly Dead].

9. Sutherland (*Transformed by the Light*) describes the four phases of the integration trajectory as follows: blocked, arrested, steady, and acceler-

ated integration. Regina Hoffman identifies five phases: shock or surprise at the content of the NDE, the need for affirmation (and the lack thereof), the impact on personal relationships, the active quest, and finally the process of integration; see R. M. Hoffman, "Disclosure Needs and Motives After Near-Death Experiences: Influences, Obstacles, and Listener Selection," *Journal of Near-Death Studies* 14 (1995): 29–48; Atwater, "Coming Back to Life"; B. Greyson, "Posttraumatic Stress Symptoms Following Near-Death Experiences," *American Journal of Orthopsychiatry* 71 (2001): 358–73.

10. R. M. Hoffman, "Disclosure Habits After Near-Death Experience: Influences, Obstacles, and Listeners Selection," *Journal of Near-Death Studies* 14 (1995): 29–48; Greyson, "Posttraumatic Stress Symptoms"; B. Greyson, "Biological Aspects of Near-Death Experiences," *Perspective in Biology and Medicine* 42, no. 1 (1998): 14–32; B. Greyson and B. Harris, "Clinical Approaches to the Near-Death Experiencer," *Journal of Near-Death Studies* 6 (1987): 41–52; I. Corbeau, "Psychische problematiek en hulpverlening na een BDE" [Psychological Problems and Support After an NDE], *Terugkeer (Tijdschrift rond bijna-dood ervaringen en zingeving)* [Return (Journal of Near-Death Experiences and Meaning)] 15, nos. 2–3 (2004): 16–22.

11. Sutherland, *Transformed by the Light.*

12. M. Grey, *Return from Death: An Exploration of the Near-Death Experience* (London: Arkana, 1985).

13. Opdebeeck, *Bijna dood* [Nearly Dead].

14. Ring, *Heading Toward Omega.*

15. C. Musgrave, "A Study of Spiritual Transformation," *Journal of Near-Death Studies* 15, no. 3 (1997): 187–201.

16. Sutherland, *Transformed by the Light,* 94.

17. J. Becker and J. de Hart, *Godsdienstige veranderingen in Nederland* [Religious Change in the Netherlands], Werkdocument 128, Sociaal Cultureel Planbureau (2006). At present, 18 percent of churchgoing people in the Netherlands are Roman Catholic, 8 percent Dutch Reformed, 7 percent Reformed, and 4 percent Muslim.

18. Pew Forum on Religion and Public Life, *U.S. Religion Landscape Survey* (2007), http://www.religions.pewforum.org; Pew Forum on Religion and Public Life, *Many Americans Mix Multiple Faiths* (2009), http://pewforum.org/newassets/images/reports/multiplefaiths/multiplefaiths.pdf; F. A. Curlin, J. D. Lantos, C. J. Roach, and S. A. Sellergren, "Religious Characteristics of U.S. Physicians," *Journal of General Internal Medicine* 20, no. 7 (2005): 629–34. In the United Kingdom only 1.0 percent of people are Hindu, 0.6 percent Sikh, 0.5 percent Jewish, and 0.3 percent Buddhist; see V. Crabtree, "Religion in the United Kingdom: Diversity, Trends and Decline" (2007), www.vexen.co.uk/UK/religion.html.

19. Sutherland, *Transformed by the Light,* 101.

20. Opdebeeck, *Bijna dood* [Nearly Dead]; P. Sartori, P. Badham, and P. Fenwick, "A Prospectively Studied Near-Death Experience with Corroborated Out-of-Body Perception and Unexplained Healing," *Journal of Near-Death Studies* 25, no. 2 (2006): 69–84.
21. Ring, *Heading Toward Omega*; Sutherland, *Transformed by the Light*.
22. Sutherland, *Transformed by the Light*, 116.
23. Sutherland, *Transformed by the Light*.
24. Corbeau, "Psychische problematiek en hulpverlening" [Psychological problems and support].
25. Corbeau, "Psychische problematiek".
26. N. E. Bush, "Is Ten Years a Life Review?" *Journal of Near-Death Studies* 10 (1991): 5–9.
27. Greyson and Harris, "Clinical Approaches."
28. Greyson, "Posttraumatic Stress Symptoms."
29. B. Greyson, "Near-Death Experiences and Antisuicidal Attitudes," *Omega* 26 (1992–1993): 81–89; Bush, "Life Review."
30. Corbeau, "Psychische problematiek en hulpverlening" [Psychological problems and support].
31. P. van Lommel et al, "Near-Death Experiences in Survivors of Cardiac Arrest: A Prospective Study in the Netherlands," *Lancet* 358 (2001): 2039–45; Ring, *Heading Toward Omega*.
32. As mentioned in the introduction, approximately 72 percent to 74 percent of the U.S. population believes in life after death; in the United Kingdom about 58 percent of the people believe in an afterlife, and 40 to 50 percent of the Dutch population believes in some form of life after death.

Chapter 4: Near-Death Experiences in Childhood

1. M. Morse and P. Perry, *Closer to the Light* (New York: Villard Books 1990).
2. P. M. H. Atwater, *The New Children and Near-Death Experiences* (Rochester, VT: Bear & Company, 2003).
3. Atwater, *New Children*.
4. Atwater, *New Children*; Morse and Perry, *Closer to the Light*.
5. Atwater, *New Children*.
6. Atwater, *New Children*.
7. Atwater, *New Children*.
8. For the Dutch study, see D. J. Bierman and R. van Wees, "Buitengewone ervaringen, andere bewustzijnsvormen en persoonlijkheid" [Extraordinary Experiences, Other Forms of Consciousness, and Personality], *Tijdschrift voor Parapsychologie* [*Journal of Parapsychology*] 60 (1993): 51–72. For the American study, see J. Palmer, "A Community Mail Survey of Psychic Experiences," *Journal of the American Society of Psychical Research* 73 (1979): 221–51.

9. K. Ring, *The Omega Project: Near-Death Experiences, UFO Encounters, and Mind at Large* (New York: William Morrow, 1992).

Chapter 5: There Is Nothing New Under the Sun

1. J. Fletcher, *The Egyptian Book of Living and Dying* (London: Duncan Baird Publishers, 2002).
2. J. Mishlove, *The Roots of Consciousness: The Classic Encyclopedia of Consciousness Studies Revised and Expanded* (Tulsa, OK: Oak Council Books, 1993).
3. J. Caesar, *The Gallic Wars*, trans. W. A. McDevitte and W. S. Bohn, http://classics.mit.edu/Caesar/gallic.6.6.html. Also see R. E. Mellor, *The Historians of Ancient Rome: An Anthology of the Major Writings*, 2nd ed. (New York: Routledge, 2004).
4. Ovid, *Metamorphoses*, trans. Samuel Garth, http://etext.virginia.edu/latin/ovid/garthl.html.
5. W. H. Huffman, *Robert Fludd and the End of the Renaissance* (London and New York: Routledge, 1988).
6. Dante Alighieri, *The Divine Comedy*, trans. A. S. Kline, http://www.poetryintranslation.com/klineasdante.htm.
7. E. Swedenborg, *Awaken from Death*, ed. James F. Lawrence (San Francisco: J. Appleseed, 1993).
8. A. Besant, *Death—and After?* (1906; repr., Wheaton, IL: Quest Books, 1998), http://www.gutenberg.org/etext/18266; R. Steiner, *Der Tod als Lebenswandlung* (Dornach, Switzerland: Rudolf Steiner Gesamtausgabe, 1917/18), translated by S. Seiler as *Death as Metamorphosis of Life* by S. Seiler (New York: Steinerbooks, 2008); A. A. Bailey, *Death: The Great Adventure* (New York: Lucis Trust, 1992); E. Byskov, *Death Is an Illusion: A Logical Explanation Based on Martinus' Worldview* (St. Paul, MN: Paragon House, 2002).
9. *The Upanishads*, trans. E. Easwaran, ed. M. N. Nagler (Tomales, CA: Nilgiri Press, 2006); J. Mascoró, *The Upanishads* (London and New York: Penguin Classics, 1965).
10. Swami Rama, *Sacred Journey: Living Purposefully and Dying Gracefully* (New Delhi: Himalayan International Institute of Yoga Science & Philosophy, 1996).
11. W. Y. Evans-Wentz, *The Tibetan Book of the Dead, or The after-death experiences on the Bardo plane, according to Lâma Kazi Dawa-Samdup's English rendering*, with foreword by Sir John Woodroffe (London: H. Milford, 1927, and London: Oxford University Press, 1971), 98ff.
12. Evans-Wentz, *Tibetan Book of the Dead*.
13. Sogyal Rinpoche, *The Tibetan Book of Living and Dying* (San Francisco: HarperSanFrancisco, 1992), 357, 364.

14. Plato, *Phaedo*, trans. Benjamin Jowett, http://philosophy.eserver.org/plato/phaedo.txt.

15. Mishlove, *Roots of Consciousness.*

16. S. P. Raphael, *Jewish Views of the Afterlife* (Northvale, NJ: Jason Aronson, 1994), 395–96.

17. L. D. Solomon, *The Jewish Book of Living and Dying* (Northvale, NJ: Jason Aronson, 1999).

18. Mishlove, *Roots of Consciousness;* Raphael, *Jewish Views.*

19. *The Holy Qur'an,* trans. Abdullah Yusuf Ali, Wordsworth Classics of World Literature (Ware, UK. Wordsworth Editions Limited, 2000).

20. B. Coppes, *Bijna dood ervaringen en wereldreligies: Getuigenis van universele waarheid* [Near-Death Experiences and World Religions: Testimony of Universal Truth] (Soesterberg, the Netherlands: Aspekt, 2006); P. Badham and L. Badham, *Death and Immortality in the Religions of the World* (New York: Paragon House, 1987); J. C. Hampe, *Sterben ist doch ganz anders: Erfahrungen mit dem eigenen Tod* (Stuttgart, Germany: Kreuz Verlag, 1975), translated as *To Die Is Gain: The Experience of One's Own Death* (Atlanta: John Knox Press, 1979); M. Fox, *Religion, Spirituality and the Near-Death Experience* (London and New York: Routledge, 2003); H. Küng, *Ewiges Leben?* [Eternal Life?] (Munich: Piper, 1982).

21. C. Zaleski, *Otherworld Journeys: Accounts of Near-Death Experience in Medieval and Modern Times* (Oxford: Oxford University Press, 1987); M. A. van der Sluijs, "Three Ancient Reports of Near-Death Experiences: Bremmer Revisited," *Journal of Near-Death Studies* 27, no. 4 (2009): 223–53.

22. Plato, *The Republic,* trans. Benjamin Jowett, http://philosophy.eserver.org/plato/republic.txt.

23. B. Colgrave and R. A. B. Mynors, eds., *Bede's Ecclesiastical History of the English People* (1969; repr., Oxford: Clarendon Press, 1991), 284–89.

24. G. G. Ritchie, *Return from Tomorrow* (Grand Rapids, MI: Zondervan, 1978).

25. F. Beaufort, "Letter to Dr. W. Hyde Wollaston," *An Autobiographical Memoir of Sir John Barrow* (London: John Murray, 1847), 398–403.

26. A. B. Boismont, *On Hallucinations,* trans. R. T. Hulme (London: Henry Renshaw, 1859).

27. F. W. H. Myers, "On Indications of Continued Terrene Knowledge on the Part of Phantasms of the Dead," *Proceedings of the Society for Psychical Research* 8 (1892): 170–252.

Chapter 6: Research into Near-Death Experiences

1. E. Kübler-Ross, *On Death and Dying* (New York: Macmillan, 1969); R. A. Moody Jr., *Life After Life* (Covington, GA: Mockingbird Books, 1975).

2. A. von St. Gallen Heim, "Notizen über den Tod durch Absturz," trans. R. Noyes and R. Kletti as "The Experience of Dying from Falls," *Jahrbuch des Schweizer Alpenclub* [Yearbook of the Swiss Alpine Club] 27 (1892): 46.

3. Data from International Association of Near-Death Studies: http://www.iands.org.

4. G. Gallup and W. Proctor, *Adventures in Immortality: A Look Beyond the Threshold of Death* (New York: McGraw-Hill, 1982); I. Schmied, H. Knoblaub, and B. Schnettler, "Todesnäheerfahrungen in Ost- und Westdeutschland: Ein empirische Untersuchung" [Near-Death Experiences in East and West Germany: An Empirical Study], in *Todesnähe: Interdisziplinäre Zugänge zu einem außergewöhnlichen Phänomen* [Near-Death: Interdisciplinary Approaches to an Extraordinary Phenomenon], ed. H. Knoblaub and H. G. Soeffner, 65–99 (Konstanz, Germany: Universitätsverlag, 1999).

5. B. Greyson, "The Incidence of Near-Death Experiences," *Medicine and Psychiatry* 1 (1998): 92–99; P. van Lommel et al., "Near-Death Experiences in Survivors of Cardiac Arrest: A Prospective Study in the Netherlands," *Lancet* 358 (2001): 2039–45; B. Greyson, "Incidence and Correlates of Near-Death-Experiences in a Cardiac Care Unit," *General Hospital Psychiatry* 25 (2003): 269–76; S. Parnia et al., "A Qualitative and Quantitative Study of the Incidence, Features and Aetiology of Near-Death Experiences in Cardiac Arrest Survivors," *Resuscitation* 48 (2001): 149–56.

6. Van Lommel et al., "Near-Death Experiences."

7. For the statistics on children, see M. Morse and P. Perry, *Closer to the Light* (New York: Villard Books, 1990). Ring's study is K. Ring, *Life at Death: A Scientific Investigation of the Near-Death Experience* (New York: Coward, McCann & Geoghegan, 1980). Sabom's study is found in M. B. Sabom, *Recollections of Death: A Medical Investigation* (New York: Harper & Row, 1982). For prospective studies among heart patients, see Greyson, "Incidence of Near-Death"; also see Greyson's 2003 study, "Incidence and Correlates." The Dutch study is Van Lommel et al., "Near-Death Experiences."

8. H. Yamamura, "Implication of Near-Death Experience for the Elderly in Terminal Care," *Nippon Ronen Igakkai Zasshi* 35, no. 2 (1998): 103–15.

9. Sabom, *Recollections of Death*; Van Lommel et al., "Near-Death Experiences."

10. G. M. Woerlee, *Mortal Minds: A Biology of the Soul and the Dying Experience* (Utrecht, the Netherlands: De Tijdstroom, 2003); B. Greyson, "Near-Death Experiences," in *Varieties of Anomalous Experiences: Examining the Scientific Evidence,* ed. E. Cardena, S. J. Lynn, and S. Krippner (Washington, DC: American Psychological Association, 2000), 315–52.

11. B. Greyson, "Biological Aspects of Near-Death Experiences," *Perspectives in Biology and Medicine* 42, no. 1 (1998): 14–32; E. D. Kelly and E. W. Kelly, "Unusual Experiences Near Death and Related Phenomena," in *Irreducible Mind: Toward a Psychology for the 21st Century*, 367–421 (Lanham: Rowman & Littlefield Publishers, 2007).

12. Woerlee, *Mortal Minds*; S. Blackmore, *Dying to Live: Science and the Near-Death Experience* (London: HarperCollins, 1993).

13. Blackmore, *Dying to Live*; Woerlee, *Mortal Minds*.

14. J. E. Whinnery and A. M. Whinnery, "Acceleration-Induced Loss of Consciousness," *Archives of Neurology* 47 (1990): 764–76.

15. T. Lempert, M. Bauer, and D. Schmidt, "Syncope and Near-Death Experience," *Lancet* 344 (1994): 829–30.

16. L. T. Meduna, *Carbon Dioxide Therapy: A Neuropsychological Treatment of Nervous Disorders* (Springfield: Charles C. Thomas, 1950).

17. Parnia et al., "Cardiac Arrest Survivors"; Ring, *Life at Death*; Greyson, "Near-Death Experiences"; P. Sartori, "The Incidence and Phenomenology of Near-Death Experiences," *Network Review (Scientific and Medical Network)* 90 (2006): 23–25.

18. K. Jansen, "Neuroscience, Ketamine and the Near-Death Experience: The Role of Glutamate and the NMDA-Receptor," in *The Near-Death Experience: A Reader*, ed. L. W. Bailey and J. Yates (New York and London: Routledge, 1996), 265–82.

19. R. Strassman, *DMT, The Spirit Molecule: A Doctor's Revolutionary Research into the Biology of Near-Death and Mystical Experiences* (Rochester, VT: Park Street Press, 2001).

20. Strassman, *DMT*.

21. A. Newberg, *Why God Won't Go Away: Brain Science and the Biology of Belief* (New York: Ballantine Books, 2002).

22. For DMT, see R. Strassman, *DMT*; for LSD, see S. Grof and J. Halifax, *The Human Encounter with Death* (New York: Dutton, 1977).

23. E. Rodin, "Comments on 'A Neurobiological Model for Near-Death Experiences,'" *Journal of Near-Death Studies* 7 (1989): 255–59.

24. W. Penfield, *The Excitable Cortex in Conscious Man* (Liverpool: Liverpool University Press, 1958); W. Penfield, *The Mystery of the Mind* (Princeton: Princeton University Press, 1975); W. Penfield, "The Role of the Temporal Cortex in Certain Psychical Phenomena," *Journal of Mental Science* 101 (1955): 451–65.

25. O. Blanke et al., "Stimulating Illusory Own-Body Perceptions: The Part of the Brain That Can Induce Out-of-Body Experiences Has Been Located," *Nature* 419 (2002): 269–70; O. Blanke et al., "Out-of-Body Experience and Autoscopy of Neurological Origin," *Brain* 127 (2004): 243–58.

26. M. A. Persinger, "Near-Death Experiences: Determining the Neuroanatomical Pathways by Experiential Patterns and Simulation in Experimental Settings," in *Healing: Beyond Suffering or Death*, ed. L. Bessette (Chabanel, Québec, Canada: Publications MNH, 1994), 277–86;

M. A. Persinger and F. Healey, "Experimental Facilitation of the Sensed Presence: Possible Intercalatation Between the Hemispheres Induced by Complex Magnetic Fields," *Journal of Nervous and Mental Diseases* 190 (2002): 533–41; P. Granqvist et al., "Sensed Presence and Mystical Experiences Are Predicted by Suggestibility, Not by the Application of Weak Complex Transcranial Magnetic Fields," *Neuroscience Letters* 379 (2005): 1–6.

27. W. B. Britton and R. R. Bootzin, "Near-Death Experiences and the Temporal Lobe," *American Psychological Society* 15, no. 4 (2004): 254–58.

28. K. R. Nelson et al., "Does the Arousal System Contribute to Near Death Experience?" *Neurology* 66, no. 1 (2006): 1003–9.

29. J. Long and J. M. Holden, "Does the Arousal System Contribute to Near-Death and Out-of-Body Experiences? A Summary and Response," *Journal of Near-Death Studies* 25, no. 3 (2007): 135–69.

30. The 1930s article is O. Pfister, "Shockdenken und Shockphantasien bei höchster Todesgefahr," *Zeitschrift für Psychoanalyse* 16 (1930): 430–55. Translated by R. Noyes and R. Kletti as "Shock Thoughts and Fantasies in Extreme Mortal Danger." For cross-cultural comparisons, see A. Kellehear, *Experiences Near Death: Beyond Medicine and Religion* (New York and Oxford: Oxford University Press, 1996).

31. G. K. Athappilly, B. Greyson, and I. Stevenson, "Do Prevailing Society Models Influence Reports of Near-Death Experiences: A Comparison of Accounts Reported Before and After 1975," *Journal of Nervous and Mental Disease* 194, no. 3 (2006): 218–33.

32. B. Greyson, "Dissociation in People Who Have Near-Death Experiences: Out of Their Bodies or Out of Their Minds?" *Lancet* 355 (2000): 460–63.

33. Woerlee, *Mortal Minds;* Blackmore, *Dying to Live.*

34. Woerlee, *Mortal Minds;* Blackmore, *Dying to Live;* Sabom, *Recollections of Death.*

35. S. Parnia and P. Fenwick, "Near-Death Experiences in Cardiac Arrest: Visions of a Dying Brain or Visions of a New Science of Consciousness. Review Article," *Resuscitation* 52 (2002): 5–11; E. W. Cook, B. Greyson, and I. Stevenson, "Do Any Near-Death Experiences Provide Evidence for the Survival of Human Personality After Death? Relevant Features and Illustrative Case Reports," *Journal of Scientific Exploration* 12 (1998): 377–406; K. Ring and S. Cooper, *Mindsight: Near-Death and Out-of-Body Experiences in the Blind* (Palo Alto, CA: William James Center/ Institute of Transpersonal Psychology, 1999).

36. Sabom, *Recollections of Death.*

37. M. M. Ghoneim and R. I. Block, "Learning and Memory During General Anaesthesia: An Update," *Anesthesiology* 87 (1997): 387–410.

38. For patients under general anesthetic, see E. R. John et al., "Invariant Reversible QEEG Effects of Anesthetics," *Consciousness and Cognition* 10 (2001): 165–83. On fMRI research among coma patients, see S. Laureys et al., "Brain Function in the Vegetative State," in *Advances in Experimental*

Medicine and Biology 550: Brain Death and Disorders of Consciousness,
ed. C. Machado and D. A. Shewmon, 229–38 (New York: Kluwer/Plenum, 2004). Also see Kelly and Kelly, *Irreducible Mind,* 367–421.

39. D. B. Cheek, "Unconscious Perception of Meaningful Sounds During Surgical Anaesthesia as Revealed Under Hypnosis," *American Journal of Clinical Hypnosis* 1 (1959): 101–13.

40. Kelly and Kelly, *Irreducible Mind.*

41. Blackmore, *Dying to Live.*

42. Kelly and Kelly, *Irreducible Mind.*

43. B. Greyson, J. M. Holden, and J. P. Mounsey, "Failure to Elicit Near-Death Experiences in Induced Cardiac Arrest," *Journal of Near-Death Studies* 25, no. 2 (2006): 85–98.

44. J. C. Saavedra-Aguilar and J. S. Gómez-Jeria, "A Neurobiological Model for Near-Death Experiences," *Journal of Near-Death Studies* 7 (1989): 205–22.

Chapter 7: The Dutch Study of Near-Death Experience

1. These facts are confirmed in a study of heart patients by Schwaninger and associates: J. Schwaninger et al., "A Prospective Analysis of Near-Death Experiences in Cardiac Arrest Patients," *Journal of Near-Death Studies* 20 (2002): 215–32. Their investigation into NDE in cardiac arrest patients, which was comparable to our study, included 174 resuscitated patients, of which 119 (68 percent) died. Of the surviving 55 patients only 30 (17 percent) could be interviewed. The other 25 patients were diagnosed as permanently brain-damaged at the time the interview was planned. Another study was conducted by Parnia and Fenwick and others: S. Parnia et al., "A Qualitative and Quantitative Study of the Incidence, Features and Aetiology of Near Death Experience in Cardiac Arrest Survivors," *Resuscitation* 48 (2001): 149–56. Their study included 220 cardiac arrest patients over a period of one year, of whom 62 percent died, and only 63 patients (28 percent) could be interviewed.

2. R. F. Hoffman, "Disclosure Habits After Near-Death Experience: Influences, Obstacles and Listeners Selection," *Journal of Near-Death Studies* 14 (1995): 29–48.

3. K. Ring, *Heading Toward Omega: In Search of the Meaning of the Near-Death Experience* (New York: William Morrow, 1984).

4. P. van Lommel et al., "Near-Death Experiences in Survivors of Cardiac Arrest: A Prospective Study in the Netherlands," *Lancet* 358 (2001): 2039–45.

5. M. J. Sauve et al., "Patterns of Cognitive Recovery in Sudden Cardiac Arrest Survivors: The Pilot Study," *Heart Lung* 25, no. 3 (1996): 172–81.

6. B. Greyson, "Incidence and Correlates of Near-Death Experiences in a Cardiac Care Unit," *General Hospital Psychiatry* 25 (2003): 269–76.

7. Hoffman, "Disclosure Habits."

8. Ring, *Heading Toward Omega.*
9. The American study is Greyson, "Incidence and Correlates"; the British is Parnia et al., "Cardiac Arrest Survivors"; see also P. Sartori, "The Incidence and Phenomenology of Near-Death Experiences," *Network Review (Scientific and Medical Network)* 90 (2006): 23–25.
10. Greyson, "Incidence and Correlates."
11. Parnia et al, "Cardiac Arrest Survivors."
12. M. B. Sabom, *Recollections of Death: A Medical Investigation* (New York: Harper & Row, 1982).
13. P. Sartori, P. Badham, and P. Fenwick, "A Prospectively Studied Near-Death Experience with Corroborated Out-of-Body Perception and Unexplained Healing," *Journal of Near-Death Studies* 25, no. 2 (2006): 69–84.

Chapter 8: What Happens in the Brain When the Heart Suddenly Stops?

1. P. van Lommel et al., "Near-Death Experiences in Survivors of Cardiac Arrest: A Prospective Study in the Netherlands," *Lancet* 358 (2001): 2044.
2. B. Greyson, "Incidence and Correlates of Near-Death Experiences in a Cardiac Care Unit," *General Hospital Psychiatry* 25 (2003): 275.
3. S. Parnia et al., "A Qualitative and Quantitative Study of the Incidence, Features and Aetiology of Near Death Experiences in Cardiac Arrest Survivors," *Resuscitation* 48 (2001): 151.
4. P. Sartori, "The Incidence and Phenomenology of Near-Death Experiences," *Network Review (Scientific and Medical Network)* 90 (2006): 23–25.
5. S. Parnia and P. Fenwick, "Near-Death Experiences in Cardiac Arrest: Visions of a Dying Brain or Visions of a New Science of Consciousness. Review article," *Resuscitation* 52 (2002): 5–11.
6. K. T. Gopalan et al., "Cerebral Blood Flow Velocity During Repeatedly Induced Ventricular Fibrillation," *Journal of Clinical Anesthesia* 11, no. 4 (1999): 290–95.
7. J. Mayer and T. Marx, "The Pathogenesis of EEG Changes During Cerebral Anoxia," in *Cardiac and Vascular Diseases/Handbook of Electroencephalography and Clinical Neurophysiology,* ed. J. H. A. van der Drift (Amsterdam: Elsevier, 1972), vol. 14A, pt. A, pp. 5–11,
8. Parnia and Fenwick, "Near-Death Experiences in Cardiac Arrest"; J. W. de Vries et al., "Changes in Cerebral Oxygen Uptake and Cerebral Electrical Activity During Defibrillation Threshold Testing," *Anesthesia Analgesia* 87 (1998): 16–20; H. Clute and W. J. Levy, "Electroencephalographic Changes During Brief Cardiac Arrest in Humans," *Anesthesiology* 73 (1990): 821–25; T. J. Losasso et al., "Electroencephalographic Monitoring of Cerebral Function During Asystole and Successful Cardiopulmonary Resuscitation," *Anesthesia Analgesia* 75 (1992): 12–19.

9. N. M. Branston et al., "Comparison of the Effects of Ischaemia on Early Components of the Somatosensory Evoked Potential in Brainstem, Thalamus, and Cerebral Cortex," *Journal of Cerebral Blood Flow Metabolism* 4, no. 1 (1984): 68–81; J. Gua, J. A. White, and H. H. Batjer, "Limited Protective Effects of Etomidate During Brainstem Ischemia in Dogs," *Journal of Neurosurgery* 82, no. 2 (1995): 278–84.

10. D. S. Smith et al., "Reperfusion Hyperoxia in the Brain After Circulatory Arrest in Humans," *Anesthesiology* 73 (1990): 12–19.

11. Mayer and Marx, "Cerebral Anoxia"; G. Buunk, J. G. van der Hoeven, and A. E. Meinders, "Cerebral Blood Flow After Cardiac Arrest," *Netherlands Journal of Medicine* 57 (2000): 106–12; Losasso et al., "Electroencephalographic Monitoring."

12. E. D. Kelly and E. W. Kelly, "Unusual Experiences Near Death and Related Phenomena," chap. 6 in *Irreducible Mind: Toward a Psychology for the 21st Century* (Lanham, MD: Rowman & Littlefield, 2007), 418.

13. M. J. Sauve et al., "Patterns of Cognitive Recovery in Sudden Cardiac Arrest Survivors: The Pilot Study," *Heart Lung* 25, no. 3 (1996): 172–81.

14. M. Fujioka et al., "Hippocampal Damage in the Human Brain After Cardiac Arrest," *Cerebrovascular Diseases* 10, no. 1 (2000): 2–7; H. C. Kinney et al., "Neuropathological Findings in the Brain of Karen Ann Quinlan: The Role of the Thalamus in the Persistent Vegetative State," *New England Journal of Medicine* 330, no. 26 (1994): 1469–75.

15. G. W. van Dijk, "Bewustzijn" [Consciousness], in *Handboek Reanimatie* [Resuscitation Handbook], 2nd rev. ed., ed. B. T. J. Meursing and R. G. van Kesteren (Utrecht, the Netherlands: Wetenschappelijke Uitgeverij Bunge, 2004), 21–25.

16. Van Dijk, "Bewustzijn" [Consciousness].

17. Van Dijk, "Bewustzijn" [Consciousness].

18. J. Herlitz et al., "Characteristics and Outcome Among Patients Suffering from In-Hospital Cardiac Arrest in Relation to the Interval Between Collapse and Start of CPR," *Resuscitation* 53, no. 1 (2000): 21–27.

19. N. A. Paradis, G. B. Martin, and M. G. Goetting, "Simultaneous Aortic Jugular Bulb, and Right Atrial Pressures During Cardiopulmonary Resuscitation in Humans: Insights into Mechanisms," *Circulation* 80 (1989): 361–68; N. A. Paradis, G. B. Martin, and J. Rosenberg, "The Effect of Standard and High Dose Epinephrine on Coronary Perfusion Pressure During Prolonged Cardiopulmonary Resuscitation," *Journal of the American Medical Association* 265 (1991): 1139–44.

20. Clute and Levy, " Electroencephalographic Changes"; Losasso et al., "Electroencephalographic Monitoring"; K. A. Hossmann and P. Kleihues, "Reversibility of Ischemic Brain Damage," *Archives of Neurology* 29, no. 6 (1973): 375–84; J. Moss and M. Rockoff, "EEG Monitoring During Cardiac Arrest and Resuscitation," *Journal of American Medical Association* 244, no. 24 (1980): 2750–51.

21. P. Safar et al., "Cerebral Resuscitation Potentials for Cardiac Arrest," *Critical Care Medicine* 30, no. 4, suppl. (2002): 140–44.

22. C. G. Coimbra, "Implications of Ischemic Penumbra for the Diagnosis of Brain Death," *Brazilian Journal of Medical and Biological Research* 32, no. 12 (1999): 1479–87.

23. M. B. Sabom, "Death: Defining the Final Frontier: The Case of Pam Reynolds," *Light and Death: One Doctor's Fascinating Account of Near-Death Experiences* (Grand Rapids, MI: Zondervan, 1998), 37–52.

24. H. Ebert, "Deep Hypothermia and Circulatory Arrest for Surgery of Complex Intracranial Aneurysms," *European Journal of Cardiothoracic Surgery* 13, no. 3 (1998): 223–29.

Chapter 9: What Do We Know About Brain Function?

1. J. E. Desmedt and D. Robertson, "Differential Enhancement of Early and Late Components of the Cerebral Somatosensory Evoked Potentials During Forced-Paced Cognitive Tasks in Man," *Journal of Physiology* 271 (1977): 761–82; P. E. Roland and L. Friberg, "Localization in Cortical Areas Activated by Thinking," *Journal of Neurophysiology* 53 (1985): 1219–43; J. C. Eccles, "The Effect of Silent Thinking on the Cerebral Cortex," *Truth Journal: International Interdisciplinary Journal of Christian Thought* 2 (1988); P. E. Roland, "Somatotopical Tuning of Postcentral Gyrus During Focal Attention in Man: A Regional Cerebral Blood Flow Study," *Journal of Neurophysiology* 46 (1981): 744–54.

2. G. W. van Dijk, "Bewustzijn" [Consciousness], in *Handboek Reanimatie* [The Resuscitation Handbook], 2nd ed., ed. B. T. J. Meursing and R. G. van Kesteren (Utrecht, the Netherlands: Wetenschappelijke Uitgeverij Bunge, 2004), 21–25.

3. C. Koch, "The Movie in Your Head," *Scientific American Mind* 16, no. 3 (2005): 8–63; J. M. Schwartz and S. Begley, *The Mind and the Brain: Neuroplasticity and the Power of Mental Force* (New York: Harper-Collins, 2002); D. Dobbs, "Fact or Phrenology?" *Scientific American Mind* 16, no. 1 (2005): 24–31.

4. M. Slob, Interview: "Waar de wetenschap niet bij kan" [Where Science Cannot Reach], *NRC Handelsblad* (January 20, 2007), education and science supplement.

5. A. I. Jack and A. Roepstorff, "Introspection and Cognitive Brain Mapping: From Stimulus-Response to Script Report," *Trends in Cognitive Science* 6, no. 8 (2002): 333–39; A. I. Jack and A. Roepstorff, "Why Trust the Subject?" *Journal of Consciousness Studies* 10, nos. 9–10 (2003): v–xx.

6. Jack and Roepstorff, "Introspection"; Jack and Roepstorff, "Why Trust the Subject?"

7. A. Freeman, "The Conscious Brain," chap. 4 in *Consciousness: A Guide to the Debates* (Santa Barbara: ABC-CLIO, 2003), 61–80.

8. A. Noë, *Out of Our Heads: Why You Are Not Your Brain, and Other Lessons from the Biology of Consciousness* (New York: Hill and Wang, 2009).

9. J. L. Saver and J. Rabin, "The Neural Substrates of Religious Experience," *Journal of Neuropsychiatry* 9, no. 3 (1997): 498–510.

10. Schwartz and Begley, *Mind and the Brain,* 18, 28.

11. H. Romijn, "About the Origin of Consciousness: A New, Multidisciplinary Perspective on the Relationship Between Brain and Mind," *Proceedings of the Koninklijke Nederlandse Akademie van Wetenschappen* 100, nos. 1–2 (1997): 181–267.

12. Electrical charges are the action potentials across the cell membrane of neurons caused by the transient and rapid alteration of voltage by an exchange of ions.

13. G. M. Edelman and G. Tononi, *A Universe of Consciousness* (New York: Basic Books, 2000).

14. The transmission of information along neurons is done via action potentials (transient alteration of voltage), which are triggered by differences in membrane potential (the electric charge or voltage difference across a cell membrane) through decreases (depolarization) and increases (hyperpolarization) in the many contacts (synapses) with other neurons. This process releases neurotransmitters in the synapses. The sum total of all the changes in membrane potentials causes constantly changing electrical fields.

15. M. Hallett, "Transcranial Magnetic Stimulation and the Human Brain," *Nature* 406 (2000): 147–50.

16. W. Penfield, *The Excitable Cortex in Conscious Man* (Liverpool: Liverpool University Press, 1958); O. Blanke et al., "Out-of-Body Experience and Autoscopy of Neurological Origin," *Brain* 127 (2004): 243–58.

17. D. Liebetanz, "Pharmacological Approach to the Mechanisms of Transcranial DC-Stimulation-Induced After-Effects of Human Motor Cortex Excitability," *Brain* 125 (2002): 2238–47.

18. Deep brain stimulation (DBS) has both an excitatory and an inhibitory effect, depending on whether the stimulation takes place in white or grey matter of the brain.

19. H. S. Mayberg et al., "Deep Brain Stimulation for Treatment-Resistant Depression," *Neuron* 45 (2005): 651–60; M. Hopkin, "Implant Boosts Activity in Injured Brain," *Nature* 448 (2007): 522; N. D. Schiff et al., "Behavioural Improvements With Thalamic Stimulation After Severe Traumatic Brain Injury," *Nature* 448 (2007): 600–603; M. N. Shadler and R. Kiani, "News and Views. Neurology: An Awakening," *Nature* 448 (2007): 539–40.

20. M. Beauregard, "Mind Does Really Matter: Evidence from Neuroimaging Studies of Emotional Self-Regulation, Psychotherapy, and Placebo Effect," *Progress in Neurobiology* 81, no. 4 (2007): 218–36.

21. M. Massimini et al., "Breakdown of Cortical Effective Connectivity During Sleep," *Science* 309, no. 5744 (2005): 2228–32.

22. Massimini et al., "Breakdown."

23. N. S. White and M. T. Alkire, "Impaired Thalamocortical Connectivity in Humans During General-Anesthetic Induced Unconsciousness," *Neuroimage* 19, no. 2, pt. 1 (2003): 401–11; M. T. Alkire and J. Miller, "General Anesthesia and the Neural Correlates of Consciousness," *Progress in Brain Research* 150 (2005): 229–44; M. T. Alkire, A. G. Hudetz, and G. Tononi, "Consciousness and Anesthesia," *Science* 322, no. 5903 (2008): 876–80.

24. T. J. Balkin et al., "The Process of Awakening: A PET Study of Regional Brain Activity Patterns Mediating the Re-establishment of Alertness and Consciousness," *Brain* 125 (2002): 2308–19.

25. J. A. Den Boer, *Neuro-filosofie: Hersenen-Bewustzijn-Vrije wil* [Neurophilosophy: The Brain-Consciousness-Free Will] (Amsterdam: Boom, 2003), 128.

26. S. Y. Berkovich, "On the Information Processing Capabilities of the Brain: Shifting the Paradigm," *Nanobiology* 2 (1993): 99–107; H. Romijn, "Are Virtual Photons the Elementary Carriers of Consciousness?" *Journal of Consciousness Studies* 9 (2002): 61–81.

27. K. Pribram, "The Neurophysiology of Remembering," *Scientific American* 220 (1969): 75; K. Lashley, "In Search of the Engram," *Psychological Mechanisms in Animal Behavior* (New York: Academic Press, 1950), 454–82.

28. R. Lewin, "Is Your Brain Really Necessary?" *Science* 210 (1980): 1232–34.

29. M. Nahm and B. Greyson, "Terminal Lucidity in Patients with Chronic Schizophrenia and Dementia: A Survey of the Literature," *Journal of Nervous and Mental Disease* 197, no. 12 (2009): 942–44.

30. Schwartz, and Begley, *Mind and the Brain.*

31. N. Doidge, *The Brain That Changes Itself: Stories of Personal Triumph from the Frontiers of Brain Science* (London: Penguin Books, 2007).

32. P. R. Huttenlocher, "Synapse Elimination and Plasticity in Developing Human Cerebral Cortex," *American Journal of Mental Deficiency* 88 (1984): 488–96.

33. M. T. Acosta, P. Montanez, and F. E. Leon-Sarmiento, "Half Brain but Not Half Function," *Lancet* 360 (2002): 643; J. Borgstein and C. Grootendorst, "Clinical Picture: Half a Brain," *Lancet* 359 (2002): 473.

34. H. S. Mayberg et al., "The Functional Neuroanatomy of the Placebo Effect," *American Journal of Psychiatry* 159 (2002): 728–37.

35. T. D. Wager, "Placebo-Induced Changes in fMRI in the Anticipation and Experience of Pain," *Science* 303 (2004): 1162–67; F. Benedetti et al., "Neurobiological Mechanisms of the Placebo Effect," *Journal of Neuroscience* 25, no. 45 (2005): 10390–402.

36. Beauregard, M. "Mind Does Really Matter: Evidence from Neuroimaging Studies of Emotional Self-Regulation, Psychotherapy, and Placebo Effect." *Progress in Neurobiology* 81, no. 4 (2007): 218–36; M. Beauregard and D. O'Leary, "Toward a Nonmaterialistic Science of Mind," chap. 6

in *The Spiritual Brain: How Neuroscience Is Revealing the Existence of the Soul* (San Francisco: HarperOne, 2007), 125–80.

37. Schwartz and Begley, *Mind and the Brain;* R. J. Davidson et al., "Alterations in Brain and Immune Function Produced by Mindfulness Meditation," *Psychosomatic Medicine* 65, no. 4 (2003): 564–70.

38. In an EEG different brain waves can be registered with increasing frequencies: delta, theta, alpha, beta, and gamma waves. Delta waves have the lowest frequency. Theta waves are said to be correlated with relaxed, meditative, and creative states, and gamma waves are commonly associated with peak concentration, optimal cognitive functioning, increased mental abilities, and increased levels of compassion and happiness. See A. Lutz et al., "Long-Term Meditators Self-Induce High-Amplitude Gamma Synchrony During Mental Practice," *Proceedings of the National Academy of Science, USA* 101, no. 46 (2004): 16369–73.

39. M. Baringa, "Buddhism and Neuroscience: Studying the Well-Trained Mind," *Science* 302 (2003): 44–46; M. Beauregard and V. Paquette, "Neural Correlates of a Mystical Experience in Carmelite Nuns," *Neuroscience Letters* 405 (2006): 186–90.

40. In split-brain patients the tissue that connects the two hemispheres and facilitates communication and coordination between those two parts (the corpus callosum) is severed to some degree.

41. R. W. Sperry, "Mental Phenomena as Causal Determinants in Brain Function," in *Consciousness of the Brain,* ed. G. G. Globus, G. Maxwell, and I. Savodnik (New York: Plenum, 1976); B. Libet et al., "Time of Conscious Intention to Act in Relation to Onset of Cerebral Activity (Readiness Potential): The Unconscious Initiation of a Freely Voluntary Act," *Brain* 106 (1983): 623–42; B. Libet, "A Testable Field Theory of Mind-Brain Interaction," *Journal of Consciousness Studies* 1 (1994): 119–26; B. Libet, *Mind Time: The Temporal Factor in Consciousness* (Cambridge, MA: Harvard University Press, 2004); M. Beauregard, P. Lévesque, and V. Paquette, "Neural Basis of Conscious and Voluntary Self-Regulation of Emotion," in *Consciousness, Emotional Self-Regulation and the Brain,* ed. M. Beauregard (Amsterdam: John Benjamins Publishing, 2004), 163–94.

42. D. Dennett, *Consciousness Explained* (Boston and London: Little, Brown, 1991); S. J. Blackmore, "There Is No Stream of Consciousness," *Journal of Consciousness Studies* 9, nos. 5–6 (2002): 17–28.

43. R. Penrose, *Shadows of the Mind* (Oxford: Oxford University Press, 1996); Beauregard and O'Leary, *Spiritual Brain;* Noë, *Out of Our Heads.*

Chapter 11: Quantum Physics and Consciousness

1. S. Hawking, *The Universe in a Nutshell* (London: Bantam Press, 2001).

2. M. Bischof, *Biophotonen—Das Licht, das unsere Zellen steuert* [Biophotons—The Light Governing Our Cells], 11th ed. (Frankfurt: Verlag

Zweitausendeins, 2001), 217. The Web site http://www.marcobischof .com contains a number of English articles on the same subject.

3. N. J. Nersessian, "Aether Or: The Creation of Scientific Concepts," *Studies in the History and Philosophy of Science* 15 (1984): 175–212.

4. F. Rutherford, G. Holton, and F. G. Watson, "Unit 4 Light and Electromagnetism," chap. 13 in *Project Physics Course* (New York: Holt, Reinhart & Winston, 1968), 14.

5. N. Bohr and J. Kalckar, eds., *Collected Works*, vol. 6, *Foundations of Quantum Physics I (1926–1932)* (Amsterdam and New York: North Holland, 1997), 91–94; B. Rosenblum and F. Kuttner, "The Observer in the Quantum Experiment," *Foundations of Physics* 32, no. 8 (2002): 1273–93; N. Bohr, *Atomic Physics and Human Knowledge* (New York: Wiley, 1958), 81; N. Bohr, *Essays 1958–1962 on Atomic Physics and Human Knowledge* (New York: Wiley, 1963), 15.

6. M. Born, *My Life and My Views: A Nobel Prize Winner in Physics Writes Provocatively on a Wide Range of Subjects* (New York: Scribner, 1968), 48.

7. E. Schrödinger, "Discussion of Probability Relations Between Separated Systems," *Cambridge Philosophical Society Proceedings* 32 (1935): 555.

8. W. Heisenberg, *Physics and Beyond* (New York: Harper & Row, 1971); R. Penrose, *Shadows of the Mind: A Search for the Missing Science of Consciousness* (Oxford: Oxford University Press, 1996).

9. J. von Neumann, *The Mathematical Foundations of Quantum Mechanics* (Princeton: Princeton University Press, 1955), quoted in A. Goswami, R. E. Reed, and M. Goswami, *The Self-Aware Universe: How Consciousness Creates the Material World* (New York: Jeremy Tarcher/Putman, 1993).

10. H. Stapp, *Mind, Matter and Quantum Mechanics*, 2nd ed. (Berlin and Heidelberg: Springer Verlag, 2004); E. Wigner, "The Problem of Measurement," *Journal of Physics* 31, no. 6 (1963); R. Nadeau and M. Kafatos, *The Non-Local Universe: The New Physics and Matters of the Mind* (New York: Oxford University Press, 1999); Goswami, Reed, and Goswami, *Self-Aware Universe*; F. A. Wolf, *The Spiritual Universe: One Physicist's Vision of Spirit, Soul, Matter and Self* (Portsmouth, NH: Moment Point Press, 1996).

11. A. Aspect, J. Dalibard, and G. Roger, "Experimental Tests of Bell's Inequality Using Varying Analyses," *Physical Review Letters* 25 (1982): 1084.

12. I. Marcikic et al., "Distribution of Time-Bin Entangled Qubits over 50 km of Optical Fiber," *Physical Review Letters* 93, no. 18 (2004): 180502-1–4; D. Greenberger, M. Horne, and A. Zeilinger, "Going Beyond Bell's Theorem," in *Bell's Theorem, Quantum Theory, and Conceptions of the Universe*, ed. M. Kafatos (Dordrecht, the Netherlands: Kluwer Academics, 1989), 73–76.

13. F. A. Wolf, *Taking the Quantum Leap: The New Physics For Nonscientists* (New York: Harper & Row, 1989), chap. 5, pp. 87–93.

14. A. Einstein, "The Concept of Space," *Nature* 125 (1930): 897–98; W. J. Moore, *Schrödinger: Life and Thought* (Cambridge and New York:

Cambridge University Press, 1989); K. C. Cole, "In Patterns, Not Particles, Physics Trust," *Los Angeles Times,* March 4, 1999.

15. N. D. Mermin, "Is the Moon There When Nobody Looks? Reality and the Quantum Theory," *Physics Today* 38, no. 4 (1985): 38–47.

16. E. D. Kelly and E. W. Kelly, *Irreducible Mind: Toward a Psychology for the 21st Century* (Lanham, MD: Rowman & Littlefield, 2007), 199–218.

17. D. Greenberger, "Remark Made During a Debate," Symposium on Fundamental Questions in Quantum Mechanics (State University of New York, Albany, 1984).

18. E. Schrödinger, *What Is Life, With Mind and Matter and Autobiographical Sketches,* Canto Edition (Cambridge: Cambridge University Press, 1944), 93.

19. L. P. Wheeler, *Josiah Willard Gibbs: The History of a Great Mind* (Woodbridge, CT: Ox Bow Press, 1998); A. Sommerfeld, "Einführung in die Quantentheorie, Oscillator und Rotator," in *Atombau und Spektrallinien* (Braunschweig, Germany: Friedrich Vieweg & Sohn, 1924), translated by Henry L. Brose as "Introduction to Quantum Theory, Oscillator and Rotator," in *Atomic Structure and Spectral Lines* (London: Methuen, 1923).

20. Penrose, *Shadows of the Mind.*

21. W. Schempp, "Quantum Holography and Neurocomputer Architectures," *Journal of Mathematical Imaging and Vision* 2 (1992): 109–64; W. Schempp, *Magnetic Resonance Imaging: Mathematical Foundations and Applications* (New York: John Wiley, 1997). Nuclear spin resonance is the principle underlying magnetic resonance imaging (MRI) for which, as for a quantum hologram, there is evidence of this nonlocal information exchange because the hydrogen nuclei in the water and the fats of the neurons must align themselves to the magnetic field on the basis of quantum-mechanical laws, with the protons of the hydrogen nuclei aligning themselves parallel or antiparallel to the inducing magnetic field.

22. G.'t Hooft, "Dimensional Reduction in Quantum Gravity," in *Salamfest,* ed. A. Ali, J. Ellis, and S. Randjbar-Daemi (Singapore: World Scientific, 1993), 284–96; B. Greene, *The Fabric of the Cosmos* (New York: Knopf, 2004).

23. H. R. Pagels, *Perfect Symmetry* (London: Joseph Publishers, 1985); E. Laszlo, *The Connectivity Hypothesis: Foundations of an Integral Science of Quantum, Cosmos, Life, and Consciousness* (Albany: State University of New York Press, 2003); E. Laszlo, *Science and the Akashic Field: An Integral Theory of Everything* (Rochester, VT: Inner Traditions, 2004).

24. R. A. Hall, *Isaac Newton, Adventurer in Thought* (Cambridge: Cambridge University Press, 1992), 27.

25. D. J. Chalmers, "Consciousness and Its Place in Nature," in *Philosophy of Mind: Classical and Contemporary Readings* (Oxford: Oxford University Press, 2002). See also http://consc.net/papers/nature.html.

26. Penrose, *Shadows of the Mind: A Search for the Missing Science of Consciousness*. (Oxford: Oxford University Press,1996).

27. D. Bohm, *Wholeness and the Implicate Order* (London: Routledge & Kegan Paul, 1980).

28. P. Weiss, *Principles of Development* (New York: Holt, 1939); A. Gurwitsch, "Über den Begriff des embryonalen Feldes" [On the Concept of the Embryonic Field], *Archiv für Entwicklungsmechanik* [Archives of Developmental Biology] 51 (1922): 383–415.

29. R. Sheldrake, *A New Science of Life* (London: Blond & Briggs, 1981); R. Sheldrake, *The Presence of the Past* (London: Fontana, 1988).

30. J. van der Greef and R. N. McBurney, "Rescuing Drug Discovery: *In Vivo* Systems Pathology and Systems Pharmacology," *Nature Reviews/ Drug Discovery* 4 (2005): 961–67.

31. Bohm, *Wholeness*.

32. A. Zeilinger, *Einsteins Schleier* [Einstein's Veil] (Munich: C. H. Beck Verlag, 2003), 103–6; G. Blatter, "Schrödinger's Cat Is Now Fat," *Nature* 406 (2000): 25–26; R. Friedman et al., "Quantum Superposition of Distinct Macroscopic States," *Nature* 406 (2000): 43–45.

33. I. Prigogine and I. Stengers, *Order Out of Chaos: Man's New Dialogue with Nature* (Boulder, CO: New Science Press, 1984); H. Fröhlich, "Coherent Excitations in Active Biological Systems," in *Modern Bioeletrochemistry*, ed. F. Gutman and H. Keyzer (New York: Plenum, 1983).

34. H. Romijn, "Are Virtual Photons the Elementary Carriers of Consciousness?" *Journal of Consciousness Studies* 9 (2002): 61–81; K. Pribram, "The Neurophysiology of Remembering," *Scientific American* 220 (1969): 75; K. Lashley, "In Search of the Engram," in *Physiological Mechanisms in Animal Behavior* (New York: Academic Press, 1950), 454–82; K. Pribram, *Languages of the Brain* (Monterey, CA: Wadsworth Publishing, 1977), 123.

35. S. Hameroff and R. Penrose, "Orchestrated Reduction of Quantum Coherence in Brain Microtubules," in *Proceedings of the International Neural Network Society, Washington DC* (Hillsdale, NJ: Erlbaum, 1995), 793–812; D. Zohar, *The Quantum Self: Human Nature and Consciousness Defined by the New Physics* (New York: William Morrow, 1990).

36. Zeilinger, *Einsteins Schleier* [Einstein's Veil].

37. Stapp, *Mind*, 266; W. James, *The Principles of Psychology* (1890; reprint, New York: Dover, 1950); W. Heisenberg, *Physics and Beyond* (New York: Harper & Row, 1971); J. von Neumann, *Mathematical Foundations of Quantum Theory* (Princeton: Princeton University Press, 1955).

38. B. Misra and E. C. G. Sudarshan, "The Zeno's Paradox in Quantum Theory," *Journal of Mathematical Physics* 18 (1977): 756–63; James, *Principles of Psychology*; Von Neumann, *Mathematical Foundations*.

Chapter 12: The Brain and Consciousness

1. D. J. Chalmers, "Consciousness and Its Place in Nature," in *Philosophy of Mind: Classical and Contemporary Readings* (Oxford: Oxford University Press, 2002); see also http://consc.net/papers/nature.html.
2. Chalmers, "Consciousness"; K. Popper and J. C. Eccles, *The Self and Its Brain* (New York: Springer, 1977).
3. D. Dennett, *Consciousness Explained* (Boston and London: Little, Brown, 1991).
4. J. C. Eccles, *Evolution of the Brain, Creation of the Self* (London and New York: Routledge, 1989), 241.
5. F. van Eeden, *Studies: Tweede Reeks* [Studies, Second Volume] (Amsterdam: W. Versluys, 1894), 321.
6. Quoted in F. D. Peat, *Infinite Potential: The Life and Times of David Bohm* (New York: Addison-Wesley, 1996).
7. F. van Eeden, *Studies: Eerste Reeks* [Studies, First Volume] (Amsterdam: W.Versluys, 1897), 226.
8. T. S. Kuhn, *The Structure of Scientific Revolutions* (Chicago: University of Chicago Press, 1962).
9. R. Strassman, *DMT, The Spirit Molecule: A Doctor's Revolutionary Research into the Biology of Near-Death and Mystical Experiences* (Rochester, VT: Park Street Press, 2001); D. Bohm and B. J. Hiley, *The Undivided Universe: An Ontological Interpretation of Quantum Physics* (London and New York: Routledge, 1995).
10. P. van Lommel, "About the Continuity of Our Consciousness," *Advances in Experimental Medicine and Biology* 550 (2004): 115–32; P. van Lommel, "Near-Death Experience, Consciousness and the Brain: A New Concept About the Continuity of Our Consciousness Based On Recent Scientific Research on Near-Death Experience in Survivors of Cardiac Arrest," *World Futures: The Journal of General Evolution* 62 (2006): 134–51. The term *phase space* has been replaced by the (more widely accepted) term *nonlocal space*, and the term *informative fields of consciousness* has been replaced by *nonlocal consciousness* because nonlocal phenomena should not really be described as a field. But the theoretical underpinnings of my vision have remained unchanged.
11. H. Walach and R. Hartmann, "Complementarity Is a Useful Concept for Consciousness Studies: A Reminder," *Neuroendocrinology Letters* 21 (2000): 221–32.
12. A visual evoked potential is a registered change in electrical activity in the EEG caused by a visual stimulation like a flash of light.
13. F. Thaheld, "Biological Nonlocality and the Mind-Brain Interaction Problem: Comments on a New Empirical Approach," *BioSystems* 2209 (2003): 1–7; J. Grinberg-Zylberbaum et al., "Human Communication and the Electrophysiological Activity of the Brain," *Subtle Energies and Energy Medicine* 3, no. 3 (1993): 25–43; J. Grinberg-Zylberbaum, M.

Deflafor, and A. Goswami, "The Einstein-Podolsky-Rosen Paradox in the Brain: The Transferred Potential," *Physics Essays* 7, no. 4 (1994): 422–28. For the replication of the results, see J. Wackermann et al, "Correlations Between Electrical Activities of Two Spatially Separated Human Subjects," *Neuroscience Letters* 336 (2003): 60–64; D. Radin, "Event-Related Electroencephalographic Correlations Between Isolated Human Subjects," *Journal of Alternative and Complementary Medicine* 10 (2004): 315–23; L. J. Standish et al., "Electroencephalographic Evidence of Correlated Event-Related Signals Between the Brain of Spatially and Sensory Isolated Subjects," *Journal of Alternative and Complementary Medicine* 10, no. 2 (2004): 307–14. For fMRI evidence of nonlocal entanglement, see L. J. Standish et al., "Evidence of Correlated Functional Magnetic Resonance Imaging Signals Between Distant Human Brains," *Alternative Therapies in Health and Medicine* 9, no. 1 (2003): 128; T. L. Richards et al., "Replicable Functional Magnetic Resonance Imaging Evidence of Correlated Brain Signals Between Physically and Sensory Isolated Subjects," *Journal of Alternative and Complementary Medicine* 11, no. 6 (2005): 955–63. For the study involving a healer, see J. Achterberg et al., "Evidence for Correlations Between Distant Intentionality and Brain Function in Recipients: An fMRI Analysis," *Journal of Alternative and Complementary Medicine* 11, no. 6 (2005): 965–71. For the recent study using laser stimulation, see R. Pizzi et al., "Non-local Correlation Between Human Neural Networks," in *Quantum Information and Computation II: Proceedings of SPIE 5436*, ed. E. Donkor, A. R. Pirick, and H. E. Brandt (SPIE—the International Society for Optical Engineering, 2004), 107–17, http://www.spiedigitallibrary.org/dbt/dbt.jsp?KEY=PSISDG&Volume=5436&Issue=1&bproc=stia&scode=SIP10.

14. R. Penrose, *Shadows of the Mind* (Oxford: Oxford University Press, 1996).

15. Strassman, *DMT, the Spirit Molecule.*

16. H. Romijn, "Are Virtual Photons the Elementary Carriers of Consciousness?" *Journal of Consciousness Studies* 9 (2002): 61–81.

17. Romijn, "Virtual Photons."

18. B. Misra and E. C. G. Sudarshan, "The Zeno's Paradox in Quantum Theory," *Journal of Mathematical Physics* 18 (1977): 756–63.

19. P. J. Marcer and W. Schempp, "Model of the Neuron Working by Quantum Holography," *Informatica* 21, no. 3 (1997): 519–34; P. J. Marcer and W. Schempp, "The Brain as a Conscious System," *International Journal of General Systems* 27, no. 11 (1998): 231–48; W. Schempp, "Quantum Holography and Neurocomputer Architectures," *Journal of Mathematical Imaging and Vision* 2 (1992): 109–64; W. Schempp, *Magnetic Resonance Imaging: Mathematical Foundations and Applications* (New York: John Wiley, 1997).

20. G. S. Engel et al., "Evidence for Wavelike Energy Transfer Through Quantum Coherence in Photosynthetic Systems," *Nature* 446 (2007): 782–86.

21. B. Julsgaard et al., "Experimental Demonstration of Quantum Memory for Light," *Nature* 432 (2004): 482–85; D. N. Matsukevich and A. Kuzmich, "Quantum State Transfer Between Matter and Light," *Science* 306 (2004): 663–66; T. Chaneliere et al., "Storage and Retrieval of Single Photons Transmitted Between Remote Quantum Memories," *Nature* 480 (2005): 833–36.

22. H. P. Hu and M. X. Wu, "Nonlocal Effects of Chemical Substances on the Brain Produced Through Quantum Entanglement," *Progress in Physics* 3 (2006): 20–26; H. P. Hu and M. X. Wu, "Photon Induced Non-Local Effects of General Anesthetics on the Brain," *NeuroQuantology* 4, no. 1 (2006): 17–31.

Chapter 13: The Continuity of the Changing Body

1. M. Ridley, *Genome: The Autobiography of a Species in 23 Chapters* (New York: HarperCollins, 2000).

2. R. N. Mantegna et al., "Linguistic Features of Non-coding DNA Sequences," *Physical Review Letters* 73 (1994): 31–69.

3. R. Robinson, "Ciliate Genome Sequence Reveals Unique Features of a Model Eukaryote," *Public Library of Science Biology* 4, no. 9 (2006).

4. J. Lederberg, "The Meaning of Epigenetics," *The Scientist* 15, no. 18 (2001): 6; M. Esteller, "How Epigenetics Affect Twins," *The Scientist* 6, no. 1 (2005): 20050707–02; G. S. Baldwin et al., "DNA Double Helices Recognize Mutual Sequence Homology in a Protein Free Environment," *Journal of Physical Chemistry B* 112, no. 4 (2008): 1060–64.

5. S. Y. Berkovich, *On the "Barcode" Functionality of the DNA, or the Phenomenon of Life in the Physical Universe* (Pittsburgh: Dorrance Publishing, 2003); Mantegna et al., "Linguistic Features."

6. E. Schrödinger, *What Is Life?* (Cambridge: Cambridge University Press, 1944); P. J. Marcer and W. Schempp, "A Mathematically Specified Template for DNA and the Genetic Code in Terms of the Physically Realisable Processes of Quantum Holography," *Proceedings of the Greenwich Symposium on Living Computers,* ed. A. M. Fedorec and P. J. Marcer (London: University of Greenwich, 1996), 45–62.

7. S. Hameroff, "Quantum Computing in DNA," *The New Frontier in Brain/Mind Science,* http://www.quantumconsciousness.org/views/Quantum ComputingInDNA.html

8. A. G. Gurwitsch, "Die Natur des spezifischen Erregers der Zellteilung" [The Nature of Specific Agents of Cell Division], *Archiv für mikroskopische Anatomie und Entwicklungsmechanik* [Archives of Microscopic Anatomy and Developmental Biology] 100 (1923–24): 11–40; M. Bischof, *Biophotonen—Das Licht, das unsere Zellen steuert* [Biophotons—The Light Directing Our Cells], 11th ed. (Frankfurt: Verlag Zweitausendeins, 2001). The web site http://www.marcobischof.com contains a number of English articles on the subject. See also F.-A. Popp and L. V.

Beloussov, eds., *Integrative Biophysics: Biophotonics* (Berlin: Springer Verlag, 2003).

9. M. Bischof, "Biophotons—the Light in Our Cells," *Journal of Optometric Phototherapy* (March 2005), 1–5; Bischof, *Biophotonen;* R. van Wijk, "Bio-photons and Bio-communication," *Journal of Scientific Exploration* 15, no. 2 (2001): 183–97.

10. In geometry, a hyperbola is a two-dimensional figure, a conic section, formed by the intersection of a cone by a plane, intersecting both halves of the cone. A nonlinear, hyperbolic reaction is specific for a quantum process. See H. Romijn, "Are Virtual Photons the Elementary Carriers of Consciousness?" *Journal of Consciousness Studies* 9 (2002): 61–81; Bischof, *Biophotonen;* Bischof, "Biophotons"; G. S. Engel et al., "Evidence for Wavelike Energy Transfer Through Quantum Coherence in Photosynthetic Systems," *Nature* 446 (2007): 782–86.

11. M. A. Gilliver et al., "Antibiotic Resistance Found in Wild Rodents," *Nature* 401 (1999): 233.

12. A. Goldbeter, "Computational Approaches to Cellular Rhythms: Review article," *Nature* 420 (2002): 238–45; A. Goldbeter et al., "From Simple to Complex Oscillatory Behavior in Metabolic and Genetic Control Networks," *Chaos* 11 (2001): 247–60; H. Fröhlich, "Coherent Excitations in Active Biological Systems," in *Modern Bioelectrochemistry,* ed. F. Gutman and H. Keyzer (New York: Plenum, 1983); A. Farhadi et al., "Evidence for Non-chemical, Non-electrical Intercellular Signalling in Intestinal Epithelial Cells," *Bioelectrochemistry* 71 (2007): 142–48.

13. B. C. Goodwin, "Developing Organisms as Self-Organizing Fields," in *Self-Organizing Systems,* ed. F. E. Yates (New York: Plenum, 1987), 167–80.

14. R. Sheldrake, *The Presence of the Past* (London: Fontana, 1988).

15. Sheldrake, *Presence of the Past.*

16. P. Weiss, *Principles of Development* (New York: Holt, 1939); A. Gurwitsch, "Über den Begriff des embryonalen Feldes" [On the concept of the embryonic field], *Archiv für Entwicklungsmechanik* [Archives of Developmental Biology] 51 (1922): 383–415; Sheldrake, *Presence of the Past;* R. Sheldrake, *A New Science of Life* (London: Blond & Briggs, 1981).

17. C. G. Jung, *Memories, Dreams, Reflections,* trans. R. and C. Winston (New York: Random House, 1961).

18. R. B. Stone, *The Secret Life of Your Cells* (West Chester, PA: Whitford Press, 1989), 82.

19. C. Backster, *Primary Perception: Biocommunication with Plants, Living Foods, and Human Cells* (Anza, CA: White Rose Millennium Press, 2003); Goldbeter, "Cellular Rhythms"; Goldbeter et al., "Oscillatory Behavior."

20. C. Sylvia and W. J. Novak, *Change of Heart* (New York: Little, Brown, 1997); P. Pearsall, *The Heart's Code* (New York: Broadway Books, 1998); P. Pearsall, G. E. Schwartz, and L. G. Russek, "Changes in Heart Transplant Recipients That Parallel the Personalities of Their Donors," *Journal of Near-Death Studies* 20, no. 3 (2002): 191–206.

Chapter 14: Endless Consciousness

1. P. Russell, *From Science to God: A Physicist's Journey into the Mystery of Consciousness* (Novato, CA: New World Library, 2002).
2. Russell, *From Science to God.*
3. D. Mesland, *Bewustzijn: De metafysische ruimte* [Consciousness: The Metaphysical Space] (Delft, the Netherlands: Eburon, 2002).
4. F. van Eeden, *Studies: Tweede reeks* [Studies: Second Volume] (Amsterdam: Versluys, 1894), 296; I. Kant, *Critique of Pure Reason,* trans. J. M. D. Meiklejohn; see http://www.gutenberg.org/etext/4280.
5. C. G. Jung, *Memories, Dreams, Reflections,* trans. R. and C. Winston (New York: Random House, 1961).
6. J. N. Ferrer, *Revisioning Transpersonal Theory: A Participatory Vision of Human Spirituality* (Albany: State University of New York Press, 2002).
7. A. H. Maslow, *Toward a Psychology of Being* (New York: Van Nostrand, 1968); S. Grof, *Beyond the Brain: Birth, Death and Transcendence in Psychotherapy* (Albany: State University of New York Press, 1985).
8. K. Wilber, *No Boundary* (Boulder, CO: Shambhala, 1981); Russell, *From Science to God.*
9. Wilber, *No Boundary.*
10. Plato, *The Republic,* trans. Benjamin Jowett, http://philosophy.eserver .org/plato/republic.txt.
11. H. Walach et al., "Spirituality: The Legacy of Parapsychology," *Archive for the Psychology of Religion* 31 (2009): 277–308.
12. W. James, *The Varieties of Religious Experience: A Study in Human Nature* (1902; New York: Mentor Books, 1958); W. James, *The Principles of Psychology,* 2 vols. (New York: Henry Holt, 1890); F. W. H. Myers, "The Subliminal Consciousness," *Proceedings of the Society for Psychical Research* 8 (1892): 298–535; E. D. Kelly and E. W. Kelly, *Irreducible Mind: Toward a Psychology for the 21st Century* (Lanham, MD: Rowman & Littlefield, 2007); F. W. H. Myers, *Human Personality and Its Survival of Bodily Death* (London: Longmans, Green, 1903); W. James, *Human Immortality: Two Supposed Objections to the Doctrine,* 2nd ed. (New York: Houghton, Mifflin, Bastion, 1900).
13. According to the *Oxford English Dictionary, endless* means "Having no end or limit of duration; unending, eternal."
14. *Akasha* is the Sanskrit word for "ether." According to ancient Indian philosophy, it is the source of the entire universe; it contains information about everything that ever was, is, and will be. See also E. Laszlo, *Science and the Akashic Field: An Integral Theory of Everything* (Rochester, VT: Inner Traditions, 2007).
15. Pew Forum on Religion and Public Life, *Many Americans Mix Multiple Faiths* (2009), http://pewforum.org/newassets/images/reports/multiple faiths/multiplefaiths.pdf.

16. For regression therapy, see J. Wade, "The Phenomenology of Near-Death Consciousness in Past-Life Regression Therapy: A Pilot Study," *Journal of Near-Death Studies* 17, no. 1 (1998): 31–53. For DMT, see Grof, *Beyond the Brain,* and R. Strassman, *DMT, The Spirit Molecule: A Doctor's Revolutionary Research into the Biology of Near-Death and Mystical Experiences* (Rochester, VT: Park Street Press, 2001).

17. M. Barbato, "Parapsychological Phenomena Near the Time of Death," *Journal of Palliative Care* 15 (1999): 30–37; E. Kübler-Ross, *On Death and Dying* (New York: Macmillan, 1969); M. De Hennezel, *La mort intime* [Intimate Death] (Paris: Éditions Robert Laffont, 1995); J. Lerma, *Into the Light: Real Life Stories About Angelic Visits, Visions of the Afterlife, and Other Pre-death Experiences* (Franklin Lakes, NJ: New Page Books, 2007); M. Callanan and P. Kelley, *Final Gifts: Understanding The Special Awareness, Needs and Communications of the Dying* (New York: Poseidon Press, 1992); P. Fenwick and E. Fenwick, *The Art of Dying: A Journey to Elsewhere* (London and New York: Continuum, 2008).

18. K. Osis and E. Haraldsson, *At the Hour of Death* (New York: Avon Books, 1977); E. W. Kelly, B. Greyson, and I. Stevenson, "Can Experiences of Near-Death Furnish Evidence of Life After Death?" *Omega* 40 (1999–2000): 13–519; I. Schmied, H. Knoblaub, and B. Schnettler, "Todesnäheerfahrungen in Ost- und Westdeutschland: Ein empirische Untersuchung" [Near-Death Experiences in East and West Germany: An Empirical Study], in *Todesnähe: Interdisziplinäre Zugänge zu einem Außergewöhnlichen Phänomen* [Near-Death: Interdisciplinary Approaches to an Extraordinary Phenomenon], ed. H. Knoblaub and H. G. Soeffner, 217–50 Konstanz, Germany: Universitätsverlag, 1999).

19. B. Guggenheim and J. Guggenheim, *Hello from Heaven: A New Field of Research—After-Death Communication—Confirms That Life And Love Are Eternal* (New York: Bantam, 1995).

20. European Value Systems Study Group, *Information Bulletin 1987* (Tilburg, the Netherlands: Tilburg University Press, 1987); E. Haraldsson and J. M. Houtkooper, "Psychic Experiences in the Multinational Human Value Study: Who Reports Them?" *Journal of the American Society for Psychical Research* 85 (1991): 145–65; E. Haraldsson, "Popular Psychology, Belief in Life After Death and Reincarnation in the Nordic Countries, Western and Eastern Europe," *Nordic Psychology* 58, no. 2 (2006), 171–80; also Pew Forum on Religion and Public Life, *Many Americans Mix Multiple Faiths* (2009), http://pewforum.org/newassets/images/reports/multiplefaiths/multiplefaiths.pdf; A. M. Greeley, "Hallucinations Among the Widowed," *Sociology and Social Research* 71, no. 4 (1987): 258–65; W. D. Rees, "The Hallucinations of Widowhood," *British Medical Journal* 4 (1971): 37–41; T. A. Rando, ed., *Parental Loss of a Child* (Champaign, IL: Research Press, 1985).

21. Guggenheim and Guggenheim, *Hello from Heaven,* 216.

22. Guggenheim and Guggenheim, *Hello from Heaven;* M. Lensink, *Toeval-lige signalen: Meer dan 100 verhalen over ervaringen rondom de dood* [Unexpected Signals: Over 100 Stories About Experiences Near Death] (Amsterdam: Schors, 2006); C. Schouterden and G. vander Linden, *"Kijk, ik ben er nog!" Met getuigenissen over tekens van overledenen* ["Look, I'm still here!" Testimonies of Signs from the Dead] (Zoetermeer, the Netherlands: Free Musketeers, 2005). The quote is from T. Waanders, *De dood en de jongen: Monument voor mijn gestorven zoon* [Death and the Boy: A Monument for My Dead Son] (Kampen, the Netherlands: Ten Have, 2006).

23. L. Halman, *The European Values Study: A Third Wave. Sourcebook of the 1999–2000 European Values Study Surveys* (Tilburg, the Nether-lands: Tilburg University, 2001); Haraldsson, "Popular Psychology." The U.S. data comes from the Association of Religion Data Archives: http:// www.thearda.com; the Pew Forum on Religion and Public Life, *U.S. Re-ligion Landscape Survey* (2007), http://www.religions.pewforum.org; and F. A. Curlin et al., "Religious Characteristics of U.S. Physicians," *Journal of General Internal Medicine* 20, no. 7 (2005): 629–34. Figures on belief in afterlife come from Halman, *European Values Study,* ques-tion 30-B. Figures on belief in reincarnation are from the Pew Forum on Religion and Public Life, *Many Americans Mix Multiple Faiths* (2009), http://pewforum.org/newassets/images/reports/multiplefaiths/multiple faiths.pdf; http:// www.religioustolerance.org/chr_poll4.htm; and Haralds-son, "Popular Psychology."

24. E. J. Larson and L. Witham, "Scientists Are Still Keeping the Faith," *Nature* 386 (1997): 235–36; E. J. Larson and L. Witham, "Leading Sci-entists Still Reject God," *Nature* 394 (1998): 313. On Europeans' belief in reincarnation, see Haraldsson, "Popular Psychology," and Halman, *European Values Study.*

25. J. Wackermann et al., "Correlations Between Electrical Activities of Two Spatially Separated Human Subjects," *Neuroscience Letters* 336 (2003): 60–64; L. J. Standish et al., "Electroencephalographic Evidence of Cor-related Event-Related Signals Between the Brain of Spatially and Sensory Isolated Subjects," *Journal of Alternative and Complementary Medicine* 10, no. 2 (2004): 307–14.

26. Waanders, *De dood en de jongen* [Death and the Boy].

27. European Value Systems Study Group, *Information Bulletin 1987;* Haraldsson and Houtkooper, "Psychic Experiences"; BBC News, "Brit-ons Report 'Psychic Powers,'" May 26, 2006, http://news.bbc.co.uk/2/ hi/uk_news/5017910.stm.

28. H. E. Puthoff, "CIA-Initiated Remote Viewing Program at Stanford Research Institute," *Journal of Scientific Exploration* 10, no. 1 (1996): 63–76; R. Targ, "A Perceptual Channel for Information Transfer over Kilometer Distances: Historical Perspectives and Recent Research," *Pro-ceedings of the IEEE* 64, no. 3 (1976): 329–54.

29. B. J. Dunne, R. G. Jahn, and R. D. Nelson, *Precognitive Remote Perception,* Princeton University Engineering Anomalies Research, School of Engineering/Applied Science, *PEAR Technical Note 83003* (Princeton: Princeton University Press, 1983); R. D. Nelson et al., "Precognitive Remote Perception: Replication of Remote Viewing," *Journal of Scientific Exploration* 10, no. 1 (1996): 109–10; S. A. Schwartz, *Opening to the Infinite: The Art and Science of Nonlocal Awareness* (Buda, TX: Nemoseen Media, 2007).

30. Schwartz, *Opening to the Infinite.*

31. Schwartz, *Opening to the Infinite.*

32. A. M. Abell, *Talks with the Great Composers* (Garmisch-Partenkirchen, Germany: G. E. Schroeder Verlag, 1964), 19–21; Kelly and Kelly, *Irreducible Mind,* chap. 7, "Genius," 423–94.

33. J. B. Alexander, "Uri's Impact on the U.S. Army," posted on http://www .urigeller.com (1996); R. Jahn and B. Dunne, *Margins of Reality: The Role of Consciousness in the Physical World* (New York: Harcourt Brace Jovanovich, 1987).

34. E. W. Davis, *Teleportation Physics Study,* Air Force Research Laboratory, Air Force Material Command, Edwards Air Force Base CA 93524–7-7048 (2006), www.fas.org/sgp/eprint/teleport.pdf; A. Zeilinger, *Einsteins Spuk* [Einstein's Phantom] (Munich: Bertelsmann, 2005).

35. L. Shuhuang et al., "Some Experiments on the Transfer of Objects Performed by Unusual Abilities of the Human Body," *Nature Journal (People's Republic of China)* 4, no. 9 (1981): 652, Defense Intelligence Agency Requirements and Validation Branch, DIA Translation LN731–8–83, Intelligence Information Report No. 6010511683 (1983); S. Kongzhi, L. Xianggao, and Z. Liangzhong, "Research into Paranormal Ability to Break Through Spatial Barriers," trans. Defense Intelligence Agency, *Chinese Journal of Somatic Science,* First Issue, 22 (1990); W. Banghui, "Evidence of the Existence of Abnormal States of Matter," trans. Defense Intelligence Agency, *Chinese Journal of Somatic Science,* First Issue, 36 (1990).

36. D. Radin, *Entangled Minds: Extrasensory Experiences in a Quantum Reality* (New York: Paraview Pocket Books, 2006); A. Goswami, R. E. Reed, and M. Goswami, *The Self-Aware Universe: How Consciousness Creates the Material World* (New York: Jeremy Tarcher/Putman, 1993); A. Goswami, *Physics of the Soul: The Quantum Book of Living, Dying, Reincarnation and Immortality* (Charlottesville, VA: Hampton Roads Publishing 2001).

Chapter 15: Some Implications of NDE Studies

1. For the general U.S. population figures, see the Association of Religion Data Archives: http://www.thearda.com/internationalData/countries/ Country_234_1.asp. For the survey of doctors, see F. A. Curlin et al.,

"Religious Characteristics of U.S. Physicians," *Journal of General Internal Medicine* 20, no. 7 (2005): 629–34; the *Nature* article on scientists is E. J. Larson and L. Witham, "Scientists Are Still Keeping the Faith," *Nature* 386 (1997): 235–36; compare with J. H. Leuba, *The Belief in God and Immortality: A Psychological, Anthropological and Statistical Study* (Boston: Sherman, French, 1926). For the figure of 7 percent, see E. J. Larson and L. Witham, "Leading Scientists Still Reject God," *Nature* 394 (1998): 313.

2. Dick Swaab, "Als we alles bewust moesten doen, zouden wij geen leven hebben" [If We Would Do Everything at a Conscious Level, It Should Be Very Difficult to Live], interview with Dick Swaab by G. Klaasen, *KRO Magazine* 47 (2009): 10–13.

3. J. Fisher, *The Case for Reincarnation* (Mississisauga, Ontario: Collins Publishers, 1984), 27.

4. I. Stevenson, *Where Reincarnation and Biology Intersect* (Westport, CT: Praeger Publishers, 1997).

5. Stevenson, *Reincarnation and Biology*.

6. Fisher, *Case for Reincarnation*.

7. Fisher, *Case for Reincarnation*.

8. P. Stark, *De hele waarheid: Verhalen uit de praktijk van orgaantransplantatie* [The Whole Truth: Real-Life Organ Transplantation Stories] (Breda, the Netherlands: Papieren Tijger, 2005).

9. E. Kompanje, "Geven en Nemen: De praktijk van postmortale orgaandonatie" [Give and Take: The Practice of Postmortem Organ Donation] (Ph.D. diss., Erasmus University, Rotterdam, 1999).

10. "Model Protocol Postmortem Organ and Tissue Donation," http://www.transplantatiestichting.nl. The protocol is available in Dutch only on this site. For more information, see also the Web site of the Eurotransplant International Foundation, http://www.eurotransplant.nl; the Web site of the NHS Evidence Health Information Resources (UK), especially the document "Organ Donation—a Protocol," http://www.library.nhs.uk/SpecialistLibrarySearch/download.aspx?resID=155999; and the Web site *The Gift of a Lifetime* (USA), especially the page "Understanding Donation," http://www.organtransplants.org/understanding/death/.

11. "Model Protocol" (in Dutch).

12. "Model Protocol" (in Dutch).

13. "Model Protocol" (in Dutch).

14. "Model Protocol" (in Dutch).

15. For organ transplantation guidelines, see "Model Protocol" (in Dutch). See also the Web site of the Eurotransplant International Foundation, http://www.eurotransplant.nl; the Web site of the NHS Evidence Health Information Resources (UK), especially the document "Organ Donation—a Protocol," http://www.library.nhs.uk/SpecialistLibrarySearch/download.aspx?resID=155999; and the Web site *The Gift of a Lifetime*

(USA), especially the page "Understanding Donation," http://www
.organtransplants.org/understanding/death/. On the process of dying,
see L. Emanuel, "Reexamining Death: The Asymptotic Model and a
Bounded Zone Definition," *Hastings Center Report* 25 (1995): 27–35;
and C. Machado and A. Shewmon, eds., *Brain Death and Disorders
of Consciousness* (New York and London: Kluwer Academic/Plenum,
2004). On the Lazarus syndrome, see J. E. Jordan, E. Dyess, and J. Cliett,
"Unusual Spontaneous Movements In Brain-Dead Patients," *Neurology* 35 (1985): 1082. On physiological changes while brain-dead, see R.
Wetzel et al., "Hemodynamic Responses in Brain Dead Organ Donor Patients," *Anesthesia and Analgesia* 64 (1985): 125–28; S. H. Pennefather,
J. H. Dark, and R. E. Bullock, "Haemodynamic Responses to Surgery
in Brain-Dead Organ Donors," *Anaesthesia* 48, no. 12 (1993): 1034–38.

16. Kompanje, *Geven en Nemen* [Give and Take]; D. Sperling, *Management
of Post-mortem Pregnancy: Legal and Philosophical Aspects* (Aldershot,
United Kingdom: Ashgate, 2006).

17. http://www.transplantatiestichting.nl.

18. See "Transplant Journey," *The Gift of a Lifetime*, http://www.organ
transplants.org. There are now 105,172 patients on the waiting list in the
United States to receive an organ transplant. In 2006 a total of 14,754
organ donors were recovered in the United States (of which 28,931 organs were transplanted), and of these organ donors 8,022 were cadaveric
donors, both brain-dead and nonbeating-heart organ donors, which represented a small increase over the total of 7,150 in 2004. Living donors
decreased from 7,004 in 2004 to 6,732 in 2006. It is estimated that about
35 percent of potential donors never become donors because family members refuse to give consent.

19. D. Sanghavi, "When Does Death Start?" *New York Times*, December
20, 2009, http://www.nytimes.com/2009/12/20/magazine/20organ-t
.html?_r=2&ref=magazine. For UK transplant numbers, see "Organ
Donation," NHS Blood and Transplant, http://www.uktransplant.org
.uk. A record number of donors were living donors; 954 people donated
a kidney or a segment of their liver or lung, representing more than half
of all donors. The highest number of nonbeating heart donor transplants
took place: 579 transplants, a 35 percent increase over 2007–2008. The
remaining 321 were brain death donors. About 7,800 patients were in
2009 listed in the United Kingdom as actively waiting for a transplant.

20. Kompanje, *Geven en Nemen* [Give and Take]; G. Lodewick, *Ik houd
mijn hart vast: Andere dimensies van orgaandonatie* [With My Heart in
My Mouth: Other Dimensions of Organ Donation] (Deventer, the Netherlands: Ankh-Hermes, 1998).

Chapter 16: Epilogue

1. D. Hammarskjöld, *Markings,* trans. Leif Sjöberg and W. H. Auden (New York: Knopf, 1964); J. Milton, *Paradise Lost* (1667), available online at http://www.literature.org/authors/milton-john/paradise-lost/.
2. E. Laszlo, *You Can Change the World: The Global Citizen's Handbook for Living on Planet Earth* (New York: Select Books, 2003).

Appendix: The Practical Significance of NDE in Health Care

1. I. Corbeau, "Psychische problematiek en hulpverlening na een bde" [Psychological Problems and Support After an NDE], *Terugkeer (Tijdschrift rond bijna-dood ervaringen en zingeving)* [Return (Journal of Near-Death Experiences and Meaning)] 15, nos. 2–3 (2004): 16–22; L. R. Derogatis, R. S. Pilman, and L. Covi, "SCL–90: An Outpatient Psychiatric Rating Scale—A Preliminary Report," *Psychopharmacology Bulletin* 9 (1973): 13–27.
2. See http://www.merkawah.nl and http://www.iands.org.
3. J. la Puma et al., "Talking to Comatose Patients," *Archives of Neurology* 45 (1988): 20–22; P. Tosch, "Patients' Recollections of Their Posttraumatic Coma," *Journal of Neuroscience Nursing* 20, no. 4 (1988): 223–28.
4. B. Greyson, "Incidence of Near-Death Experiences Following Attempted Suicide," *Suicide and Life-Threatening Behavior* 16, no. 1 (1986): 40–45; B. Greyson, "Near-Death Experiences Precipitated by Suicide Attempt: Lack of Influence of Psychopathology, Religion, and Expectations," *Journal of Near-Death Studies* 9 (1991): 183–88.
5. B. Greyson, "Near-Death Experiences and Anti-suicidal Attitudes," *Omega* 26 (1992–1993): 81–89.
6. H. Stolp, "De Gouden Vogel: Dagboek van een stervende jongen" (Rotterdam, the Netherlands: Lemniscaat, 1987), translated as *The Golden Bird* (New York: Dial Books for Young Readers, 1987); E. Elsaesser-Valarino, *Talking with Angel: About Illness, Death and Survival* (Edinburgh: Floris Books, 2005).
7. J. Lerma, *Into the Light: Real Life Stories About Angelic Visits, Visions of the Afterlife, and Other Pre-death Experiences* (Franklin Lakes, NJ: New Page Books, 2007); M. Callanan and P. Kelley, *Final Gifts: Understanding the Special Awareness, Needs and Communications of the Dying* (New York: Poseidon Press, 1992); M. Nahm and B. Greyson, "Terminal Lucidity in Patients with Chronic Schizophrenia and Dementia: A Survey of the Literature," *Journal of Nervous and Mental Disease* 197, no. 12 (2009): 942–44.
8. Lerma, *Into the Light;* Callanan and Kelley, *Final Gifts.*
9. E. Kübler-Ross, *On Death and Dying* (New York: Macmillan, 1969); E. Kübler-Ross, *Death, the Final Stage of Growth* (Englewood Cliffs, NJ: Prentice-Hall, 1975).

10. B. Guggenheim and J. Guggenheim, *Hello from Heaven: A New Field of Research—After-Death Communication—Confirms That Life and Love Are Eternal* (New York: Bantam Books, 1995); M. Lensink, *Toevallige Signalen: Meer dan 100 verhalen over ervaringen rondom de dood* [Unexpected Signals: Over 100 Stories About Experiences Near Death] (Amsterdam: Schors, 2006); T. Waanders, *De dood en de jongen: Monument voor mijn gestorven zoon* [Death and the Boy: A Monument for My Dead Son] (Kampen, the Netherlands: Ten Have, 2006).

11. F. A. Curlin et al., "Religious Characteristics of U.S. Physicians," *Journal of General Internal Medicine* 20, no. 7 (2005): 629–34.

12. Kübler-Ross, *On Death and Dying;* Kübler-Ross, *Death;* M. de Hennezel, *La mort intime* [Intimate Death] (Paris: Éditions Robert Laffont, 1995); M. de Hennezel and J.-Y. Leloup, *L'art de mourir* [The Art of Dying] (Paris: Éditions Robert Laffont, 1997).

13. E. Brandt, "Minder euthanasie, meer meldingen" [Less Euthanasia, More Reports], *Medisch Contact* 62, no. 19 (2007): 804–7.

14. B. van Schuijlenburg, *Waar was de patiënt? Omgang met mensen met een bijnadood-ervaring in de hulpverlening* [Where Was the Patient? Supporting People with a Near-Death Experience] (Assen, the Netherlands: Van Gorcum, 1994).

Glossary

AKASHA: The Sanskrit word for ether. According to ancient Indian philosophy, Akasha is the source of the entire universe; it contains information about everything that ever was, is, and will be.

ANESTHESIA: General anesthesia, causing a patient to lose consciousness.

ANEURYSM: A balloonlike bulge at a weakened spot in a blood vessel, not unlike a protruding bubble on the inside of a bicycle tire.

ANOXIA: Total loss of oxygen supply.

ANTIGEN: An antigen is a molecule capable of producing antibodies and triggering an immune response.

BLACK HOLE: A region in space where the gravitational field is so powerful that it absorbs everything in its vicinity, including light.

BOSE-EINSTEIN CONDENSATE: A (quantum) physics system in which the many parts that form an organized system not only behave like a whole but actually become a whole.

BRAIN DEATH: Coma with indications of irreparable damage to the cerebral cortex and brain stem.

BRAIN STEM: The oldest part of the brain; it connects the cerebrum with the cerebellum and the spinal cord and governs vital functions such as heartbeat, respiration, and blood pressure.

CAUSALITY: Because time is unidirectional, the order of cause and effect is fixed.

CEREBRAL CORTEX: The largest part of the brain, spread across frontal lobe, temporal bone, parietal and occipital lobes, each with its specific functions.

CES: Cranial Electrotherapy Stimulation; see also TDCS.

CLAIRAUDIENCE: The sensation of knowing what somebody is thinking.

CLAIRSENTIENCE: The sensation of knowing what somebody is feeling.

CLAIRVOYANCE: The sensation of knowing what is happening at a great distance.

CLINICALLY DEAD: Unconsciousness caused by the loss of heartbeat and respiration. Unless patients are resuscitated within five to ten minutes, they will die.

COGNITIVE FUNCTIONS: The ability to understand, think, reason, and remember.

COHERENT: Two waves are coherent if they oscillate regularly and have a certain relationship that enables interference.

COMA: A state of deep unconsciousness, precipitated by a range of different disorders; also known as "apparent death."

COMPLEMENTARITY: Light behaves like either a particle or a wave depending on the experiment design, but never like both at the same time.

COMPLEX: Composite and complicated.

CONCEPTION: Fertilization.

CONTINUITY: There is never any discontinuity; everything happens orderly in time and space.

CORNEAL REFLEX: The blinking of the eye at touch.

CORRELATION: Interrelation or connection.

CORTEX: See cerebral cortex.

DBS: Deep brain stimulation through the implanting of deep electrodes in hyperactive or underactive areas of the brain.

DEATHBED VISION: The experience of nonlocal consciousness during the terminal phase.

DECOHERENCE: The leaking of information in living quantum systems.

DEFIBRILLATION: The administration of an electric shock to the chest of a cardiac arrest patient during resuscitation (ventricular fibrillation).

DELUSION: An incorrect interpretation of a correct perception.

DENDRITES: The branched projections of neurons, which connect synapses with other neurons.

DEPERSONALIZATION: The phenomenon of identity loss, coupled with feelings of detachment, alienation, and a sense of unreality.

DIFFERENTIATION: The process of becoming more specific by developing more specialized functions or by transforming into a more heterogeneous and detailed state.

DISSOCIATION: The escape from the frightening reality of a trauma, or a disruption in identity, memory, or consciousness.

DMT: Dimethyltryptamine, a psychoactive substance produced by the pineal gland. It causes a state of mind that bears a striking resemblance to an NDE.

DNA: Deoxyribonucleic acid. The human DNA consists of 23 pairs of chromosomes and contains approximately 30,000 genes made up of more than 3 billion base pairs. About 5 percent of DNA encodes protein and is called exon.

DONOR PREVENTION: The prevention of brain death.

DUALISM: Based on the premise of two opposing or neighboring, irreducible basic principles.

ECG: Electrocardiogram; the registration of the heart's electrical activity.

ECHO: Ultrasound analysis.

EDEMA: Swelling.

EEG: Electroencephalogram; the registration of the brain's electrical activity.

ELECTRON: An elementary particle that forms part of an atom or moves freely in space. It carries a negative electric charge. The electron's antiparticle is called a positron.

EMPATHY: The ability to imagine and sympathize with the feelings or thoughts of others.

EMPIRICAL RESEARCH: All research activities or experiments that draw on direct or indirect observations and that are not based on a predetermined theoretical foundation.

ENHANCED INTUITIVE SENSITIVITY: The experience of verifiably correct information about events or persons or about future events that could not have been perceived via the senses or obtained via memories. See also Intuition.

ENTANGLEMENT: See nonlocal entanglement.

EPIGENETICS: The study of reversible changes in gene function, brought about by external factors without causing changes in the DNA sequence in the cell nucleus.

EPILEPSY: An illness with attacks (insults, seizures, or convulsions) caused by brain disorders and sometimes coupled with unconsciousness.

EPIPHENOMENALISM: The view according to which all mental phenomena, processes, or states are mere side effects or by-products (epiphenomena) of cerebral processes.

EPIPHYSIS: The epiphysis or pineal gland is an endocrine gland in the brain. An endocrine gland produces hormones.

EPS: Electrophysiological study, in which arrhythmias are diagnosed and induced with the help of catheters in the heart.

EUTHANASIA: Literally: good death. The practice of deliberately ending another person's life, at their request and according to strict guidelines. Active euthanasia involves the administration of, for example, a lethal injection, usually in a patient in the final stages of a malignant disease or AIDS, precipitating an accelerated and unnatural death.

EXISTENTIAL CRISIS: A crisis accompanied by feelings of anxiety, panic, and powerlessness.

FARADAY CAGE: An isolated cagelike construction of conductive material that blocks all electromagnetic radiation.

FIELD: A field is an immaterial and invisible area that extends throughout the whole of space and that is capable of exerting an influence on our visible, material world.

FMRI: Functional magnetic resonance imaging.

FOURIER TRANSFORM: This is used to transform a linear system (time) into wave functions or frequencies or vice versa.

GENE: A gene carries the information for a specific hereditary property in a cell.

GENOME: The genome of an organism comprises one complete set of chromosomes and describes the combination of all hereditary factors.

GLUON: An elementary particle that is responsible for transmitting a powerful nuclear force. The electrical charge of gluons equals 0, their spin equals 1.

HALLUCINATION: A sensory perception that is experienced as real by the hallucinating person but that does not correspond with reality.

HIPPOCAMPUS: Connecting brain structure, located inside the medial temporal lobe and belonging to the limbic system.

HOLISTIC SYSTEM: A system based on the premise that everything is connected to everything else, which is why a problem can never be examined or solved in isolation.

HOLOGRAM: A spatial three-dimensional image on a flat surface, which originates with the help of coherent laser light, giving rise to interference patterns. Each part contains the entire image.

HOSPICE: A home or unit in a hospital or nursing home that specializes in terminal care and where terminally ill patients can be cared for until their death.

HYPERBOLA: In geometry, a two-dimensional figure, a conic section, formed by the intersection of a cone by a plane, intersecting both halves of the cone. A nonlinear, hyperbolic reaction is specific for a quantum process.

HYPERESTHESIA: Hypersensitivity.

HYPNOSIS: An artificially created state of consciousness characterized by relaxation and concentration but without waking consciousness.

HYPOPHYSIS: The hypophysis, or the pituitary gland, is a gland in the middle of the head, below the brain, secreting hormones. The hypophysis fulfills an important role in hormone regulation.

HYPOTHALAMUS: Brain structure below the thalamus, which forms part of the limbic system and regulates blood pressure, respiration, emotions, and so forth.

HYPOTHERMIA: Cooling of the body's temperature below normal.

HYPOXIA: Oxygen shortage.

ICD: Implantable cardioverter-defibrillator, implanted in patients with recurrent life-threatening arrhythmias who do not, or not sufficiently, respond to medication.

ILLUSION: An apparent reality, or a false sense of reality.

IMMATERIAL: Mental, incorporeal, spiritual.

IMMUNE SYSTEM: The body's defense system, aimed at combating foreign intruders or cells that are changing inside the body.

INSTANTANEOUS: Immediate, that is, much faster than the speed of light.

INTERACTIONISM: The scientific study of all processes whereby individuals and groups interact with one another.

INTERFACE: Place of resonance.

INTERFERENCE: Phenomenon whereby overlapping coherent waves create a specific pattern.

INTUITION: The experience of verifiably correct information about events or people or about future events that could not have been perceived via the senses or obtained from memory. It is a manifestation of nonlocal consciousness, a form of direct knowledge that has no rational basis. It is coupled with a sense of inner certainty and familiarity. It can consist of telepathy, clairvoyance, clairsentience, clairaudience (see separate entries), or prognostic dreams.

INTUITIVE, REMOTE VIEWING: Also known as nonlocal perception. It refers to people's ability to perceive remotely (nonlocally) objects that are either selected at random or merely indicated by coordinates on a map. A form of clairvoyance (see separate entry).

ISOTOPE ANALYSIS: The imaging of organs or blood vessels through the injection of minute quantities of a radioactive substance.

JUNK DNA: Nonprotein encoding DNA, or intron; its function remains unknown.

LIMBIC SYSTEM: Part of the brain that encompasses the hippocampus, hypothalamus, and amygdala; it plays a role in emotions and memories.

LOCALITY: Concept that objects are influenced only by direct (local) contact. The concept excludes any remote influence.

LSD: Psychoactive substance, not found in nature, that causes an experience of consciousness bearing some resemblance to an NDE.

LUCID DREAM: A dream in which the dreamer is aware of the fact that he or she is dreaming.

LUMINESCENCE: The radiation that sometimes follows a change in atomic energy level. It occurs when energy-rich electrons leap into a less energy-rich orbit around the atom's nucleus. The electron emits the released energy as a photon (electromagnetic radiation). If enough energy is released, the radiation emitted can be within the wavelength visible to man.

MATERIALISM: The philosophy that reduces all reality to matter.

MBCT: Mindfulness based cognitive therapy; this combination of cognitive therapy and meditation with mindfulness has produced clear clinical improvement of certain syndromes or disorders.

MEASURING PROBLEM: See quantum measuring problem.

MEDITATION: The practice of concentrated, silent, and inner attention or concentration on an object.

MEG: Magnetoencephalogram; the registration of the brain's magnetic activity.

MEMBRANE POTENTIAL: The electric charge or voltage difference across a cell membrane.

METAPHYSICS: The school of philosophy that does not study material reality as we perceive it through the senses but instead studies reality that transcends matter.

MITOCHONDRIA: The cell's power stations.

MORPHOGENETIC FIELDS: Fields with formative information.

MORTUARY: A place where the body of a deceased person can be stored for a few days until the funeral or cremation takes place.

MRI: See fMRI.

NEAR-DEATH EXPERIENCE OR NDE: A special state of consciousness arising during an imminent or actual period of physical, psychological, or emotional death.

NEUROPLASTICITY: A constant process of adaptation in the cerebral cortex caused by the fact that our mental, intellectual, and physical activities influence both the number and the location of the connections (synapses) between neurons.

NEUROTRANSMITTERS: Chemical substances in the brain that transmit and regulate electrical signals between neurons (brain cells).

NEUTRON: A subatomic particle without electrical charge, found in atomic nuclei.

NONLINEAR: Not occurring according to linear principles. A nonlinear, hyperbolic reaction is specific for a quantum process.

NONLOCAL CONNECTION: Quantum phenomenon whereby spatially separated particles possess properties that are connected beyond time and place; also known as entanglement.

NONLOCAL PERTURBATION: See telekinesis.

NONLOCAL SPACE: A space where time and place play no role, where everything is instantaneously and continuously connected. The nonlocal space represents a hidden reality that exerts a constant influence on our material world.

NONPERIODIC: Unpredictable. Quantum-mechanical processes in living matter are nonperiodic. See also nonstatistical.

NONSTATISTICAL: Unpredictable or chaotic quantum-mechanical processes originating in the nonlocal space for living organisms. See also nonperiodic.

OBJECTIVITY: Independent of people's opinions; sometimes used to imply neutrality or impartiality. The opposite of subjectivity.

PALLIATIVE: Palliative care is aimed at alleviating or easing pain, especially when recovery is not or no longer possible.

PANPSYCHISM: Also known as idealism: according to this model, all material, physical systems contain a form of subjective consciousness at an elementary or fundamental level.

PARADIGM: A philosophical principle, framework, or view with a set of rules and regulations.

PARANORMAL: Phenomena that cannot be explained with our "normal" physical laws and prevailing concepts.

PARASYMPATHIC: The parasympathetic nervous system is part of the autonomous nervous system and influences the internal organs; it is most active when the body is at rest, when it stimulates metabolism, cell generation, and the buildup of our body's reserves.

PARKINSON'S DISEASE: A neurological disorder caused by the loss of brain cells that secrete the substance dopamine, which is necessary for motor control.

PERIMORTEM EXPERIENCE: The sensation of somebody's presence at a moment when it is not yet known that he or she has died.

PERIODICITY: Recurring at regular intervals. Processes in dead matter are periodic.

PERMISSION: Consent.

PET-SCAN: Positron emission tomography; a brain imaging technique using radioactive isotopes.

PHARMACOLOGY: The study of drug action.

PHASE SPEED: The speed at which a wave propagates. The phase speed increases in proportion to the wavelength.

PHENOMENAL: Based on subjective perception in the mind.

PHENOMENON: Observable occurrence.

PHOBIA: A mental disorder whereby people, usually for unknown reasons, develop a pathological fear of specific things or situations.

PHOTON: Particle, carrier of electromagnetic force: a "bundle" of light.

PHYSIOLOGY: The study of the normal vital functions and phenomena related to humans, animals, and plants.

PINEAL GLAND: The pineal gland, or epiphysis, is an endocrine gland in the brain. An endocrine gland produces hormones.

PLACEBO EFFECT: The self-healing effect brought about by somebody's beliefs and expectations, with demonstrable effect in brain research, among other fields.

POSTHUMOUS: After life; when somebody has died.

POSTMORTEM: After death.

POSTMORTEM EXPERIENCE: The sensation of contact with (the consciousness of) a deceased loved one in the weeks, months, or years after his or her death.

PROBABILITY WAVE: In quantum mechanics, a wave indicating the probability with which a particle will be found in a particular location. Also known as wave function.

PROSPECTIVE STUDY: This kind of study approaches all consecutive patients with a predefined diagnosis and therefore has much greater scientific value than a retrospective study.

PROTONS: A subatomic particle with a positive electric charge. A proton's charge is equal, though opposite, to the electron's charge.

PSYCHOKINESIS: See telekinesis.

QUANTUM MEASURING PROBLEM: The problem posed by the fact that the countless possibilities implied in a probability wave are lost as soon as a result is measured. Without perception, a quantum object has no definitive location in time and space, nor any of the fixed properties that classical physics tends to ascribe to objects.

QUANTUM MECHANICS/QUANTUM PHYSICS: Theory describing the world of atoms and subatomic particles.

QUANTUM SPIN CORRELATION: Also known as nuclear spin resonance or quantum spin coherence; coherence or rhythmic oscillation of the spin of all the smallest particles, such as protons, neutrons, and electrons, in the cell nucleus. See also Spin.

QUANTUM SUPERPOSITION: Occurs during the coherent organization of photons and the self-organization of biological processes.

QUANTUM ZENO EFFECT: Following a series of rapidly successive perceptions in a quantum system, the effect of the perception appears to freeze and the ever-changing system appears to come to a standstill.

QUARKS: Elementary or subatomic particles and fundamental constituents of matter. Quarks have a spin, and each quark has an antiparticle.

REDUCTIONISM: The act of reducing or tracing back to more fundamental properties.

REGRESSION THERAPY: Regression involves going back in time, to a time even before birth. Also known as reincarnation therapy.

REINCARNATION: The belief that (aspects of) the consciousness of a living being does not disappear after death but is reborn in another living being.

REMOTE VIEWING: Also known as nonlocal perception. People are capable of intuitive, remote (nonlocal) perception of objects that are randomly selected or merely indicated by coordinates on a map. A form of clairvoyance (see separate entry).

REM SLEEP: Rapid-eye-movement sleep; the period during sleep when people dream; it is coupled with increased brain activity.

RESISTANCE: The capacity of a living organism to withstand a certain influence.

RESONANCE: Vibration with the same frequency and phase.

RESUSCITATION: Attempts at reviving someone with a cardiac or respiratory arrest through external heart massage, respiration, and the administration of an electric shock (defibrillation).

RETROSPECTIVE STUDY: The study of past events, often involving a randomly and unknown selection of patients, which renders the results of this type of study less reliable.

SELF-ORGANIZING SYSTEM: This kind of system gives rise to patterns or structures that interact with their surroundings without being directly caused by external factors.

SHOCK: Low blood pressure.

SPECT-SCAN: Single photon emission computed tomography, an imaging technique based on the injection of radioactive isotopes, using single photon emission.

SPIN: Quantum-mechanical property of elementary particles undergoing a rotation (intrinsic momentum). A spin is a fundamental property of nature, like electrical charge or mass. All the smallest particles, such as protons, neutrons, and electrons, possess a spin, which is either positive or negative and which is always a multiple of ½.

SPIRITUALITY: The search for meaning through inner experience.

STATISTICAL PROCESSES: (Predictable, ordered, regular) quantum-mechanical processes that originate in the nonlocal space of dead matter.

STRING THEORY: Theory based on one-dimensional oscillating energy lines (open or closed strings).

SUBJECTIVITY: The personal opinion or personal viewpoint of an individual.

SUPERHOLOGRAM: A multidimensional hologram.

SUPERPOSITION: A calculation cannot determine the location of a particle (such as a photon). If the location of a particle is not known, it is "in superposition." This means that the particle is surrounded by a kind of cloud of possible locations. See also probability wave or wave function.

SYMPATHETIC: The sympathetic nervous system is part of the autonomous nervous system and is most active during exertion and stress; it regulates the release of energy, influences organs so that the body can function, slows digestion.

SYNAPSE: The junction between neurons.

SYNCHRONICITY: The not strictly causal or supposedly accidental concurrence of events.

SYNESTHESIA: The cross-influencing of sensory impressions.

TDCS: Transcranial direct current stimulation, the administration, under anesthesia, of an electrical shock to seriously depressed patients. See also CES.

TELEKINESIS: The power to influence or move matter nonlocally with the mind. Also known as psychokinesis or nonlocal perturbation.

TELEPATHY: Nonlocal connection between people's consciousness; contact with people at a great distance.

TELEPORTATION: The moving of material objects through the mind (mind power).

TERMINAL PATIENTS: Patients in the terminal phase of an intractable illness.

TES: Transcranial electrical stimulation; the stimulation of brain functions via an electrical field aimed at the skull.

THALAMUS: An important part of the brain; just like the hypothalamus, it is located in the midbrain. It functions like a kind of relay station between various parts of the brain.

TMS: Transcranial magnetic stimulation; the stimulation of brain functions via a magnetic field aimed at the skull.

TRANCE: A dissociative phenomenon whereby someone experiences another level of consciousness, is less sensitive to external stimuli, but is neither unconscious nor asleep. The sense of self can be diminished.

TRANSCENDENCE: Rising above.

TRANSFORMATION: Change, metamorphosis.

TRANSMISSION: Passing on.

TRANSPERSONAL: Aspects in somebody's consciousness that go beyond the personal or the ego.

TRAUMA: Physical or psychological damage.

UNCERTAINTY PRINCIPLE: According to the Heisenberg uncertainty principle in quantum mechanics, there is a fundamental limit to the accuracy with which certain complementary physical properties (wave-particle) can be measured.

UNIVERSE: The entire space-time continuum in which we live, together with all matter and energy, is known as the universe or cosmos.

VACUUM: A volume of space that is essentially empty of matter but still can possess a huge amount of energy and information.

VEGETATIVE STATE: A form of coma with spontaneous respiration and brain-stem reflexes.

VENTRICULAR FIBRILLATION: The chaotic flailing of the heart, resulting in a cardiac arrest that can be treated only by defibrillation (electric shock).

VERIFIABLE: Refers to the fact that it should be possible to check the accuracy of a claim or result.

VIBRATION: Pulsation with a certain frequency; also known as oscillation or the periodic, repeated inversion of the direction of movement.

VIRTUAL: That which is seemingly real or a possibility.

WAVE FUNCTION: See probability wave.

WAVELENGTH: Inversely proportional to frequency. The symbol for frequency is Hertz (Hz), and 1 Hz corresponds with a period of 1 second.

ZERO POINT ENERGY: A vacuum is not empty; at absolute zero, −273.15 degrees Celsius, it is full of energy (a "plenum"), and at microlevel it undergoes constant quantum fluctuations, which create new quanta "from nothing," which promptly disappear again. These quantum fluctuations are also known as the vacuum's zero point energy. It can give rise to virtual particles (with antiparticles) that instantly destroy each other again.

Bibliography

Web Sites

www.pimvanlommel.nl
www.iands.org: IANDS United States, International Association of Near-Death Studies.
www.nderf.org: Near-Death Experience Research Foundation.
www.merkawah.nl: Dutch branch of IANDS.
www.transpantatiestichting.nl: Model protocol postmortem organ and tissue donation.

Literature

Abell, A. M. *Talks with the Great Composers*. Garmisch-PartenKirchen, Germany: G. E. Schroeder Verlag, 1964.

Achterberg, J., K. Cooke, T. Richards, L. Standish, L. Kozak, and J. Lake. "Evidence for Correlations Between Distant Intentionality and Brain Function in Recipients: An fMRI Analysis." *Journal of Alternative and Complementary Medicine* 11, no. 6 (2005): 965–71.

Acosta, M. T., P. Montanez, and F. E. Leon-Sarmiento. "Half Brain but Not Half Function." *Lancet* 360 (2002): 643.

Alexander, J. B. "Uri's Impact on the U.S. Army." 1996. http://www.urigeller.com.

Ali, A. A., J. Ellis, and S. Randjbar-Daemi, eds. *Salamfest*. Singapore: World Scientific, 1993.

Alkire, M. T., A. G. Hudetz, and G. Tononi. "Consciousness and Anesthesia." *Science* 322, no. 5903 (2008): 876–80.

Alkire, M. T., and J. Miller. "General Anesthesia and the Neural Correlates of Consciousness." *Progress in Brain Research* 150 (2005): 229–44.

Aspect, A., J. Dalibard, and G. Roger. "Experimental Tests of Bell's Inequality Using Varying Analyses." *Physical Review Letters* 25 (1982): 1084.

Association of Religion Data Archives. http://www.thearda.com.

Athappilly, G. K., B. Greyson, and I. Stevenson. "Do Prevailing Society Models Influence Reports of Near-Death Experiences: A Comparison of Accounts Reported Before and After 1975." *Journal of Nervous and Mental Disease* 194, no. 3 (2006): 218–33.

Atwater, P. M. H. *Coming Back to Life: The After-Effects of the Near-Death Experience.* Rev. ed. New York: Citadel, 2001.

———. *The New Children and Near-Death Experiences.* Rochester, VT: Bear & Company, 2003.

Backster, C. "Primary Perception: Biocommunication with Plants, Living Foods, and Human Cells." Anza, CA: White Rose Millennium Press, 2003.

Badham, P., and L. Badham. *Death and Immortality in the Religions of the World.* New York: Paragon House, 1987.

Bailey, A. A. *Death: The Great Adventure.* New York: Lucis Trust, 1992.

Bailey, L. W., and J. Yates, eds. *The Near-Death Experience: A Reader.* New York: Routledge, 1996.

Baldwin, G. S., N. J. Brooks, R. E. Robson, A. Wynveen, A. Goldar, S. Leikin, J. M. Seddon, and A. A. Kornyshev. "DNA Double Helices Recognize Mutual Sequence Homology in a Protein Free Environment." *Journal of Physical Chemistry B* 112, no. 4 (2008): 1060–64.

Balkin, T. J., A. R. Braun, N. J. Wesensten, K. Jeffries, M. Varga, P. Baldwin, G. Belenky, and P. Herscovitch. "The Process of Awakening: A PET Study of Regional Brain Activity Patterns Mediating the Re-Establishment of Alertness and Consciousness." *Brain* 125 (2002): 2308–19.

Banghui, W. "Evidence of the Existence of Abnormal States of Matter." Translated by the Defense Intelligence Agency. *Chinese Journal of Somatic Science* First Issue, 36 (1990).

Barbato, M. "Parapsychological Phenomena Near the Time of Death." *Journal of Palliative Care* 15 (1999): 30–37.

Baringa, M. "Buddhism and Neuroscience: Studying the Well-Trained Mind." *Science* 302 (2003): 44–46.

BBC News. "Britons Report 'Psychic Powers.'" http://news.bbc.co.uk/2/hi/uk_news/5017910.stm.

Beaufort, F. "Letter to Dr. W. Hyde Wollaston." *An Autobiographical Memoir of Sir John Barrow.* London: John Murray, 1847.

Beauregard, M. "Mind Does Really Matter: Evidence from Neuroimaging Studies of Emotional Self-Regulation, Psychotherapy, and Placebo Effect." *Progress in Neurobiology* 81, no. 4 (2007): 218–36.

———, ed. *Consciousness, Emotional Self-Regulation and the Brain.* Amsterdam: John Benjamins Publishing, 2004.

Beauregard, M., and D. O'Leary. *The Spiritual Brain: How Neuroscience Is Revealing the Existence of the Soul.* San Francisco: HarperOne, 2007.

Beauregard, M., and V. Paquette. "Neural Correlates of a Mystical Experience in Carmelite Nuns." *Neuroscience Letters* 405 (2006): 186–90.

Beauregard, M., P. Lévesque, and V. Paquette. "Neural Basis of Conscious and Voluntary Self-Regulation of Emotion." In *Consciousness, Emotional Self-Regulation and the Brain,* edited by M. Beauregard, 163–94. Amsterdam: John Benjamins Publishing, 2004.

Becker, J., and J. de Hart. *Godsdienstige veranderingen in Nederland* [Religious Change in the Netherlands]. Werkdocument 128. The Hague, the Netherlands: Sociaal Cultureel Planbureau, 2006.

Benedetti, F., H. S. Mayberg, T. D. Wager, C. S. Stohler, and J.-K. Zubieta. "Neurobiological Mechanisms of the Placebo Effect." *Journal of Neuroscience* 25, no. 45 (2005): 10390–402.

Bergson, H. *L'Énergie Spirituelle.* In *Bibliothèque de Philosophie Contemporaine fondée par Félix Alcan.* 58th ed. Paris: Presses Universitaire de France, 1955. Translated by H. W. Carr as *Mind-Energy: Lectures and Essays.* London: Macmillan, 1920.

Berkovich, S. Y. *On the "Barcode" Functionality of the DNA, or the Phenomenon of Life in the Physical Universe.* Pittsburgh: Dorrance Publishing, 2003.

———. "On the Information Processing Capabilities of the Brain: Shifting the Paradigm." *Nanobiology* 2 (1993): 99–107.

Besant, A. *Death—and After?* 1906; reprint, Wheaton, IL: Quest Books, 1998. http://www.gutenberg.org/etext/18266.

Bessette, L., ed. *Healing: Beyond Suffering or Death.* Chabanel, Quebec, Canada: Publications MNH, 1994.

Bierman, D. J., and R. Van Wees. "Buitengewone ervaringen, andere bewustzijnsvormen en persoonlijkheid" [Extraordinary Experiences, Other Forms of Consciousness, and Personality]. *Tijdschrift voor Parapsychologie* [Journal for Parapsychology] 60 (1993): 51–72.

Bischof, M. *Biophotonen—Das Licht, das unsere Zellen steuert* [Bio-Photons—The Light Directing Our Cells]. 11th ed. Frankfurt: Verlag Zweitausendeins, 2001. The Web site http://www.marcobischof.com contains a number of English articles on the subject.

———. "Biophotons—The Light in our Cells." *Journal of Optometric Phototherapy* (2005), 1–5.

Blackmore, S. *Dying to Live: Science and the Near-Death Experience.* London: Grafton, 1993.

———. "There Is No Stream of Consciousness." *Journal of Consciousness Studies* 9, nos. 5–6 (2002): 17–28.

Blanke, O., T. Landis, L. Spinelli, and M. Seeck. "Out-of-Body Experience and Autoscopy of Neurological Origin." *Brain* 127 (2004): 243–58.

Blanke, O., S. Ortigue, T. Landis, and M. Seeck. "Stimulating Illusory Own-Body Perceptions: The Part of the Brain That Can Induce Out-of-Body Experiences Has Been Located." *Nature* 419 (2002): 269–70.

Blatter, G. "Schrödinger's Cat Is Now Fat." *Nature* 406 (2000): 25–26.

Boer, J. A. den. *Neuro-filosofie: Hersenen—Bewustzijn—Vrije wil* [Neurophilosophy: The Brain—Consciousness—Free Will]. Amsterdam: Boom, 2003.

Bohm, D. *Wholeness and the Implicate Order.* London: Routledge & Kegan Paul, 1980.

———, and B. J. Hiley. *The Undivided Universe: An Ontological Interpretation of Quantum Physics.* London and New York: Routledge, 1995.

Bohr, N. *Atomic Physics and Human Knowledge.* New York: Wiley, 1958.

———. *Essays 1958–1962 on Atomic Physics and Human Knowledge.* New York: Wiley, 1963.

———and J. Kalckar, eds. *Collected Works.* Vol. 6, *Foundations of Quantum Physics I (1926–1932).* Amsterdam and New York: North Holland, 1997.

Boismont, A. B. *On Hallucinations.* Translated by R. T. Hulme. London: Henry Renshaw, 1859.

Borgstein J., and C. Grootendorst. "Clinical Picture: Half a Brain." *Lancet* 359 (2002): 473.

Born, M. *My Life and My Views: A Nobel Prize Winner in Physics Writes Provocatively on a Wide Range of Subjects.* New York: Scribner, 1968.

Brandt, E. "Minder euthanasie, meer meldingen" [Less Euthanasia, More Reports]. *Medisch Contact* 62, no. 19 (2007): 804–7.

Branston, N. M., A. Ladds, L. Symon, and A. D. Wang. "Comparison of the Effects of Ischaemia on Early Components of the Somatosensory Evoked Potential in Brainstem, Thalamus, and Cerebral Cortex." *Journal of Cerebral Blood Flow Metabolism* 4, no. 1 (1984): 68–81.

Britton, W. B., and R. R. Bootzin. "Near-Death Experiences and the Temporal Lobe." *American Psychological Society* 15, no. 4 (2004): 254–58.

Bush, N. E. "Is Ten Years a Life Review?" *Journal of Near-Death Studies* 10 (1991): 5–9.

Buunk, G., J. G. van der Hoeven, and A. E. Meinders. "Cerebral Blood Flow After Cardiac Arrest." *Netherlands Journal of Medicine* 57 (2000): 106–12.

Byskov, E. *Death Is an Illusion: A Logical Explanation Based on Martinus' Worldview.* St. Paul, MN: Paragon House, 2002.

Caesar, J. *The Gallic Wars.* Translated by W. A. McDevitte and W. S. Bohn. http://classics.mit.edu/Caesar/gallic.6.6.html.

Callanan, M., and P. Kelley. *Final Gifts: Understanding the Special Awareness, Needs and Communications of the Dying.* New York: Poseidon Press, 1992.

Cardena, E., S. J. Lynn, and S. Krippner, eds. *Varieties of Anomalous Experiences: Examining the Scientific Evidence.* Washington, DC: American Psychological Association, 2000.

Chalmers, D. J. "Consciousness and Its Place in Nature." In *Philosophy of Mind: Classical and Contemporary Readings,* edited by D. J. Chalmers. Oxford: Oxford University Press, 2002. See also http://consc.net/papers/nature.htlm.

———. "Facing Up to the Problem of Consciousness." *Journal of Consciousness Studies* 3, no. 1 (1995): 200.

———. "The Puzzle of Conscious Experience." *Scientific American* 271 (1995): 80–86.

Chanelière, T., D. N. Matsukevich, S. D. Jenkins, S. Y. Lan, T. A. B. Kennedy, and A. Kuzmich. "Storage and Retrieval of Single Photons Transmitted Between Remote Quantum Memories." *Nature* 438 (2005): 833–36.

Cheek, D. B. "Unconscious Perception of Meaningful Sounds During Surgical Anaesthesia as Revealed Under Hypnosis." *American Journal of Clinical Hypnosis* 1 (1959): 101–13.

Clute, H., and W. J. Levy. "Electroencephalographic Changes During Brief Cardiac Arrest in Humans." *Anesthesiology* 73 (1990): 821–25.

Coimbra, C. G. "Implications of Ischemic Penumbra for the Diagnosis of Brain Death." *Brazilian Journal of Medical and Biological Research* 32, no. 12 (1999): 1479–87.

Cole, K. C. "In Patterns, Not Particles, Physics Trust." *Los Angeles Times,* March 4, 1999.

Colgrave, B., and R. A. B. Mynors, eds. *Bede's Ecclesiastical History of the English People.* 1969. Reprint, Oxford: Clarendon Press, 1991.

Cook, E. W., B. Greyson, and I. Stevenson. "Do Any Near-Death Experiences Provide Evidence for the Survival of Human Personality After Death? Relevant Features and Illustrative Case Reports." *Journal of Scientific Exploration* 12 (1998): 377–406.

Coppes, B. *Bijna Dood Ervaringen en Wereldreligies: Getuigenis van universele waarheid* [Near-Death Experiences and World Religions: Testimony of Universal Truth]. Soesterberg, Netherlands: Aspekt, 2006.

Corbeau, I. "Psychische problematiek en hulpverlening na een BDE" [Psychological Problems and Support after an NDE]. *Terugkeer: Tijdschrift rond bijna-dood ervaringen en zingeving* [Return: Journal of Near-Death Experiences and Meaning]15, nos. 2–3 (2004): 16–22.

Crabtree, V. "Religion in the United Kingdom: Diversity, Trends and Decline." 2007. www.vexen.co.United Kingdom /United Kingdom /religion.html.

Curlin, F. A., J. D. Lantos, C. J. Roach, and S. A. Sellergren. "Religious Characteristics of U.S. Physicians." *Journal of General Internal Medicine* 20, no. 7 (2005): 629–34.

Davidson, R. J., J. Kabat-Zinn, J. Schumacher "Alterations in Brain and Immune Function Produced by Mindfulness Meditation." *Psychosomatic Medicine* 65, no. 4 (2003): 64–70.

Davis, E. W. *Teleportation Physics Study.* Air Force Research Laboratory, Air Force Material Command, Edwards Air Force Base CA 93524–7048, 2006. http://www.fas.org/sgp/eprint/teleport.pdf.

Dennett, D. *Consciousness Explained.* Boston: Little, Brown, 1991.

Derogatis, L. R., R. S. Pilman, and L. Covi. "SCL–90: An Outpatient Psychiatric Rating Scale—A Preliminary Report." *Psychopharmacology Bulletin* 9 (1973): 13–27.

Desmedt, J. E., and D. Robertson. "Differential Enhancement of Early and Late Components of the Cerebral Somatosensory Evoked Potentials During Forced-Paced Cognitive Tasks in Man." *Journal of Physiology* 271 (1977): 761–82.

Dijk, G. W. Van. "Bewustzijn" [Consciousness]. In *Handboek Reanimatie* [The Resuscitation Handbook], edited by B. T. J. Meursing and R. G. van Kesteren, 21–25. 2nd rev. ed. Utrecht, Netherlands: Wetenschappelijke Uitgeverij Bunge, 2004.

Dobbs, D. "Fact or Phrenology?" *Scientific American Mind* 16, no. 1 (2005): 24–31.

Doidge, N. *The Brain That Changes Itself: Stories of Personal Triumph from the Frontiers of Brain Science.* London: Penguin Books, 2007.

Donkor, E., A. R. Pirick, and H. E. Brandt, eds. *Quantum Information and Computation II: Proceedings of SPIE 5436.* 2004.

Drift, Ed van der. *Cardiac and Vascular Diseases Handbook of Electroencephalography and Clinical Neurophysiology.* Amsterdam: Elsevier, 1972.

Dunne, B. J., R. G. Jahn, and R. D. Nelson. *Precognitive Remote Perception.* Princeton Engineering Anomalies Research, School of Engineering/Applied Science. *PEAR Technical Note 83003.* 1983.

Ebert, H., A. Brawanski, A. Philipp, R. Behr, O. W. Ullrich, C. Keyl, and D. E. Birnbaum. "Deep Hypothermia and Circulatory Arrest for Surgery of Complex Intracranial Aneurysms." *European Journal of Cardiothoracic Surgery* 13, no. 3 (1998): 23–229.

Eccles, J. C. "The Effect of Silent Thinking on the Cerebral Cortex." *Truth Journal: International Interdisciplinary Journal of Christian Thought* 2 (1988).

——. *Evolution of the Brain, Creation of the Self.* London and New York: Routledge, 1989.

Edelman, G. M., and G. Tononi. *A Universe of Consciousness.* New York: Basic Books, 2000.

Eeden, F. van. *Studies: Eerste Reeks* [Studies: First Volume]. 3rd ed. Amsterdam: W. Versluys, 1897.

——. *Studies: Tweede Reeks* [Studies: Second Volume]. Amsterdam: W. Versluys, 1894.

Einstein, A. "The Concept of Space." *Nature* 125 (1930): 897–98.

Elsaesser-Valarino, E. *Talking with Angel: About Illness, Death and Survival.* Edinburgh: Floris Books, 2005.

Emanuel, L. "Reexamining Death: The Asymptotic Model and a Bounded Zone Definition." *Hastings Center Report* 25 (1995): 27–35.

Engel, G. S., T. R. Calhoun, E. L. Read, T.-K. Ahn, T. Manal, Y.-C. Cheng, R. E. Blankenship, and G. R. Fleming. "Evidence for Wavelike Energy Transfer Through Quantum Coherence in Photosynthetic Systems." *Nature* 446 (2007): 782–86.

Esteller, M. "How Epigenetics Affect Twins." *The Scientist* 6, no. 1 (2005): 20050707–02.

European Value Systems Study Group. *Information Bulletin 1987.* Tilburg, the Netherlands: Tilburg University Press, 1987.

Evans Bush, N. "Afterward: Making Meaning After a Frightening Near-Death Experience." *Journal of Near-Death Studies* 21, no. 2 (2002): 99–133.

Evans-Wentz, W. Y. *The Tibetan Book of the Dead, or the After-Death Experiences on the Bardo Plane, According to Lâma Kazi Dawa-Samdup's English Rendering*. 1927. Reprint, London: Oxford University Press, 1971. See also http://www.summum.us/mummification/tbotd/book2.shtml.

Farhadi, A., C. Forsyth, A. Banan, M. Shaik, P. Engen, J. Z. Fields, and A. Keshavarzian. "Evidence for Non-Chemical, Non-Electrical Intercellular Signalling in Intestinal Epithelial Cells." *Bioelectrochemistry* 71 (2007): 142–48.

Fenwick, P., and E. Fenwick. *The Art of Dying: A Journey to Elsewhere*. New York: Continuum, 2008.

———. *The Truth in the Light: An Investigation of over 300 Near-Death Experiences*. New York: Berkley Books, 1997.

Ferrer, J. N. *Revisioning Transpersonal Theory: A Participatory Vision of Human Spirituality*. Albany: State University of New York Press, 2002.

Fisher, J. *The Case for Reincarnation*. Mississisauga, Ontario: Collins Publishers, 1984.

Fletcher, J. *The Egyptian Book of Living and Dying*. London: Duncan Baird Publishers, 2002.

Fox, M. *Religion, Spirituality and the Near-Death Experience*. New York: Routledge, 2003.

Freeman, A. *Consciousnes:; A Guide to the Debates*. Santa Barbara: ABC-CLIO, 2003.

Friedman, R., et al. "Quantum Superposition of Distinct Macroscopic States." *Nature* 406 (2000): 43–45.

Fröhlich, H. "Coherent Excitations in Active Biological Systems." In *Modern Bioelectrochemistry*, edited by F. Gutman and H. Keyzer. New York: Plenum, 1983.

Fujioka, M., K. Nishio, S. Miyamoto, K. I. Hiramatsu, T. Sakaki, K. Okuchi, T. Taoka, and S. Fujioka. "Hippocampal Damage in the Human Brain After Cardiac Arrest." *Cerebrovascular Diseases* 10, no. 1 (2000): 2–7.

Gallup, G., and W. Proctor. *Adventures in Immortality: A Look Beyond the Threshold of Death*. New York: McGraw-Hill, 1982.

Ghoneim, M. M., and R. I. Block. "Learning and Memory During General Anaesthesia: An Update." *Anesthesiology* 87 (1997): 387–410.

Gilliver, M. A., M. Bennett, M. Begon, S. M. Hazel, and C. A. Hart. "Antibiotic Resistance Found in Wild Rodents." *Nature* 401 (1999): 233.

Globus, G. G., G. Maxwell, and I. Savodnik, eds. *Consciousness of the Brain*. New York: Plenum, 1976.

Goldbeter, A. "Computational Approaches to Cellular Rhythms: Review Article." *Nature* 420 (2002): 238–45.

———, D. Gonze, G. Houart, J.-C. Leloup, J. Halloy, and G. Dupont. "From Simple to Complex Oscillatory Behavior in Metabolic and Genetic Control Networks." *Chaos* 11 (2001): 247–60.

Goodwin, B. C. "Developing Organisms as Self-Organizing Fields." In *Self-Organizing Systems,* edited by F. E. Yates, 167–80. New York: Plenum, 1987.

Gopalan, K. T., J. Lee, S. Ikeda, and C. M. Burch. "Cerebral Blood Flow Velocity During Repeatedly Induced Ventricular Fibrillation." *Journal of Clinical Anesthesia* 11, no. 4 (1999): 290–95.

Goswami, A. *Physics of the Soul: The Quantum Book of Living, Dying, Reincarnation, and Immortality.* Charlottesville, VA: Hampton Roads Publishing, 2001.

———, R. E. Reed, and M. Goswami. *The Self-Aware Universe: How Consciousness Creates the Material World.* New York: Jeremy Tarcher/Putman, 1993.

Granqvist, P., M. Fredrikson, P. Unge, A. Hagenfeldt, S. Valind, D. Larhammar, and M. Larsson. "Sensed Presence and Mystical Experiences Are Predicted by Suggestibility, Not by the Application of Weak Complex Transcranial Magnetic Fields." *Neuroscience Letters* 379 (2005): 1–6.

Greef, J. van der, and R. N. McBurney. "Rescuing Drug Discovery: In Vivo Systems Pathology and Systems Pharmacology." *Nature Reviews/Drug Discovery* 4, no. 4 (2005): 961–67.

Greeley, A. M. "Hallucinations Among the Widowed." *Sociology and Social Research* 71, no. 4 (1987): 258–65.

Greenberger, D., M. Horne, and A. Zeilinger. "Going Beyond Bell's Theorem." In *Bell's Theorem, Quantum Theory, and Conceptions of the Universe,* edited by M. Kafatos, 73–76. Dordrecht, the Netherlands: Kluwer Academics, 1989.

Greene, B. *The Fabric of the Cosmos.* New York: Knopf, 2004.

Grey, M. *Return from Death: An Exploration of the Near-Death Experience.* London: Arkana, 1985.

Greyson, B. "Biological Aspects of Near-Death Experiences." *Perspectives in Biology and Medicine* 42, no. 1 (1998): 14–32.

———. "Dissociation in People Who Have Near-Death Experiences: Out of Their Bodies or Out of Their Minds?" *Lancet* 355 (2000): 460–463.

———. "Incidence and Correlates of Near-Death Experiences in a Cardiac Care Unit." *General Hospital Psychiatry* 25 (2003): 269–76.

———. "The Incidence of Near-Death Experiences." *Medicine and Psychiatry* 1 (1998): 92–99.

———. "Incidence of Near-Death Experiences Following Attempted Suicide." *Suicide and Life-Threatening Behavior* 16, no. 1 (1986): 40–45.

———. "The Near-Death Experience Scale: Construction, Reliability and Validity." *Journal of Nervous and Mental Disease* 171 (1983): 369–75.

———. "Near-Death Experiences." In *Varieties of Anomalous Experiences: Examining the Scientific Evidence,* edited by E. Cardena, S. J. Lynn, and S. Krippner, 315–52. Washington, DC: American Psychological Association, 2000.

———. "Near-Death Experiences and Antisuicidal Attitudes." *Omega* 26 (1992–93): 81–89.

———. "Near-Death Experiences and Personal Values." *American Journal of Psychiatry* 140 (1983): 618–20.

———. "Near-Death Experiences Precipitated by Suicide Attempt: Lack of Influence of Psychopathology, Religion, and Expectations." *Journal of Near-Death Studies* 9 (1991): 183–88.

———. "Posttraumatic Stress Symptoms Following Near-Death Experiences." *American Journal of Orthopsychiatry* 71 (2001): 358–73.

———. "Reduced Death Threat in Near-Death Experiences." *Death Studies* 16 (1992): 533–46.

———, and B. Harris. "Clinical Approaches to the Near-Death Experiencer." *Journal of Near-Death Studies* 6 (1987): 41–52.

Greyson, B., J. M. Holden, and J. P. Mounsey. "Failure to Elicit Near-Death Experiences in Induced Cardiac Arrest." *Journal of Near-Death Studies* 25, no. 2 (2006): 85–98.

Grinberg-Zylberbaum, J., M. Deflafor, and A. Goswami. "The Einstein-Podolsky-Rosen Paradox in the Brain: The Transferred Potential." *Physics Essays* 7, no. 4 (1994): 422–28.

Grinberg-Zylberbaum, J., M. Deflafor, M. E. Sanchez-Arellano, and J. A. Guevara. "Human Communication and the Electrophysiological Activity of the Brain." *Subtle Energies and Energy Medicine* 3, no. 3 (1993): 25–43.

Grof, S. *Beyond the Brain: Birth, Death, and Transcendence in Psychotherapy.* Albany: State University of New York Press, 1985.

———, and J. Halifax. *The Human Encounter with Death.* New York: Dutton, 1977.

Groth-Marnat, G., and R. Summers. "Altered Beliefs, Attitudes, and Behaviors Following Near-Death Experiences." *Journal of Human Psychology* 38 (1998): 110–25.

Gua, J., J. A. White, and H. H. Batjer. "Limited Protective Effects of Etomidate During Brainstem Ischemia in Dogs." *Journal of Neurosurgery* 82, no. 2 (1995): 278–84.

Guggenheim, B., and J. Guggenheim. *Hello from Heaven: A New Field of Research—After-Death Communication—Confirms That Life and Love Are Eternal.* New York: Bantam Books, 1995.

Gurwitsch, A. G. "Die Natur des spezifischen Erregers der Zellteilung" [The Nature of Specific Agents of Cell Division]. *Archiv für mikroskopische Anatomie und Entwicklungsmechanik* [Archives of Microscopic Anatomy and Developmental Biology] 100 (1923–24): 11–40.

———. "Über den Begriff des embryonalen Feldes" [On the Concept of the Embryonic Field]. *Archiv für Entwicklungsmechanik* [Archives of Developmental Biology] 51 (1922): 383–415.

Gutman F., and H. Keyzer, eds. *Modern Bioelectrochemistry.* New York: Plenum, 1983.

Hall, R. A. *Isaac Newton, Adventurer in Thought*. Cambridge: Cambridge University Press, 1992.

Hallett, M. "Transcranial Magnetic Stimulation and the Human Brain." *Nature* 406 (2000): 147–50.

Halman, L. *The European Values Study: A Third Wave. Sourcebook of the 1999–2000 European Values Study Surveys*. Tilburg, the Netherlands: Evs, WORC, Tilburg University, 2001.

Hameroff, S. "Quantum Computing in DNA." *New Frontier in Brain/Mind Science*. http://www.quantumconsciousness.org/views/QuantumComputing InDNA.html.

———, and R. Penrose. "Orchestrated Reduction of Quantum Coherence in Brain Microtubules." *Proceedings of the International Neural Network Society*, 793–812. Hillsdale, NJ: Erlbaum, 1995.

Hammarskjöld, D. *Markings*. Translated by Leif Sjöberg and W.H. Auden. London: Faber and Faber, 1964.

Hampe, J. C. *To Die Is Gain: The Experience of One's Own Death*. Translated by Margaret Kohl. London: Darton, Longman & Todd, 1975.

Haraldsson, E. "Popular Psychology, Belief in Life After Death and Reincarnation in the Nordic Countries, Western and Eastern Europe." *Nordic Psychology* 58, no. 2 (2006): 171–80.

Haraldsson, E., and J. M. Houtkoper. "Psychic Experiences in the Multinational Human Value Study: Who Reports Them?" *Journal of the American Society for Psychical Research* 85 (1991): 145–65.

Hawking, S. *The Universe in a Nutshell*. London: Bantam Press, 2001.

Heim, A. von St. Gallen. "The Experience of Dying from Falls." Translated by R. Noyes and R. Kletti. *Jahrbuch des Schweizer Alpenclub* [Yearbook of the Swiss Alpine Club] 27 (1892): 327–37.

Heisenberg, W. *Physics and Beyond*. New York: Harper & Row, 1971.

———. *Physics and Philosophy*. New York: Harper & Row, 1958.

Hennezel, M. de. *La mort intime* [Intimate Death]. Paris: Éditions Robert Laffont, 1995.

———, and J.-Y. Leloup. *L'art de mourir* [The Art of Dying]. Paris: Éditions Robert Laffont, 1997.

Herlitz, J., A. Bang, B. Alsen, and S. Aune. "Characteristics and Outcome Among Patients Suffering from In-Hospital Cardiac Arrest in Relation to the Interval Between Collapse and Start of CPR." *Resuscitation* 53, no. 1 (2000): 21–27.

Hoffman, R. F. "Disclosure Habits After Near-Death Experience: Influences, Obstacles, and Listeners Selection." *Journal of Near-Death Studies* 14 (1995): 29–48.

Holden, J. M. "Veridical perception in near-death experiences." In J. M. Holden, B. Greyson, and D. James, eds., *The Handbook of Near-Death Experiences*. Santa Barbara, CA: Praeger/ABC-CLIO, 2009; pp. 185–211.

Hooft, G. 't. "Dimensional Reduction in Quantum Gravity." In *Salamfest,* edited by A. A. Ali, J. Ellis, and S. Randjbar-Daemi, 284–96. Singapore: World Scientific, 1993.

Hopkin, M. "Implant Boosts Activity in Injured Brain." *Nature* 448 (2007): 522.

Hossmann, K. A., and P. Kleihues. "Reversibility of Ischemic Brain Damage." *Archives of Neurology* 29, no. 6 (1973): 375–84.

Hu, H. P., and M. X. Wu. "Nonlocal Effects of Chemical Substances on the Brain Produced Through Quantum Entanglement." *Progress in Physics* 3 (2006): 20–26.

———. "Photon Induced Non-Local Effects of General Anesthetics on the Brain." *NeuroQuantology* 4, no. 1 (2006): 17–31.

Huffman, W. H. *Robert Fludd and the End of the Renaissance.* New York: Routledge, 1988.

Huttenlocher, P. R. "Synapse Elimination and Plasticity in Developing Human Cerebral Cortex." *American Journal of Mental Deficiency* 88 (1984): 488–96.

Jack, A. I., and A. Roepstorff. "Introspection and Cognitive Brain Mapping: From Stimulus-Response to Script Report." *Trends in Cognitive Science* 6, no. 8 (2002): 333–39.

———. "Why Trust the Subject?" *Journal of Consciousness Studies* 10, nos. 9–10 (2003): v–xx.

Jahn, R., and B. Dunne. *Margins of Reality: The Role of Consciousness in the Physical World.* New York: Harcourt Brace Jovanovich, 1987.

James, W. *Human Immortality: Two Supposed Objections to the Doctrine.* 2nd ed. New York: Houghton, Mifflin, Bastion, 1900.

———. *The Principles of Psychology.* 2 vols. New York: Henry Holt, 1890.

———. *The Varieties of Religious Experience: A Study in Human Nature.* 1902. New York: Mentor Books, 1958.

Jansen, K. "Neuroscience, Ketamine and the Near-Death Experience: The Role of Glutamate and the NMDA-Receptor." In *The Near-Death Experience: A Reader,* edited by L. W. Bailey and J. Yates, 265–82. New York: Routledge, 1996.

Jeans, J. H. *The Mysterious Universe.* New York: Macmillan, 1930.

John, E. R., L. S. Prichep, W. Kox, P. Valdés-Sosa, J. Bosch-Bayard, E. Aubert, M. Tom, F. diMichele, and L. D. Gugino. "Invariant Reversible QEEG Effects of Anesthetics." *Consciousness and Cognition* 10 (2001): 165–83.

Jordan, J. E., E. Dyess, and J. Cliett. "Unusual Spontaneous Movements in Brain-Dead Patients." *Neurology* 35 (1985): 1082.

Julsgaard, B., J. Sherson, J. I. Cirac, J. Fiurasek, and E. S. Polzik. "Experimental Demonstration of Quantum Memory for Light." *Nature* 432 (2004): 482–85.

Jung, C. G. *Memories, Dreams, Reflections.* Edited by Aniela Jaffé, translated by R. and C. Winston. New York: Random House, 1961.

Kant, I. *Critique of Pure Reason.* Translated by J. M. D. Meiklejohn. 1781. Hamburg: Meiner, 1998. http://www.gutenberg.org/etext/4280.

Kellehear, A. "Culture, Biology, and the Near-Death Experience: A Reappraisal." *Journal of Nervous and Mental Disease* 181 (1993): 148–56.

———. *Experiences Near Death: Beyond Medicine and Religion.* Oxford and New York: Oxford University Press, 1996.

Kelly, E. D., and E. W. Kelly. *Irreducible Mind: Toward a Psychology for the 21st Century.* Lanham, MD: Rowman & Littlefield, 2007.

Kelly, E. W., B. Greyson, and I. Stevenson. "Can Experiences Near-Death Furnish Evidence of Life After Death?" *Omega* 40 (1999–2000): 13–519.

Kennedy, D., and C. Norman. "What We Don't Know." *Science* 309, no. 5731 (2005): 75.

Kerkhoffs, J. *Droomvlucht in coma* [Dream Flight in Coma]. Melick, the Netherlands: Marga Genot Melick, 1994.

Kinney, H. C., J. Korein, A. Panigraphy, P. Dikkes, and R. Goode. "Neuropathological Findings in the Brain of Karen Ann Quinlan: The Role of the Thalamus in the Persistent Vegetative State." *New England Journal of Medicine* 330, no. 26 (1994): 1469–75.

Knoblaub, H., and H. G. Soeffner, eds. *Todesnähe: Interdisziplinäre Zugänge zu einem außergewöhnlichen Phänomen* [Near-Death: Interdisciplinary Approaches to an Extraordinary Phenomenon]. Konstanz, Germany: Universitätsverlag, 1999.

Koch, C. "The Movie in Your Head." *Scientific American Mind* 16 (2005): 3, 58–63.

Kompanje, E. *Geven en Nemen: De praktijk van postmortale orgaandonatie* [Give and Take: The Practice of Postmortem Organ Donation]. Doctoral thesis, Erasmus University, Rotterdam, the Netherlands, 1999.

Kongzhi, S., L. Xianggao, and Z. Liangzhong. "Research into Paranormal Ability to Break Through Spatial Barriers." Translated by the Defense Intelligence Agency. *Chinese Journal of Somatic Science.* First Issue, 22 (1990).

Korthals Altes, A. *Uit coma* [Out of Coma]. The Hague, the Netherlands: Mirananda, 2002.

Kübler-Ross, E. *Death, the Final Stage of Growth.* Englewood Cliffs, NJ: Prentice-Hall, 1975.

———. *On Death and Dying.* New York: Macmillan, 1969.

Kuhn, T. S. *The Structure of Scientific Revolutions.* Chicago: University of Chicago Press, 1962.

Küng, H. *Eternal Life? Life After Death as a Medical, Philosophical, and Theological Problem.* Translated by E. Quinn. London: Collins, 1984.

Larson, E. J., and L. Witham. "Leading Scientists Still Reject God." *Nature* 394 (1998): 313.

———. "Scientists Are Still Keeping the Faith." *Nature* 386 (1997): 235–36.

Lashley, K. "In Search of the Engram." Society of Experimental Biology Symposium 4. *Psychological Mechanisms in Animal Behavior.* New York: Academic Press, 1950.

Laszlo, E. *The Connectivity Hypothesis: Foundations of an Integral Science of Quantum, Cosmos, Life, and Consciousness.* Albany: State University of New York Press, 2003.

———. *Science and the Akashic Field: An Integral Theory of Everything.* 2nd ed. Rochester, VT: Inner Traditions, 2007.

———. *You Can Change the World: The Global Citizen's Handbook for Living on Planet Earth.* New York: Select Books, 2003.

Laureys, S., M.-E. Faymonville, X. de Tiège, P. Peigneux, J. Berré, G. Moonen, S. Goldman, and P. Maquet. "Brain Function in the Vegetative State." In *Advances in Experimental Medicine and Biology 550: Brain Death and Disorders of Consciousness,* edited by C. Machado and D. A. Shewmon, 229–38. New York: Kluwer/Plenum, 2004.

Lederberg, J. "The Meaning of Epigenetics." *The Scientist* 15, no. 18 (2001): 6.

Lempert, T., M. Bauer, and D. Schmidt. "Syncope and Near-Death Experience." *Lancet* 344 (1994): 829–30.

Lensink, M. *Toevallige Signalen: Meer dan 100 verhalen over ervaringen rondom de dood* [Unexpected Signals: over 100 Stories About Experiences near Death]. Amsterdam: Schors, 2006.

Lerma, J. *Into the Light: Real Life Stories About Angelic Visits, Visions of Afterlife, and Other Pre-Death Experiences.* Franklin Lakes, NJ: New Page Books, 2007.

Leuba, J. H. *The Belief in God and Immortality: A Psychological, Anthropological and Statistical Study.* Boston: Sherman, French, 1926.

Lewin, R. "Is Your Brain Really Necessary?" *Science* 210 (1980): 1232–34.

Libet, B. *Mind Time: The Temporal Factor in Consciousness.* Cambridge: Harvard University Press, 2004.

———. "A Testable Field Theory of Mind-Brain Interaction." *Journal of Consciousness Studies* 1 (1994): 119–26.

———, C. A. Gleason, E. W. Wright, and D. K. Pearl. "Time of Conscious Intention to Act in Relation to Onset of Cerebral Activity (Readiness Potential): The Unconscious Initiation of a Freely Voluntary Act." *Brain* 106 (1983): 623–42.

Liebetanz, D., M. A. Nitsche, F. Tergau, and W. Paulus. "Pharmacological Approach to the Mechanisms of Transcranial DC-Stimulation-Induced After-Effects of Human Motor Cortex Excitability." *Brain* 125 (2002): 2238–47.

Lodewick, G. *Ik houd mijn hart vast: Andere dimensies van orgaandonatie* [With My Heart in My Mouth: Other Dimensions of Organ Donation]. Deventer, Netherlands: Ankh-Hermes, 1998.

Lommel, P. van. "About the Continuity of Our Consciousness." *Advances in Experimental Medicine and Biology* 550 (2004): 115–132.

———. "Near-Death Experience, Consciousness and the Brain: A New Concept About the Continuity of Our Consciousness Based on Recent Scientific Research on Near-Death Experience in Survivors of Cardiac Arrest." *World Futures: The Journal of General Evolution* 62 (2006): 134–51.

————, R. van Wees, V. Meyers, and I. Elfferich. "Near-Death Experiences in Survivors of Cardiac Arrest: A Prospective Study in the Netherlands." *The Lancet* 358 (2001): 2039–45.

Long, J., and J. M. Holden. "Does the Arousal System Contribute to Near-Death and Out-of-Body Experiences? A Summary and Response." *Journal of Near-Death Studies* 25, no. 3 (2007): 135–69.

Losasso, T. J., D. A. Muzzi, F. B. Meyer, and F. W. Sharbrough. "Electroencephalographic Monitoring of Cerebral Function During Asystole and Successful Cardiopulmonary Resuscitation." *Anesthesia Analgesia* 75 (1992): 12–19.

Lutz, A., L. L. Greischar, N. B. Rawlings, M. Ricard, and R. J. Davidson. "Long-Term Meditators Self-Induce High-Amplitude Gamma Synchrony During Mental Practice." *Proceedings of the National Academy of Science, USA* 101, no. 46 (2004): 16369–73.

Machado, C., and A. Shewmon, eds. *Brain Death and Disorders of Consciousness.* New York and London: Kluwer Academic/Plenum, 2004.

Mantegna, R. N., S. V. Buldyrev, A. L. Goldberger, S. Havlin, C. K. Peng, M. Simons, and H. E. Stanley. "Linguistic Features of Non-Coding DNA Sequences." *Physical Review Letters* 73 (1994): 31–69.

Marcer, P. J., and W. Schempp. "The Brain as a Conscious System." *International Journal of General Systems* 27, no. 11 (1998): 231–48.

————. "A Mathematically Specified Template for DNA and the Genetic Code in Terms of the Physically Realisable Processes of Quantum Holography." *Proceedings of the Greenwich Symposium on Living Computers,* edited by A. M. Fedorec and P. J. Marcer, 45–62. London: University of Greenwich, 1996.

————. "Model of the Neuron Working by Quantum Holography." *Informatica* 21, no. 3 (1997): 519–34.

Marcikic, I., H. de Rietmatten, W. Tittel, H. Zbinden, M. Legré, and N. Gisin. "Distribution of Time-Bin Entangled Qubits over 50 Km of Optical Fiber." *Physical Review Letters* 93, no. 18 (2004): 180502-1–4.

Mascoró, J. *The Upanishads.* New York: Penguin, 1965.

Maslow, A. H. *The Psychology of Science.* New York: Harper & Row, 1966.

————. *Toward a Psychology of Being.* New York: Van Nostrand, 1968.

Maso, I. *Argumenten voor een inclusieve wetenschap* [Arguments in Favor of an Inclusive Science]. Paper presented at the conference Science, Worldview, and Us, Brussels, Belgium, June 2003.

Massimini, M., F. Ferrarelli, R. Huber, S. K. Esser, H. Singh, and G. Tononi. "Breakdown of Cortical Effective Connectivity During Sleep." *Science* 309, no. 5744 (2005): 2228–32.

Matuksevich, D. N., and A. Kuzmich. "Quantum State Transfer Between Matter and Light." *Science* 306 (2004): 663–66.

Mayberg, H. S., A. M. Lozano, V. Voon, H. E. McNeely, D. Seminowicz, C. Hamani, J. M. Schwalb, and S. H. Kennedy. "Deep Brain Stimulation for Treatment-Resistant Depression." *Neuron* 45 (2005): 651–60.

Mayberg, H. S., J. A. Silva, S. K. Brannan, J. L. Tekell, R. K. Mahurin, S. McGinnis, and P. A. Jerabek. "The Functional Neuranatomy of the Placebo Effect." *American Journal of Psychiatry* 159 (2002): 728–37.

Mayer, J., and T. Marx. "The Pathogenesis of EEG Changes During Cerebral Anoxia." In *Cardiac and Vascular Diseases Handbook of Electroencephalography and Clinical Neurophysiology,* edited by J. H. A. van der Drift, 5–11. Amsterdam: Elsevier, 1972.

Meduna, L. T. *Carbon Dioxide Therapy: A Neuropsychological Treatment of Nervous Disorders.* Springfield, IL: Charles C. Thomas, 1950.

Mellor, R., ed. *The Historians of Ancient Rome: An Anthology of the Major Writings.* 2nd ed. New York: Routledge, 2004.

Mermin, N. D. "Is the Moon There When Nobody Looks? Reality and the Quantum Theory." *Physics Today* 38, no. 4 (1985): 38–47.

Mesland, D. *Bewustzijn: De metafysische ruimte* [Consciousness: The Metaphysical Space]. Delft, the Netherlands: Eburon, 2002.

Meursing, B. T. J., and R. G. van Kesteren, eds. *Handboek Reanimatie* [The Resuscitation Handbook]. 2nd ed. Utrecht, the Netherlands: Wetenschappelijke Uitgeverij Bunge, 2004.

Mishlove, J. *The Roots of Consciousness: The Classic Encyclopedia of Consciousness Studies.* Rev. ed. Tulsa, OK: Oak Council Books, 1993.

Misra, B., and E. C. G. Sudarshan. "The Zeno's Paradox in Quantum Theory." *Journal of Mathematical Physics* 18 (1977): 756–63.

Moody, R. A., Jr. *Life After Life.* Covington, GA: Mockingbird Books, 1975.

Moore, W. J. *Schrödinger: Life and Thought.* Cambridge: Cambridge University Press, 1989.

Morse, M. *Transformed by the Light.* New York: Villard Books, 1990.

———, and P. Perry. *Closer to the Light.* New York: Villard Books, 1990.

Moss, J., and M. Rockoff. "EEG Monitoring During Cardiac Arrest and Resuscitation." *Journal of the American Medical Association* 244, no. 24 (1980): 2750–51.

Musgrave, C. "A Study of Spiritual Transformation." *Journal of Near-Death Studies* 15, no. 3 (1997): 187–201.

Myers, F. W. H. *Human Personality and Its Survival of Bodily Death.* London: Longmans, Green, 1903.

———. "On Indications of Continued Terrene Knowledge on the Part of Phantasms of the Dead." *Proceedings of the Society for Psychical Research* 8 (1892): 170–252.

———. "The Subliminal Consciousness." *Proceedings of the Society for Psychical Research* 8 (1892): 298–535.

Nadeau, R., and M. Kafatos. *The Non-Local Universe: The New Physics and Matters of the Mind.* Oxford and New York: Oxford University Press, 1999.

Nahm, M., and B. Greyson. "Terminal Lucidity in Patients with Chronic Schizophrenia and Dementia: A Survey of the Literature." *Journal of Nervous and Mental Disease* 197, no. 12 (2009): 942–44.

Nelson, K. R., M. Mattingly, S. A. Lee, and F. A. Schmitt. "Does the Arousal System Contribute to Near Death Experience?" *Neurology* 66, no. 1 (2006): 1003–9.

Nelson, R. D., B. J. Dunne, Y. H. Dobyns, and R. G. Jahn. "Precognitive Remote Perception: Replication of Remote Viewing." *Journal of Scientific Exploration* 10, no. 1 (1996): 109–10.

Nersessian, N. J. "Aether Or: The Creation of Scientific Concepts." *Studies in the History and Philosophy of Science* 15 (1984): 175–212.

Neumann, J. von. *Mathematical Foundations of Quantum Theory*. Princeton: Princeton University Press, 1955.

Newberg, A. *Why God Won't Go Away: Brain Science and the Biology of Belief*. New York: Ballantine Books, 2002.

Noë, A. *Out of Our Heads: Why You Are Not Your Brain, and Other Lessons from the Biology of Consciousness*. New York: Hill and Wang, 2009.

Opdebeeck, A. *Bijna dood: Leven met bijna-doodervaringen* [Nearly Dead: Living with Near-Death Experiences]. Tielt, Belgium: Uitgeverij Terra-Lannoo, 2001.

Osis, K., and E. Haraldsson. *At the Hour of Death*. New York: Avon Books, 1977.

Ovid. *Metamorphoses*. Translated by Samuel Garth. http://etext.virginia.edu/latin/ovid/garthl.html.

Owen, A. M., M. R. Coleman, M. Boly, M. H. Davis, S. Laureys, and J. D. Pickard. "Detecting Awareness in the Vegetative State." *Science* 313 (2006): 1402.

Pagels, H. R. *Perfect Symmetry*. London: Joseph Publishers, 1985.

Palmer, J. "A Community Mail Survey of Psychic Experiences." *Journal of the American Society of Psychical Research* 73 (1979): 221–51.

Paradis, N. A., G. B. Martin, and M. G. Goetting. "Simultaneous Aortic Jugular Bulb, and Right Atrial Pressures During Cardiopulmonary Resuscitation in Humans: Insights into Mechanisms." *Circulation* 80 (1989): 361–68.

Paradis, N.A., G. B. Martin, and J. Rosenberg. "The Effect of Standard and High Dose Epinephrine on Coronary Perfusion Pressure During Prolonged Cardiopulmonary Resuscitation." *Journal of the American Medical Association* 265 (1991): 1139–44.

Parnia, S., and P. Fenwick. "Near-Death Experiences in Cardiac Arrest: Visions of a Dying Brain or Visions of a New Science of Consciousness." Review article. *Resuscitation* 52 (2002): 5–11.

Parnia, S., D. G. Waller, R. Yeates, and P. Fenwick. "A Qualitative and Quantitative Study of the Incidence, Features and Aetiology of Near Death Experiences in Cardiac Arrest Survivors." *Resuscitation* 48 (2001): 149–56.

Pearsall, P. *The Heart's Code*. New York: Broadway Books, 1998.

———, G. E. Schwartz, and L. G. Russek. "Changes in Heart Transplant Recipients That Parallel the Personalities of Their Donors." *Journal of Near-Death Studies* 20, no. 3 (2002): 191–206.

Peat, F. D. *Infinite Potential: The Life and Times of David Bohm*. New York: Addison-Wesley, 1996.

Penfield, W. *The Excitable Cortex in Conscious Man*. Liverpool: Liverpool University Press, 1958.

———. *The Mystery of the Mind*. Princeton: Princeton University Press, 1975.

———. "The Role of the Temporal Cortex in Certain Psychical Phenomena." *Journal of Mental Science* 101 (1955): 451–65.

Pennefather, S. H., J. H. Dark, and R. E. Bullock. "Haemodynamic Responses to Surgery in Brain-Dead Organ Donors." *Anaesthesia* 48, no. 12 (1993): 1034–38.

Penrose, R. *Shadows of the Mind: A Search for the Missing Science of Consciousness*. Oxford: Oxford University Press, 1996.

Persinger, M. A. "Near-Death Experiences: Determining the Neuroanatomical Pathways by Experiential Patterns and Simulation in Experimental Settings." In *Healing: Beyond Suffering or Death*, edited by L. Bessette, 277–86. Chabanel, Quebec, Canada: Publications MNH, 1994.

———, and F. Healey. "Experimental Facilitation of the Sensed Presence: Possible Intercalatation Between the Hemispheres Induced by Complex Magnetic Fields." *Journal of Nervous and Mental Diseases* 190 (2002): 533–41.

Pew Forum on Religion and Public Life. *Many Americans Mix Multiple Faiths*. 2009. http://pewforum.org/newassets/images/reports/multiple faiths/multiplefaiths.pdf.

———. *U.S. Religion Landscape Survey*. 2007. http://religions.pewforum .org/reports.

Pfister, O. "Shockdenken und Shockphantasien bei höchster Todesgefahr." *Zeitschrift für Psychoanalyse* 16 (1930): 430–55. Translated by R. Noyes and R. Kletti as "Shock Thoughts and Fantasies in Extreme Mortal Danger."

Pizzi, R., A. Fantasia, F. Gelain, D. Rossetti, and A. Vescovi. "Non-Local Correlation Between Human Neural Networks." In *Quantum Information and Computation II: Proceedings of SPIE 5436*, edited by E. Donkor, A. R. Pirick, and H. E. Brandt, 107–17. 2004. http://www.spiedigital library.org/dbt/dbt.jsp?KEY=PSISDG&Volume=5436&Issue=1&bproc =stia&scode=SIP10.

Plato. *Phaedo*. Translated by Benjamin Jowett. http://philosophy.eserver .org/plato/phaedo.txt.

———. *The Republic*. Translated by Benjamin Jowett. http://philosophy .eserver.org/plato/republic.txt.

Popp, F.-A., and L. V. Beloussov, eds. *Integrative Biophysics Biophotonics*. Berlin: Springer Verlag, 2003.

Popper, K., and J. C. Eccles. *The Self and Its Brain*. New York: Springer, 1977.

Pribram, K. *Languages of the Brain*. Monterey, CA: Wadsworth, 1977.

———. "The Neurophysiology of Remembering." *Scientific American* 220 (1969): 75.

Prigogine, I., and I. Stengers. *Order Out of Chaos: Man's New Dialogue with Nature.* Boulder, CO: New Science Press, 1984.

Puma, J. la, D. L. Schriedermayer, A. E. Gulyas, and M. Siegler. "Talking to Comatose Patients." *Archives of Neurology* 45 (1988): 20–22.

Puthoff, H. E. "CIA-Initiated Remote Viewing Program at Stanford Research Institute." *Journal of Scientific Exploration* 10, no. 1 (1996): 63–76.

———, and R. Targ. "A Perceptual Channel for Information Transfer over Kilometres Distances: Historical Perspectives and Recent Research." *Proceedings of the IEEE* 64, no. 3 (1976): 329–54.

Qur'an. Translated by Abdullah Yusuf Ali. Wordsworth Classics of World Literature. Ware, UK: Wordsworth Editions Limited, 2000.

Radin, D. *Entangled Minds: Extrasensory Experiences in a Quantum Reality.* New York: Simon & Schuster, 2006.

———. "Event-Related Electroencephalographic Correlations Between Isolated Human Subjects." *Journal of Alternative and Complementary Medicine* 10 (2004): 315–23.

Rando, T. A., ed. *Parental Loss of a Child.* Champaign, IL: Research Press, 1985.

Raphael, S. P. *Jewish Views of the Afterlife.* Northvale, NJ: Jason Aronson, 1994.

Rees, W. D. "The Hallucinations of Widowhood." *British Medical Journal* 4 (1971): 37–41.

Richards, T. L., L. Kozak, L. C. Johnson, and L. J. Standish. "Replicable Functional Magnetic Resonance Imaging Evidence of Correlated Brain Signals Between Physically and Sensory Isolated Subjects." *Journal of Alternative and Complementary Medicine* 11, no. 6 (2005): 955–63.

Ridley, M. *Genome: The Autobiography of a Species in 23 Chapters.* New York: HarperCollins, 2000.

Ring, K. *Heading Toward Omega: In Search of the Meaning of the Near-Death Experience.* New York: Quill William Morrow, 1984.

———. *Life at Death: A Scientific Investigation of the Near-Death Experience.* New York: Coward, McCann & Geoghegan, 1980.

———. *The Omega Project: Near-Death Experiences, UFO-Encounters, and Mind at Large.* New York: William Morrow, 1992.

———, and S. Cooper. *Mindsight: Near-Death and Out-of-Body Experiences in the Blind.* Palo Alto, CA: William James Center/Institute of Transpersonal Psychology, 1999.

Ring, K., and E. Elsaesser-Valarino. *Lessons from the Light: What We Can Learn from the Near-Death Experience.* New York and London: Insight Books, Plenum, 1998.

Ritchie, G. G. *Return from Tomorrow.* Grand Rapids, MI: Zondervan, 1978.

Rivas, T. *Een gesprek met TG over de man met het gebit* [A Conversation with T.G. About the Man with the Dentures]. *Terugkeer* 19, no. 3 (2008): 12–20.

Robinson, R. "Ciliate Genome Sequence Reveals Unique Features of a Model Eukaryote." *Public Library of Science Biology* 4, no. 9 (2006). http://www.ncbi.nlm.nih.gov/pmc/articles/PMC1557400/.

Rodin, E. "Comments on 'A Neurobiological Model for Near-Death Experiences.'" *Journal of Near-Death Studies* 7 (1989): 255–59.

Roland, P. E. "Somatotopical Tuning of Postcentral Gyrus During Focal Attention in Man: a Regional Cerebral Blood Flow Study." *Journal of Neurophysiology* 46 (1981): 744–54.

——, and L. Friberg. "Localization in Cortical Areas Activated by Thinking." *Journal of Neurophysiology* 53 (1985): 1219–43.

Romijn, H. "About the Origin of Consciousness: A New, Multidisciplinary Perspective on the Relationship Between Brain and Mind." *Proceedings of the Koninklijke Nederlandse Akademie van Wetenschappen* 100, nos. 1–2 (1997): 181–267.

——. "Are Virtual Photons the Elementary Carriers of Consciousness?" *Journal of Consciousness Studies* 9 (2002): 61–81.

Rosenblum, B., and F. Kuttner. "The Observer in the Quantum Experiment." *Foundations of Physics* 32, no. 8 (2002): 1273–93.

Russell, P. *From Science to God: A Physicist's Journey into the Mystery of Consciousness.* Novato, CA: New World Library, 2002.

Rutherford, F., G. Holton, and F. G. Watson. *Project Physics Course.* New York: Holt, Reinhart & Winston, 1968.

Saavedra-Aguilar, J. C., and J. S. Gómez-Jeria. "A Neurobiological Model for Near-Death Experiences." *Journal of Near-Death Studies* 7 (1989): 205–22.

Sabom, M. B. *Light and Death: One Doctor's Fascinating Account of Near-Death Experiences.* Grand Rapids, MI: Zondervan, 1998.

——. *Recollections of Death: A Medical Investigation.* New York: Harper & Row, 1982.

Safar, P., W. Behringer, B. W. Bottiger, and F. Sterz. "Cerebral Resuscitation Potentials for Cardiac Arrest." *Critical Care Medicine* 30, no. 4 suppl. (2002): 140–44.

Sanghavi, D. "When Does Death Start?" *New York Times,* December 20, 2009. http://www.nytimes.com/2009/12/20/magazine/20organ-t.html?_r=2&ref=magazine

Sartori, P. "The Incidence and Phenomenology of Near-Death Experiences." *Network Review* 90 (2006): 23–25.

——, P. Badham, and P. Fenwick. "A Prospectively Studied Near-Death Experience with Corroborated Out-of-Body Perception and Unexplained Healing." *Journal of Near-Death Studies* 25, no. 2 (2006): 69–84.

Sauve, M. J., J. A. Walker, S. M. Massa, R. A. Winkle, M. M. Scheinman. "Patterns of Cognitive Recovery in Sudden Cardiac Arrest Survivors: The Pilot Study." *Heart Lung* 25, no. 3 (1996): 172–81.

Saver, J. L., and J. Rabin. "The Neural Substrates of Religious Experience." *Journal of Neuropsychiatry* 9, no. 3 (1997): 498–510.

Schempp, W. *Magnetic Resonance Imaging: Mathematical Foundations and Applications.* New York: John Wiley, 1997.

———. "Quantum Holography and Neurocomputer Architectures." *Journal of Mathematical Imaging and Vision* 2 (1992): 109–64.

Schiff, N. D, J. T. Giacino, K. Kalmar, J. D. Victor, K. Baker, M. Gerber, B. Fritz, B. Eisenberg, J. O'Connor, E. J. Kobylarz, S. Farris, A. Machado, C. McCagg, F. Plum, J. J. Fins, and A. R. Rezai. "Behavioural Improvements with Thalamic Stimulation After Severe Traumatic Brain Injury." *Nature* 448 (2007): 600–603.

Schmied, I., H. Knoblaub, and B. Schnettler. "Todesnäheerfahrungen in Ost- und Westdeutschland: Eine empirische Untersuchung" [Near-Death Experiences in East and West Germany: An Empirical Study]. In *Todesnähe: Interdisziplinäre Zugänge zu einem außergewöhnlichen Phänomen* [Near-Death: Interdisciplinary Approaches to an Extraordinary Phenomenon], edited by H. Knoblaub and H. G. Soeffner. Konstanz, Germany: Universitätsverlag, 1999.

Schouterden, C., and G. van der Linden. *"Kijk, ik ben er nog!" Met getuigenissen over tekens van overledenen* ["Look, I'm Still Here!" Testimonies of Signs from the Dead]. Zoetermeer, the Netherlands: Free Musketeers, 2005.

Schrödinger, E. "Discussion of Probability Relations Between Separated Systems." *Cambridge Philosophical Society Proceedings* 32 (1935): 555.

———. *What Is Life; With Mind and Matter and Autobiographical Sketches.* Canto Edition. 1944; Cambridge: Cambridge University Press, 1992.

Schuijlenburg, B. van. *Waar was de patiënt? Omgang met mensen met een bijna-dood ervaring in de hulpverlening* [Where Was the Patient? Supporting People with a Near-Death Experience]. Assen, the Netherlands: Van Gorcum, 1994.

Schwaninger, J., P. R. Eisenberg, K. B. Schechtman, and A. N. Weiss. "A Prospective Analysis of Near-Death Experiences in Cardiac Arrest Patients." *Journal of Near-Death Studies* 20 (2002): 215–32.

Schwartz, J. M., and S. Begley. *The Mind and the Brain: Neuroplasticity and the Power of Mental Force.* New York: HarperCollins, 2002.

Schwartz, S. A. *Opening to the Infinite: The Art and Science of Nonlocal Awareness.* Buda, TX: Nemoseen Media, 2007.

Shadler, M. N., and R. Kiani. "News and Views. Neurology: An Awakening." *Nature* 448 (2007): 539–40.

Sheldrake, R. *A New Science of Life.* London: Blond & Briggs, 1981.

———. *The Presence of the Past.* London: HarperCollins, 1988.

Shuhuang, L., et al. "Some Experiments on the Transfer of Objects Performed by Unusual Abilities of the Human Body." *Nature Journal (People's Republic of China)* 4, no. 9 (1981): 652. Defense Intelligence Agency Requirements and Validation Branch, DIA Translation LN731–83, Intelligence Information Report No. 6010511683 (1983).

Slob, M. Interview: "Waar de wetenschap niet bij kan" [Where Science Cannot Reach]. *NRC Handelsblad,* January 20, 2007.

Sluijs, M. A., van der. "Three Ancient Reports of Near-Death Experiences: Bremmer Revisited." *Journal of Near-Death Studies* 27, no. 4 (2009): 223–53.

Smit, R. "Corroboration of the Dentures Anecdote Involving Veridical Perception in a Near-Death Experience." *Journal of Near-Death Studies* 27, no. 1 (2008): 47–61.

Smith, D. S., W. Levy, M. Maris, and B. Chance. "Reperfusion Hyperoxia in the Brain After Circulatory Arrest in Humans." *Anesthesiology* 73 (1990): 12–19.

Sogyal Rinpoche. *The Tibetan Book of Living and Dying*. San Francisco: HarperSanFransisco, 1992.

Solomon, L. D. *The Jewish Book of Living and Dying*. Northvale, NJ: Jason Aronson, 1999.

Sommerfeld, A. *Atombau und Spektrallinien*. Braunschweig, Germany: Friedrich Vieweg & Sohn, 1919. Translated by H. L. Brose as *Atomic Structure and Spectral Lines*. Methuen, 1923.

Sperling, D. *Management of Post-Mortem Pregnancy: Legal and Philosophical Aspects*. Aldershot, UK: Ashgate, 2006.

Sperry, R. W. "Mental Phenomena As Causal Determinants in Brain Function." In *Consciousness of the Brain,* edited by G. G. Globus, G. Maxwell, and I. Savodnik. New York: Plenum, 1976.

Standish, L. J., L. C. Johnson, L. Kozak, and T. L. Richards. "Evidence of Correlated Functional Magnetic Resonance Imaging Signals Between Distant Human Brains." *Alternative Therapies in Health and Medicine* 9, no. 1 (2003): 128.

Standish, L. J, L. Kozak, L. C. Johnson, and T. Richards. "Electroencephalographic Evidence of Correlated Event-Related Signals Between the Brain of Spatially and Sensory Isolated Subjects." *Journal of Alternative and Complementary Medicine* 10, no. 2 (2004): 307–14.

Stapp, H. *Mind, Matter and Quantum Mechanics*. 2nd ed. Berlin: Springer Verlag, 2004.

Stark, P. *De hele waarheid: Verhalen uit de praktijk van orgaantransplantatie* [The Whole Truth: Real-Life Organ Transplantation Stories]. Breda, the Netherlands: Papieren Tijger, 2005.

Steiner, R. *Der Tod als Lebenswandlung*. Dornach, Switzerland: Rudolf Steiner Gesamtausgabe, 1917/18. Translated by S. Seiler as *Death as Metamorphosis of Life*. New York: Steinerbooks, 1961.

Stevenson, I. *Where Reincarnation and Biology Intersect*. Westport, CT: Praeger, 1997.

Stolp, H. *De Gouden Vogel: Dagboek van een stervende jongen*. Rotterdam, the Netherlands: Lemniscaat, 1987. Translated as *The Golden Bird*. New York: Dial Books, 1987.

Stone, R. B. *The Secret Life of Your Cells*. West Chester, PA: Whitford Press, 1989.

Strassman, R. *DMT, the Spirit Molecule: A Doctor's Revolutionary Research into the Biology of Near-Death and Mystical Experiences*. Rochester, VT: Park Street Press, 2001.

Sutherland, C. *Transformed by the Light: Life After Near-Death Experiences*. Sydney: Bantam Books, 1992.

Swaab, D. "Als we alles bewust moesten doen, zouden wij geen leven hebben: Interview met Dick Swaab" [If We Would Do Everything at a Conscious Level, It Should Be Very Difficult to Live: Interview with Dick Swaab]. By G. Klaasen. *KRO Magazine* 47 (2009): 10–13.

Swami Rama. *Sacred Journey: Living Purposefully and Dying Gracefully*. New Delhi: Himalayan International Institute of Yoga Science & Philosophy, 1996.

Swedenborg, E. *Awaken from Death*. Edited by James F. Lawrence. San Francisco: J. Appleseed, 1993.

Sylvia, C., and W. J. Novak. *Change of Heart*. New York: Little, Brown, 1997.

Thaheld, F. "Biological Non-Locality and the Mind-Brain Interaction Problem: Comments on a New Empirical Approach." *BioSystems* 2209 (2003): 1–7.

Tosch, P. "Patient's Recollections of Their Posttraumatic Coma." *Journal of Neuroscience Nursing* 20, no. 4 (1988): 223–28.

Vries, J.W. de, P. F. A. Bakker, G. H. Visser, J. C. Diephuis, and A. C. van Huffelen. "Changes in Cerebral Oxygen Uptake and Cerebral Electrical Activity During Defibrillation Threshold Testing." *Anesthesia Analgesia* 87 (1998): 16–20.

Waanders, T. *De dood en de jongen: Monument voor mijn gestorven zoon* [Death and the Boy: A Monument for My Dead Son]. Kampen, the Netherlands: Ten Have, 2006.

Wackermann, J., C. Seiter, H. Keibel, and H. Walach. "Correlations Between Electrical Activities of Two Spatially Separated Human Subjects." *Neuroscience Letters* 336 (2003): 60–64.

Wade, J. "The Phenomenology of Near-Death Consciousness in Past-Life Regression Therapy: A Pilot Study." *Journal of Near-Death Studies* 17, no. 1 (1998): 31–53.

Wager, T. D., J. K. Rilling, E. E. Smith, A. Sokolik, K. L. Casey, R. J. Davidson, S. M. Kosslyn, R. M. Rose, and J. D. Cohen. "Placebo-Induced Changes in fMRI in the Anticipation and Experience of Pain." *Science* 303 (2004): 1162–67.

Walach, H., R. Hartmann. "Complementarity Is a Useful Concept for Consciousness Studies: A Reminder." *Neuroendocrinology Letters* 21 (2000): 221–32.

Walach, H., N. Kohls, N. von Stillfried, T. Hinterberger, and S. Schmidt. "Spirituality: The Legacy of Parapsychology." *Archive for the Psychology of Religion* 31: (2009): 277–308.

Weiss, P. *Principles of Development*. New York: Holt, 1939.

Wetzel, R., N. Setzer, J. L. Stiff, and M. C. Rogers. "Hemodynamic Responses in Brain Dead Organ Donor Patients." *Anesthesia and Analgesia* 64 (1985): 125–28.

Whinnery, J. E., and A. M. Whinnery. "Acceleration-Induced Loss of Consciousness." *Archives of Neurology* 47 (1990): 764–76.

White, N. S., and M. T. Alkire. "Impaired Thalamocortical Connectivity in Humans During General-Anesthetic Induced Unconsciousness." *Neuroimage* 19, no. 2, pt. 1 (2003): 401–11.

Wigner, E. "The Problem of Measurement." *Journal of Physics* 31 (1963): 6.

Wijk, R. van. "Bio-Photons and Bio-Communication." *Journal of Scientific Exploration* 15, no. 2 (2001): 183–97.

Wilber, K. *No Boundary*. Boulder, CO: Shambhala, 1981.

Woerlee, G. M. *Mortal Minds: A Biology of the Soul and the Dying Experience*. Utrecht, the Netherlands: De Tijdstroom, 2003.

Wolf, F.A. *The Spiritual Universe: One Physicist's Vision of Spirit, Soul, Matter and Self*. Portsmouth, NH: Moment Point Press, 1996.

———. *Taking the Quantum Leap: The New Physics for Non-Scientists*. New York: Harper & Row, 1989.

Yamamura, H. "Implication of Near-Death Experience for the Elderly in Terminal Care." *Nippon Ronen Igakkai Zasshi* 35, no. 2 (1998): 103–15.

Zaleski, C. *Otherworld Journeys: Accounts of Near-Death Experience in Medieval and Modern Times*. Oxford: Oxford University Press, 1987.

Zeilinger, A. *Einsteins Schleier* [Einstein's Veil]. Munich: C. H. Beck Verlag, 2003.

———. *Einsteins Spuk* [Einstein's Ghost]. Munich: Bertelsmann, 2005.

Zohar, D. *The Quantum Self: Human Nature and Consciousness Defined by the New Physics*. New York: William Morrow, 1990.

Index

Page numbers of illustrations appear in italics.

afterlife, *28*, 354; belief in a form of, xiv, 55, 69, 83, 93, 149, 316–17, 358; scientists and, 329–30, 355
Akasha field, 211, 308
Altes, Alison Korthals, xii
Aspect, Alain, 238, 275, 294
Atwater, P. M. H., 48, 72–73, 75
auras, 62
autoscopy/autoscopic experience, 13, 14
Awaken from Death (Swedenborg), 85

Bach, Richard, 1
Backster, Cleve, 298–99
Bailey, Alice, 85
Beaufort, Adm. Francis, 101
Beauregard, Mario, 198, 201
Bede (The Venerable Bede), 99
Bergson, Henri, 317
Berkovich, Simon, 194, 288–89
Besant, Annie, 85
Besso, Michele, 229
Betz, W., 153
Biophotonen (Bischof), 290
biophotons, 283, 290–91, 294
birth, x, xiii; nonlocal consciousness preceding, 82, 346; as passing from one state of consciousness to another, xx, 318, 325; transpersonal theory and, 305; tunnel experience as memory of, 130
Bischof, Marco, 290
Blackmore, Susan, 115, 127, 201
Blanke, Olaf, 121–22, 190
Bohm, David, 247, 249, 264, 266
Bohr, Niels, 223, 235, 249, 254, 289
Boismont, Brierre de, 101
border, perception of, during NDE, 12, 14, 15, 33, 39–40, 102, 143, 144, 173

Born, Max, 236
Bosch, Jeroen, *28*
Bragg, Sir William Lawrence, 281
Brahms, Johannes, 322
brain, x–xii, xxii, 160–61, *179*; during anesthesia, 128; carbon dioxide, NDEs, and, 116–17; cardiac arrest and, viii, 161–64, 262; chemical reactions, NDEs and, 117–18; computer vs., 200–202; consciousness and, ix, xvi, xvii, xxii, 9, 26, 106, 113, 160, 177, 178–79, 183–86, 195, 202, 223, 251, 252–53, 257–79, 307, 328, 331; consciousness relationship, six theories, 257–60; consciousness research using TMS, 191–93; cooling, 168, 340; death, xii, 141, 163, 166, 171, 328–29, 334–43; electrical stimulation, 120–22, 190; electromagnetic activity and brain function, 120–22, 188–93, 268–69, 272–73; facts, 202; flat EEG and, 340; fMRI, *180*, 180–81, 183, 191, *197*, 200; Fourier transform in, 275–76; heart stopping, effect of, 164–65; holographic hypothesis, 194; information transfer via quantum spin, 274–77, 292; interface of nonlocal consciousness and, 270–71; measurable activity, 178, *180*, 180–81, 183, 185, 198, 200; meditation and, 199–200; memory and, 106, 183–86, 193–95, 251–52, 279; memory and cell death, 186, 279, 281; memory loss, causes, 195; NDE during impairment, 8, 111, 112, 133, 164, 169–76, 202; NDE while not impaired, 111; neurons and electromagnetic fields, 186–88, *187*; neuroplasticity, 195–97, *197*, 200, 253, 259, 266, 279; oxygen deficiency